IEE CIRCUITS AND SYSTEMS SERIES 3

Series Editors: Dr D. G. Haigh
Dr R. Soin

ANALOGUE-DIGITAL ASICs

circuit techniques, design tools and applications

Other volumes in this series:

ANALOGUE-DIGITAL ASICs

circuit techniques, design tools and applications

Edited by
**R. S. Soin,
F. Maloberti
and J. Franca**

Peter Peregrinus Ltd. on behalf of the Institution of Electrical Engineers

Published by: Peter Peregrinus Ltd., London, United Kingdom

© 1991: Peter Peregrinus Ltd.

Peter Peregrinus Ltd.,
Michael Faraday House,
Six Hills Way, Stevenage,
Herts. SG1 2AY, United Kingdom

British Library Cataloguing in Publication Data

A CIP catalogue record for this book
is available from the British Library.

ISBN 0 86341 259 9

Printed in England by Short Run Press Ltd., Exeter

Table of Contents

List of Illustrations

Foreword

This volume arose from a one day course on mixed analogue–digital ASICs, held at ECCTD'89 (European Conference on Circuit Theory and Design) in Brighton, England. Following the success of the course in attracting a large number of attendees, the course organisers were invited to produce a multi–authored volume on the subject. The editors have extended the original request both in the quality and the number of contributions and contributors than were present at ECCTD'89. In view of the very strong interest in this area of electronics, a much updated version of the course, based on this book, is being offered at ECCTD'91 in Copenhagen, Denmark and at the the ASIC'91 conference in Rochester, New York. Although the need for mixed–signal electronics has been present for some time, there continue to be strong developments in technologies, design methods, design techniques and CAD tools to help produce more integrated solutions to this problem. The present book will serve as a useful foundation for practising engineers and postgraduate students working in this field.

David Haigh (Series Editor)
London, August, 1991.

Acknowledgements

All the work leading to the production of a camera ready copy of this book was carried out at GenRad in Fareham. We would like to extend our sincerest gratitude to the management for the use of GenRad's excellent facilities and their commitment to the project.

The editors would like to express, in the strongest possible terms, their appreciation to Alan Gladwell for his superhuman efforts in completing the drawings and the text.

On behalf of Randeep Soin we would also like to extend our thanks to his many friends and colleagues, in particular Chris Pitchford, John Palmer, Youri Koulagenko Korsak, Simon Helliwell and James Ball, for their help with the proof reading and checking.

Finally we would like to thank our families for their patience, tolerance and support in the face of many days and hours of absence from domestic duties and responsibilities while preparing this volume.

In the final days of preparation of this book, one of the editors (RSS) was greatly saddened to learn of the passing away of his uncle, Dr. Om Prakash Madhok. We would like to mark the memory of his life and his work, tending the sick in three continents, over a working life extending to more than forty years.

Randeep Singh Soin
Jose Franca
Franco Maloberti

List of Contributers

J. E. Franca
C. Leme
J. Vital
M.H Martins
A.S Steiger Garcao
N.Horta
**Instituto Superior Tecnico, Lisbon,
Portugal**

F. Maloberti
**University of Pavia
Italy**

P. L. O'Leary
**Jonnaeum Research
Graz
Austria**

P. Senn
**CNET
Grenoble
France**

M. Steyaert
W. Sansen
**Katholieke Universitiet, Leuven
Belgium**

S. J. Morris
**Dialog Semiconductor
Swindon
U.K**

P. Walker
N. Morris
R. Cottrell
**LSI Logic,
Sidcup, Kent
U.K**

K. G. Nichols
A. D. Brown
**University of Southampton
U.K**

P .F. Kilty
R. S. Soin
**GenRad Ltd.,
Fareham
U.K**

D. G. Haig
**University College, University of
London**

N. Battersby
C. Toumazou
**Imperial College, University of London
U.K**

B. Singh
**Phoenix VLSI Consultants,
Towcester,
U.K**

P. Shepeherd
**University of Bath
U.K**

J. B. Hughes
**Philips Research Labs.
Redhill, Surrey
U.K**

A. Rueda
J. L. Huertas
**Univesidad de Sevilla
Spain**

C. Caillon
**SGS–Thomson
Grenoble
France**

Chapter 1

Introduction

1.1 Analogue – digital signal processing

Digital techniques have a number of well known advantages over their analogue counterparts for electronic signal processing, which are briefly outlined in Table 1.1. Digital signals are represented by only two states and are hence more immune to noise than analogue signals which can have a continium of values. Also, the nature of the signals they process renders analogue designs more problematic to evaluate and simulate, because they require greater precision in modelling than digital circuits, where more abstract models are often sufficient to evaluate performance. As digital circuits are easier to program and adjust to particular applications, they can be more readily produced in generic form, while analogue circuits often have to be individually customised because of the greater difficulty of programming.

Table 1.1 Analogue counterparts for electronic signal processing

Characteristics	Analogue design	Digital design
Signal amplitude	Continuum of values for amplitude (and time)	Two amplitude states
Design	Customised	Standardised
Methodology component values	Continuum of values	Components with values
Model requirements	Requires precise modeling capability	Only requires a precise large signal model
Programmability	Hard to change design	Easily programmable by software
Design Level	Design at the circuit level	Designed at the systems level
Use of CAD tools	Difficult to use with CAD tools	Amenable to CAD tools

As a consequence of these potential advantages, the main thrust of (very) large scale integration of electronic circuits has been in digital solutions. Therefore, much more effort has been devoted to the development of digital, rather than analogue, circuits, tools and applications. Nevertheless, electronic systems will always require some form of analogue signal processing for two reasons. Firstly, the real world is analogue; that is signals in the physical world are continuous both in amplitude and in time, and hence analogue techniques are required in the acquisition and conditioning of such signals before they can be processed by digital electronic systems. Also, signals which are subsequently output from digital electronic systems must be in analogue form. A typical example [1] of this is shown in Figure 1.1. A second important reason why analogue techniques will remain important is that of bandwidth. In fact the effective bandwidth which can be processed with analogue techniques is higher than with digital techniques as is readily observed from Figure 1.2, which shows the signal bandwidth requirements for signal processing in various application areas. For this reason analogue techniques will always be required for high bandwidth, high performance applications.

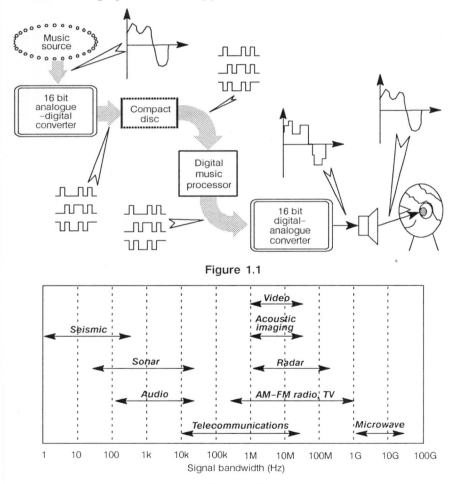

Figure 1.1

Figure 1.2

In view of the above discussion, modern electronic signal processing systems will include some form of analogue, as well as digital, processing, as illustrated in Figure 1.3. Here the physical signal is sensed, perhaps by a

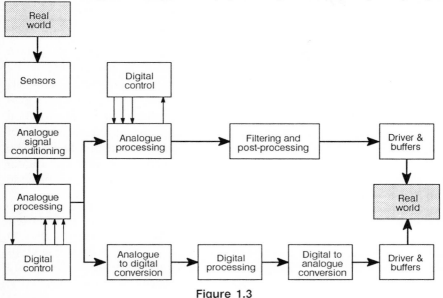

Figure 1.3

transducer, and captured as an analogue electrical signal. The output of the sensor may go through some signal conditioning, such as filtering to remove unwanted background noise signals. This may be followed by some further analogue processing. The signal could then follow one of two, or perhaps even both routes. In one the signal could go through a suitable chain of analogue signal processing functions before conversion to a form suitable for its final physical function. Alternatively, the signal could be converted to digital form with an analogue to digital converter, then be processed digitally, before being converted to an analogue form suitable for driving a physical transducer. Despite its acute relevance in todays electronic systems, it should be clear that the need to combine analogue and digital functionality has been felt for a long time. For reasons of performance, reliability and economy, it has generally been desirable to integrate as much functionality in a single integrated circuit as possible. This, however, has been hindered by technology limitation of high–performance analogue circuit designs. In the early days of bipolar ICs, components such as amplifiers, comparators and regulators could be integrated (fabricated), while discrete, precision, passive components had to be employed to meet even modest performance requirements. Hence, typically, analogue and digital functions were realised separately, and exchanged signals at the board level. In an effort to increase the size and number of functions which could be integrated on a common substrate, the processing technologies were optimised for purely digital circuit techniques, and more rarely for combined digital and analogue circuit techniques.

Newer processing technologies such as analogue CMOS and BiCMOS have enabled good quality components suitable for analogue circuits to be fabricated on the same substrate as digital circuit components. This has been coupled with the developments of powerful analogue signal processing methodologies, and circuit techniques such as switched capacitor, and now switched current, which can exploit the types of components which are becoming available in newer technologies. The relevance of these techniques is greatly enhanced by the fact that digital feedback circuitry can be used to obtain adaptive capabilities and provide greater precision in the performance of analogue circuitry.

It is instructive to consider further technical and some commercial reasons why the two types of signal processing should be mixed on common substrates. Firstly, the ability to implement entire systems on single chips will provide the designer with greater flexibility in the way he partitions and designs the system, leading to technically superior solutions. Secondly, single–chip systems possess smaller component counts and hence need fewer interconnections, which can lead to improved reliability. Thirdly, many applications, for example mobile applications, have to meet stringent power, weight and size restrictions, which can best be met with more integrated and compact solutions. Finally, a reduction in the number of components reduces production and inventory costs. Such factors may be very important in consumer applications where production runs may be very large.

Driven by the emerging market for mixed analogue–digital ICs, there has also been an increased availability from various vendors of powerful cell libraries, compilation techniques, CAD tools and cell–based design methodologies geared to analogue–digital designs. Such factors are likely to cause a rapid growth in this sector of the market in the years ahead.

1.2 Outline of the Book

Analogue–digital ICs encompass a variety of scientific and technological disciplines, such as processing technology, circuit techniques, functional building blocks, applications and computer–aided design. All such disciplines are addressed in the four separate parts of this book as outlined in figures 1.4 and 1.5.

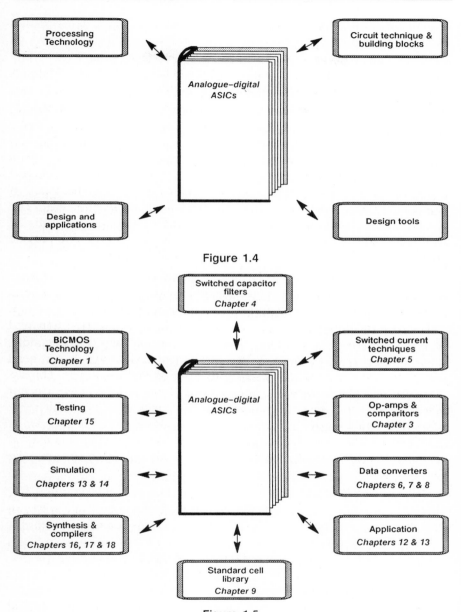

Figure 1.4

Figure 1.5

As it is the fundamental vehicle for IC fabrication, process technology is addressed first, in chapter 2, which considers one particular BiCMOS technology. BiCMOS offers more accurate analogue devices and better noise immunity at the expense of somewhat greater complexity and cost than pure CMOS technology, which nevertheless is also widely used to implement analogue functions. This can in fact be seen in the next part of the book, which addresses a variety of circuit techniques and functional building blocks for designing analogue–digital ICs with both CMOS and BiCMOS technologies. This spans from chapter 3 to chapter 9.

In the treatment of circuit techniques the emphasis is on analogue. The reason for this emphasis is that the technologies which are available are primarily geared to a digital environment, with some modifications to accommodate analogue functionality. Since analogue techniques have to be adapted to an established digital environment, problem areas remain with analogue circuits. On the other hand, standard, well–documented techniques exist for design and evaluation of digital circuits.

Chapter 3 concentrates on the design of the work–horses of analogue techniques, operational amplifiers and comparators. There is a continuing need for high–performance amplifiers with high open–loop gains, high–gain bandwidths and low power consumption. High–accuracy functions such as inverting amplifiers, summing amplifiers, integrators and buffers, can be implemented using such operational amplifiers and a few extra passive components. Thus, complex analogue blocks such as SC filters, and very sensitive amplifiers for medical and other applications, can be implemented. Analogue to digital converters, on the other hand, require comparators, which must meet high speed and resolution requirements for high performance A/D converters. The chapter discusses several of these circuit techniques, both for continuous and discrete time applications, and also addresses their noise performance, which can be a severe problem in analogue circuits, especially when digital circuits are fabricated on the same substrate. Switched capacitor (SC) filters have played an important part in the development of mixed–signal integrated circuits, particularly as they are inherently suitable for fabrication in MOS technology, and rely on capacitor ratios and reference clock frequencies to define time constants, with an accuracy which can be as good as 0.1%. Chapter 4 gives a brief review of SC filters, followed by a discussion of the various building blocks which go to make up such filters, and of optimum switching schemes for higher performance and for high–order filter design. The chapter ends with a number of case studies.

For the implementation of SC techniques, special steps have to be introduced into standard digital processes to fabricate components such as accurate linear capacitors. In contrast to this situation, chapter 5 introduces switched–current techniques as a method of implementing analogue, sampled–data signal processing without the need to employ precision capacitors on chip. This makes switched–current techniques more suited to standard digital processing technology. The chapter discusses the basic principle behind this approach and gives various examples of basic building blocks, such as memory cells and integrators. The final part of the chapter also shows how Finite Impulse Response (FIR) filters and complex blocks such as biquadratic sections may be obtained with this technique.

The next three chapters of the book concentrate on the class of functional building blocks for data converters. Data converters are paramount in mixed systems, since the quality of the signal conversion can be crucial to the overall performance of the system. In the past, converters were stand–alone ICs, and hence the processing technology most appropriate for their implementation could be selected. Design techniques based on Bipolar and JFET were preferred, as these provided better analogue performance in terms of accuracy, noise and speed. Now, in the case of mixed–signal ASICs, data converters are an integral part of the system, and therefore the choice of technology cannot be dictated by the requirements of the converters alone. Instead, global considerations such as area–consumption and power–dissipation have to be carefully accounted for. Indeed, the decision on technology is often restricted to CMOS, a technology not optimum for convertors. Chapter 6 gives a general introduction to data converters. It defines and illustrates the various static and dynamic performance parameters which characterise data converters, discusses basic design considerations and gives examples of various implementation techniques. The chapter concludes with a discussion on the speed and resolution limits of data converters.

The next two chapters concentrate on advanced and some non–conventional solutions to the data conversion problem. Chapter 7 describes Over–sampling Converters. For low to medium signal bandwidth applications Over–sampling Converters offer an attractive solution to the requirements of high performance, while not demanding special tuning of digital processing technology to make good analogue components. Over–sampling Converters owe their designation to the fact that the ratio of the adopted sampling frequency to the Nyquist rate, commonly referred to as the over–sampling factor, must be high. Over–sampling factors of several hundred are not rare in high–quality digital audio applications. The main beneficial effect of the over–sampling technique is that the power of the error associated with the quantisation noise is spread out over a frequency band much wider than that of of the specific signal of interest. The out of band quantisation noise can be filtered by purely digital means while not affecting the input signal. Thus the quantisation noise power is reduced and the effective resolution of the converter is increased. This chapter first provides an intuitive description of over–sampling conversion, followed by a discussion of the basic sigma delta scheme as well as some of most well known derivatives.

Another popular A/D conversion technique, based on the successive approximation algorithm, is addressed in chapter 8. Such converters require more accurate analogue circuitry than their over–sampling counterparts, but since they usually need simpler digital circuitry they can lead to potentially more economically attractive solutions with respect to silicon area and power consumption. In successive approximation A/D converters the conversion cycle is carried out in a number of steps equal to the required number of bits, and hence the conversion speed reduces as the resolution increases. Specialised calibration circuitry can be added to the basic architecture to push the bit–resolution from the typical untrimmed technology limit of 8 to 10 bits up to as high as 16 bits. Self–calibrated successive approximation converters can therefore cover a broad range of product specifications with respect to both conversion resolution and speed.

The complimentary digital to analogue (D/A) function is also addressed in chapter 8, where some specialised algorithmic converters are considered to achieve specific target specifications for power, area and added functionality. Such functionality concerns not only the traditional programmability of the resolution and speed of conversion, but also a further dimension of filtering associated with the conversion function. This is implemented by a mixed analogue–digital building block, which finds interesting application areas in complete D/A interface systems, such as in the case of data transmission telephone lines. In conjunction with the description of the architectures and circuit techniques of the various A/D and D/A converters presented in the chapter, some integrated design examples are also illustrated by means of prototype ICs.

The next part of the book, extending from chapter 9 through to 12, concentrates on the design and applications of mixed–signal ASICs. Chapter 9 focuses on a commercial BiCMOS standard cell library providing great flexibility for designing mixed analogue–digital IC applications. The chapter presents brief descriptions of the BiCMOS process employed, the cell libraries available and the CAD tools and systems, and finally provides two application examples. The success of designing with this cell library as with any other, depends crucially on the CAD system(figure 1.6), and in particular on the simulators available. The solution adopted here is a tight coupling between a multi–level digital simulator (Mozart) and an advance analogue modelling and simulation system, (ELDO/FAS [3]), which provides a powerful and flexible modelling language and can efficiently perform circuit and behavioural level simulation.

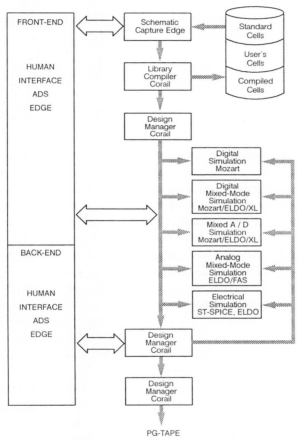

Figure 1.6 ADS environment
(XL and FAS are options of the ELDO simulator)

In order to discuss certain practical aspects of mixed analogue–digital design, chapter 10 deals first with the pernicious effects of process variations and then introduces appropriate design techniques to overcome these effects. In particular, the chapter discusses methods of layout to improve element matching, and overcome problems associated with boundary effects , noise, power supply and substrate coupling. Two optimised layout structures are shown at the end of the chapter.

The next two chapters are concerned with application examples from various market segments. Chapter 11 starts by giving brief descriptions of some chips designed by a silicon house for external customers. It then discusses four case–studies from the telecommunications and automotive markets to illustrate some basic techniques, design considerations and results. Chapter 12 is devoted to mixed–signal ASICs in one very important area, that of high definition television. New standards and formats for transmission have been defined and agreed and the next major challenge is to develop efficient circuit solutions to launch this service in the consumer market. Strategic system aspects of HDTV are discussed to indicate the areas of need of integrated circuit solution. The chapter also provides one example of such a "video processor", implemented in an advanced CMOS technology

The final part of the book deals with CAD and supporting tools. Chapters 13 and 14 discuss simulation where a fundamental problem exists in that these ASICs use different signal and circuit level abstractions. Whereas the logic simulator sees digital circuits represented in terms of gates and processing of binary valued signals, the circuit simulator models circuits in terms of much more detailed electrical components and computes continuous signal values. There is a great need to develop combined simulators which can handle both types of signal and circuit abstraction. Chapter 13 discusses various factors in the development of such mixed–mode simulators. However, there is also need for faster higher–level, or "behavioural", mixed–mode simulators to perform system simulation. This area is discussed in chapter 14, which deals with macro models, special–purpose languages and event–driven analogue models. The chapter concludes with a discussion of some future developments and gives some hints on extensions of VHDL for analogue.

Chapter 15 describes the important area of testing mixed analogue–digital circuits. For mixed–signal, as for digital ASICs, the testing of the final product is largely left to the designer himself. ASIC vendors may perform some preliminary tests before packaging, perhaps by using a designer's test pattern. However, the design of test routines is very much the responsibility of the designer. Although concepts such as design for testability have emerged, these have primarily been in the digital field, where many tools and techniques have been developed. However, the diversity of analogue functions available means that performance can be measured in many different ways, and it is very time–consuming to go through a process of measurement of analogue performance functions. While arguing that there is no universal solution, the chapter also discusses a number of approaches, such as partitioning and use of DSP (Digital Signal Processing), logical decomposition, and the use of artificial– and knowledge–based systems.

Chapter 16, "Towards high–level synthesis", is the first of three chapters on synthesis of mixed analogue–digital or pure analogue circuits for mixed–signal applications. While synthesis tools are becoming available for digital circuits, as with many other CAD tools the mixed–signal and analogue counterparts are not as fully developed. Earlier systems allowed limited analogue synthesis based on a few standard topologies, dedicated to circuits with regular structures, like SC circuits and some classes of data converters. In recent years several knowledge–based systems have been proposed aimed at rendering automated design of analogue digital circuits more efficient and capable of adapting to rapid changes in technology. This chapter describes a particular computer–assisted design environment, which employs high–level symbolic tools capable of analysing and supporting the synthesis of system architectures defined in terms of specific functional blocks, whose architectures in turn are defined in terms of circuit components.

Chapter 17, on the other hand, is more specific and deals with synthesis of data converters. The final chapter, "Automated analogue design", reviews the whole area of analogue compilers and module generators with a view to efficiently automate the design of analogue circuits for mixed–signal applications.

1.3 References

[1]   Allen, P. E., Chapter 18, "Future of Analogue Integrated Circuit Design.", in "Analogue IC design: the current mode approach", Toumazou, C., Lidgey, F.J., and Haig, D., (Eds.), pub. Peter Peregrinus, 1990

[2]   Morris, S. G., "CMOS Cell based designs for Mixed–Signal applications.", Silicon Design, Nov. 1–2, 1989.

[3]   ELDO User Manual, "ANACAD Computer Systems" Ulm, Germany

Chapter 2

Technology and Modeling aspects of an advanced BiCMOS ASIC process

P. Walker and N. Morris
LSI Logic (Europe) Limited

2.1 Introduction

One of the main trends in silicon technology at present is the integration of CMOS and bipolar processes to form a single production process (BiCMOS) – this growth in BiCMOS is supported by marketing forecasts which show its share of the total integrated circuit market increasing into the 1990s. The impetus for this largely comes from the expected proliferation in mixed analogue digital integrated circuit. Many of these BiCMOS processes are very complex and contain transistor structures that rival some of the more advanced structures used in conventional bipolar or CMOS technologies, with the result that complex, high performance circuits are now being produced. Like any advanced process, however, this can present significant problems, since complex chips require complex software tools in order to predict their performance accurately and speedily – clearly a requirement for fast turn–around ASICs (Application Specific Integrated Circuits).

This chapter describes LSI Logic's analogue BiCMOS technology and attempts to highlight the importance of modelling and simulation at all stages of chip manufacture and development – in particular, process and device modelling are discussed as ways to improve the overall manufacturing process of a designed circuit. At its lowest level this means modelling the manufacturing steps to enable a more stable process to be developed and manufactured, whilst at its highest level it means statistically modelling the single transistor performance to enable the reliable prediction of circuit performance. This is especially crucial in an industry where there is a large cost attached to a batch of silicon wafers – for example, if a final device (prior to packaging) costs $2, there are 500 devices on one silicon wafer and there are 24 wafers in one batch, there is a potential batch revenue of $24000. Clearly, any improvements in yield to achieve the full potential value are important, especially when the profit margin for some designs can be quite narrow.

2.2 LSI Logic analogue BiMOS technology

2.2.1 Background

The analogue BiCMOS technology from LSI Logic originally started development at Bell Northern Research in Harlow, UK (formerly STL) in the early 1980s when a programme was initiated to combine poly-silicon emitter bipolar transistors (BJTs) with an advanced CMOS process. In 1987, LSI Logic and STC signed agreements giving LSI the rights to the process; the process was subsequently developed by LSI in Sidcup, UK between 1987 and 1990.

The first LAD (LSI Analogue–Digital) BiCMOS product was a series of ASIC arrays called the LAD310 series (Figure 2.1).These are divided into a digital area comprising a 'sea of gates' and an area containing analogue elements such as BJTs, MOS transistors (MOSFETs), capacitors and resistors (see Figure 2.2),allowing the combination of digital and analogue functions on a fast turn–around ASIC. Careful consideration has been given to electrical isolation between the two regions to ensure no interference or crosstalk problems occur.

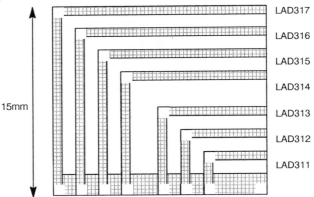

Figure 2.1 The LAD310 Series of arrays

The main market area at which the LAD product has been aimed is telecommunications, because of the product's high speed performance, low cost and fast turn–around times. High frequency performance is achieved as the core BJT used in the array has a cut–off frequency (f_T) of 5GHz; low cost is achieved by low process complexity; fast turn–around comes courtesy of an array product. In fact, the product is very suitable for any application requiring high frequency operation (several 100MHz) and more than 5000 logic gates. The only realistic alternative for these applications is often multi–chip modules.

Technology and Modeling issues for an advanced BiCMOS ASIC process

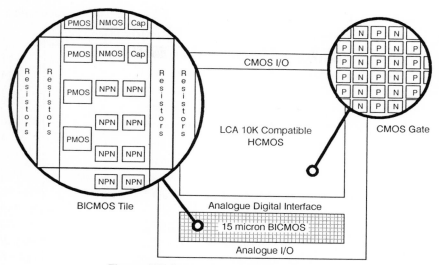

Figure 2.2 Construction of an LAD array

The BICMOS process has shown its versatility more recently since it has been successfully used for an LAD ECL design and a full–custom RF design, operating at frequencies over 1GHz.

2.2.2 Process technology

One of the key aspects of the technology is that the CMOS performance is fully compatible with LSI Logic's LCA10K HCMOS process, allowing it to make full use of existing digital library functions. This is important since the existing library functions have been thoroughly proven on many HCMOS designs. This did, in fact, require a good deal of further engineering work, since the HCMOS process and the BiCMOS process are built on different starting materials.

The doping profiles for the CMOS process have also been engineered so that they are compatible with those needed for high performance BJTs – the result of this is that only three extra masks are required to produce the full BiCMOS process. When this is combined with the patented 'collector shunt' structure for the reduction of collector resistance, the whole process is not a great deal more complex to manufacture than the original CMOS process – this is highlighted by Table 1 which compares the two basic process flows.

Table 2.1 Comparison of process flows for CMOS and BiCMOS processes

CMOS Flow	BiCMOS Flow
Well formation	Well formation
Island definition	Island definition
	Shunt implant
N-Field implants	N-Field implant
Field oxidation	Field oxidation
Threshold voltage adjust implant	Threshold voltage adjust implant
Gate oxidation	Gate oxidation
	Polysilicon 1 deposition
	Base implant
	Emitter cut
Polysilicon deposition	Polysilicon 2 deposition
Polysilicon dope	Emitter implant
Polysilicon definition	Polysilicon definition
Contact implants (N+ and P+)	Contact implants (N+ and P+)
First layer dielectric deposition	First layer dielectric deposition
Planarization	Planarization
Contact cut	Contact cut
Metal 1 deposition and definition	Metal 1 deposition and definition

There now follows a step–by–step guide to how the complete devices are manufactured, including cross–sections of the process architecture and typical mask layouts. A diagram is associated with each of the following sub–sections, starting with:

2.2.2.1 Well formation (Figure 2.3)

The starting material is lightly–doped p–type (P–) silicon on a heavily doped p–type (P+) base layer. The presence of a heavily doped layer is a standard method used to suppress unwanted parasitic transistor action (latch–up) between MOSFETs.

As a first step, the silicon is oxidised and the n–well regions are defined in photoresist using the mask shown in Figure 2.3a. Implantation using phosphorus is then performed, before n–well processing is completed by a thermal cycle ('drive') in a furnace to diffuse the phosphorus.

P–well definition follows using a mask that is the complement of the n–well – it is generated by oversizing the n–well features and then reversing the field (see figure 3a). Following p–well definition and a boron implantation step, a furnace diffusion forms the final p–well profile.

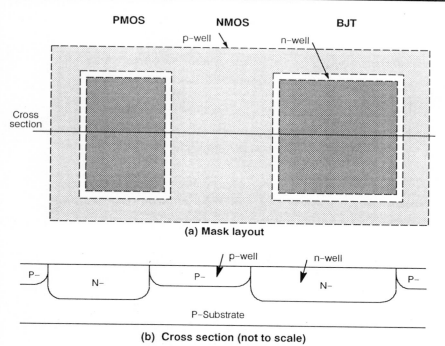

(b) **Cross section (not to scale)**

Figure 2.3 Well formation

The two well profiles are shown in Figure 2.3b.

2.2.2.2 Island definition and field region implants (Figure 2.4)

The next stage in the process is to deposit silicon nitride on top of a thermally grown layer of silicon dioxide. The island (active device area) mask is then defined (in photoresist) and the silicon nitride etched to form the nitride 'caps' shown in figure 4b – the areas still covered by nitride will not oxidise and will become the active device regions whilst the areas between them are termed the 'field isolation regions'.

Next the shunt mask is defined to create a region for an n–type implant in the field region between collector and base islands. The implant and subsequent drive conditions for the shunt region have been carefully chosen to provide the optimum electrical characteristics for the BJT. This was done using a combination of careful measurement, statistical experimental design and process modelling.

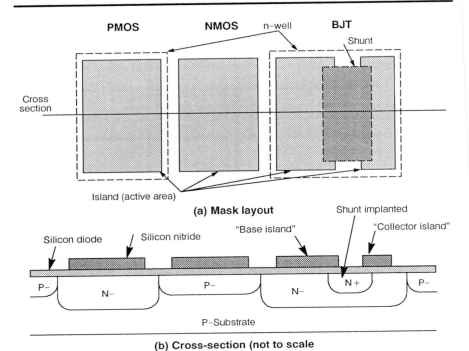

(a) Mask layout

(b) Cross-section (not to scale

Figure 2.4 Island defination and field implants

After shunt processing, the field regions receive a boron implant (not shown) to prevent leakage across these regions due to the turning on of parasitic MOSFETs. This implant is termed 'n–field', the mask being generated from the n–well mask, in a similar manner to the p–well mask.

2.2.2.3 Field oxidation – Island formation (Figure 2.5)

Field oxidation simply consists of a furnace treatment that is performed to grow silicon dioxide in the field regions of the wafer to provide isolation between active areas. Oxide growth is inhibited in areas that are capped with nitride, but there is some growth at the edges of nitride regions that produces a characteristic 'bird's beak' shape on the edge of the oxide profile.

Following oxidation, the nitride layer is removed.

2.2.2.4 Gate oxidation and base implant (Figure 2.6)

The next stage of processing is very critical to final MOSFET performance. First of all there is an implant to produce the required threshold voltages for the MOSFETs (not shown). This is followed by the gate oxidation which

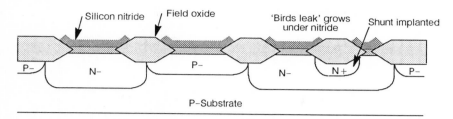

Figure 2.5 Field oxidation (not to scale)

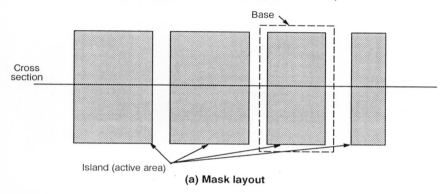

(a) Mask layout

(b) Cross-section (not to scale)

Figure 2.6 Gate oxidation and base implant

grows a thin (25 nm) layer of silicon dioxide to act as the dielectric between the gate contact and the silicon. The pre–growth cleans and the actual growth of this thin oxide are very important, as they are critical for the overall performance of the MOSFET – threshold voltage, drive current, breakdown voltage and leakage are all affected by this step. This is followed by a poly–silicon deposition which protects the gate oxide during the subsequent processing steps.

Finally, as shown in Figure 2.6, the base region of the BJT is implanted with boron and diffused.

2.2.2.5 Emitter formation (Figure 2.7)

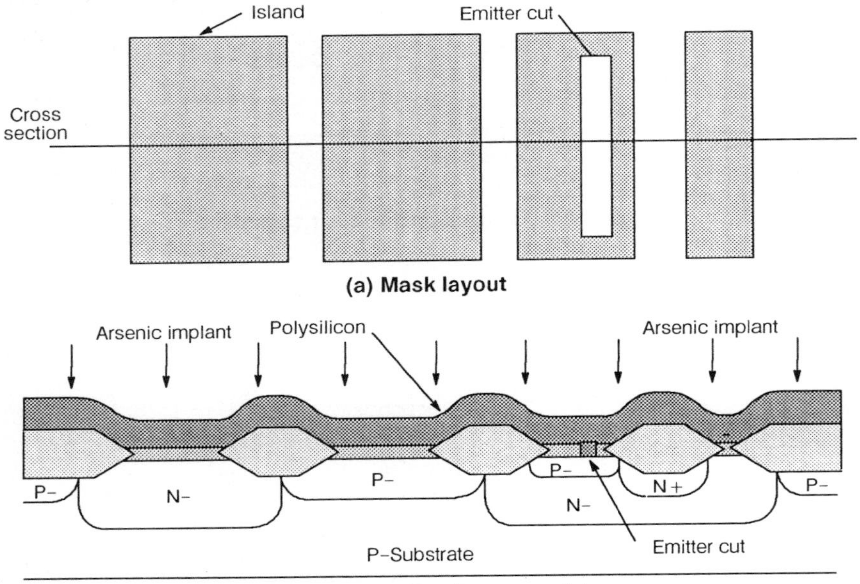

(a) Mask layout

(b) Cross-section (not to scale)

Figure 2.7 Emitter formation

This stage in processing is a critical stage in fabricating the BJT. The basic steps involve definition of the emitter using the mask shown in Figure 2.7a, an emitter etch through the thin poly-silicon with arsenic and oxide layers (formed in previous gate processing), the deposition of a thick layer of poly-silicon and finally the implantation of the poly-silicon. The critical step is the clean, prior to the deposition of the thick layer of poly-silicon. Control of this clean is important, as the presence of a reasonable thickness of oxide ($\sim$2nm) can produce a very large BJT current gain (500 or more). This occurs because the presence of an oxide layer inhibits the back–injection of holes from the base region into the emitter region, which in turn reduces the base current (current gain is ratio of collector to base current). Since this value of current gain is too high for most applications, removal of the interfacial oxide is important.

Figure 2.7b shows the emitter cut through the thin oxide and the poly-silicon implant (emitter implant).

2.2.2.6 Gate definition (Figure 2.8)

The next main processing step is the definition of the poly-silicon layer, which defines the gate size and the emitter contact. Following the gate etch, the contact implants are performed – phosphorus implants for the N+ contacts to NMOS (n–channel MOS) sources/drains and BJT collectors and a boron implant for the P+ contacts to PMOS (p–channel MOS) sources/drains and BJT bases.

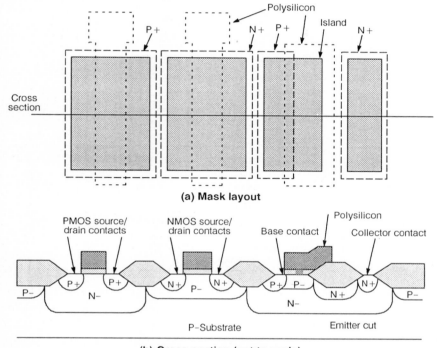

(a) Mask layout

(b) Cross-section (not to scale)

Figure 2.8 Gate definition

The relevant masks are shown in Figure 2.8a and a profile is shown in Figure 2.8b.

2.2.2.7 Contact cut (Figure 2.9)

The next step involves the deposition of first layer dielectric (doped silicon dioxide) over the surface of the wafer which is then given a heat treatment to produce a planar (flat) surface on which to deposit metal – if the surface is not planar, cracking and/or thinning of metal can occur. The heat treatment also diffuses arsenic from the poly-silicon into the silicon base region to form the emitter. Following this, the contact mask is defined (Figure 2.9a) and the contacts etched through the oxide down to the silicon surface.

Figure 2.9b shows the final profile prior to metallization.

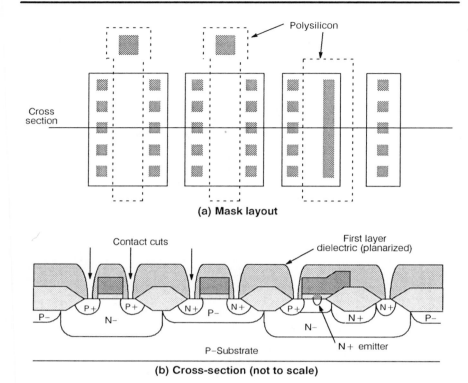

(a) **Mask layout**

(b) **Cross-section (not to scale)**

Figure 2.9 Contact cut

2.2.2.8 Metal 1 definition (Figure 2.10)

The final step shown in these diagrams is the deposition and definition of the first layer of metal. This metal layer is a combination of aluminium, silicon and copper – the aluminium dominates the combination (>98%) and is present for its high conductivity; the silicon is included as it helps prevent 'junction spiking' where aluminium reacts with silicon and 'spikes' can grow into the silicon surface and possibly short out the contact implants; the copper is present as it helps prevent electromigration (metal diffusion when stressed with high currents).

The complete BiCMOS process includes two layers of metal.

The final profiles for the PMOS and NMOS FETs and the BJT are shown in Figure 2.10b.

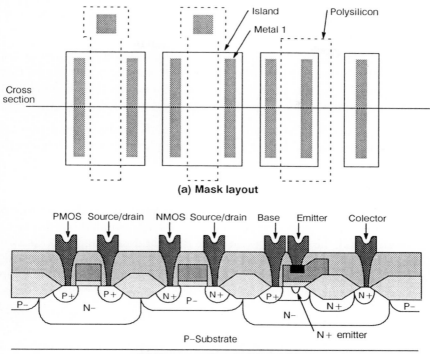

(a) Mask layout

(b) Cross-section (not to scale)

Figure 2.10 Metal 1 definition

2.2.3 Summary

Whilst this section has been specific to LSI Logic's analogue BiCMOS technology, the processing techniques are representative of many similar CMOS, bipolar and BiCMOS processes. The main area in which this process differs from others is in the BJT shunt implant for the reduction of collector resistance. Most other comparable processes use a high conductivity 'buried layer' to reduce collector resistance, and although this can produce a slightly lower resistance than the shunt structure, buried layers are more costly to manufacture and make a process significantly more complex. Therefore, the shunt structure is ideal for ASIC applications where fast turn–around and high yield are issues, since it maintains the relatively low complexity associated with the CMOS process.

The next section provides a brief overview of the process simulation tools which have been used to develop LSI Logic's analogue BiCMOS process.

2.3 Process simulation

2.3.1 Background

In order to continually develop new processing technologies (such as the BiCMOS process described previously), and in order to accurately predict the behaviour of discrete devices and circuits, much use has been made of simulation and modelling tools for both process and device design.

In this section, a brief overview of some trends in device and process simulation is provided, along with specific examples of process modelling as applied to the BiCMOS technology.

2.3.2 Modeling

In the previous section, the basic building blocks of the BiCMOS process were described in general terms. However, for the detailed development work (calculating the doping profiles, the implant doses, etc.), a calibrated process model was developed, using commercially available software. This was then used to investigate any potential changes to the process, to develop improvements to the device performance and to study future modifications.

Currently, there exists a wide range of simulation tools to assist in this type of process development. Initially evolving from some relatively simple models (capable, for example, of predicting simple ion–implanted and diffused impurity profiles), these tools have now reached the stage where they can model thin film deposition, photo–lithographic (mask) processing, etching and oxidation in a wide variety of processing ambients.

Although there are an increasing number of process simulators being written for the commercial sector, the most commonly used are the still the SUPREM[1] family of models, originally developed at Stanford University, California [1] and subsequently marketed on a commercial basis by a number of software vendors.

2.3.2.1 One–dimensional process modeling

For simple one–dimensional forms of analysis, where depth profiles are the main consideration, the process simulator SUPREM3 is now a widely used, well calibrated, process development tool.

SUPREM3 is a general purpose simulator, allowing most of the major effects (implantation, oxidation, diffusion, etc.) to be modelled with a degree of accuracy which is suitable for most applications. The results generally take the form of a depth dependent impurity concentration, in

[1] Stanford University Process Engineering Models

which the positions of any junctions or interfaces can be easily visualised. Inclusion of poly-silicon layers, oxides and nitrides play an integral part of the modelling capabilities and information can also be obtained on the number of active impurities present in a given layer (and the net chemical concentrations if desired).

In the case of the analogue BiCMOS process, SUPREM3 has been used extensively to model all areas of the process, from the initial n–well formation through to the final device. The poly-silicon emitter transistor is one particular aspect of the technology that has been modelled in some detail. Figure 2.11 shows a one–dimensional depth profile of a typical

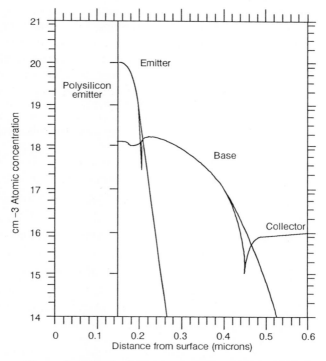

Figure 2.11 Typical Bipolar depth profile (SUPREM3)

(BiCMOS) bipolar structure, in which the arsenic profile from the poly-silicon emitter is seen to have diffused through the base region as the result of high–temperature processing. By using SUPREM3 to tailor the implant conditions, it was possible to modify this structure to give an optimized device which could be used in a much wider range of applications.

Once a calibrated process model is obtained, calculations of other key physical parameters (such as the basewidth and the peak base doping) can be easily made, and the model can also be used to give some indication of the inherent control and stability of the process.

In the latter case, it is possible to explore the effects that minor variations in processing (e.g. small changes in temperature and time) would have on the final device structure so that, in this way, one can optimize the device structure so that it becomes relatively insensitive to small process perturbations. In addition, the shunt processing conditions were also optimized using this type of analysis.

When used in conjunction with a process–based device simulator such as PISCES, which is capable of accepting a one–dimensional (SUPREM3–type) doping profile, the facility to provide direct feedback to the process designer exists, again making the development cycle much shorter.

The obvious drawback to SUPREM3 is that it only provides analysis in one dimension. However, recent years have seen the emergence of a number of two and three–dimensional simulators, some brief details of which are given in the next section.

2.3.2.2 Simulations in two- and three-dimensions

In order to provide a more detailed overview of modern device structures, recent emphasis has been placed on developing two– and three–dimensional extensions to the simple one–dimensional process simulators.

There is already a two dimensional version of SUPREM3 available (SUPREM4) and various other two–dimensional codes have recently appeared [e.g. TITAN (France) and COMPOSITE (Germany)]. However, due to the nature of some of the models used in these codes, a certain degree of expertise is required before accurate simulations are achieved.

This is especially true in cases where the structures have small lateral dimensions (e.g. for short channel MOSFETs). In these cases, the lateral diffusion of impurities (for example, under the gate region of the device) may be enhanced under certain processing conditions and the resulting impurity profile then becomes a critical factor for accurate device design. Care should therefore be taken to ensure that the simulation conditions (e.g. the type and accuracy of the models used) reflect the degree and complexity of the analysis being performed.

Nevertheless, despite the additional care required in use (and the additional computing resources required), two–dimensional simulators of this type do provide a worthwhile tool for studying a much wider range of processing conditions and for developing a greater understanding of the physical processes involved.

As an example, simulation of oxidation where there is a nitride mask present (i.e. the formation of a 'bird's beak' profile) is an easily visualised process in two–dimensions (see Figure 2.12), as is the lateral spread of

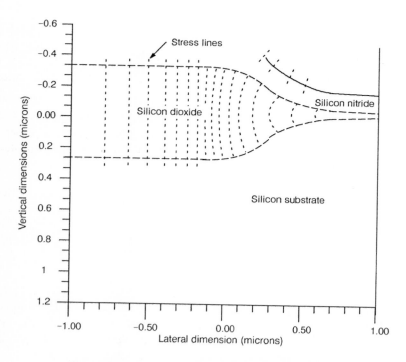

Figure 2.12 Typical 'Bird's Beak' profile under

implanted and diffused impurity profiles. The capability to simulate simple forms of photo-lithography and masking is also possible in a two–dimensional simulator such as SUPREM4, although there are a now a much wider range of commercial simulators which have specialist applications in process topography and thin film processing (see, for example, SSAMPLE for 2–D topography, SOLID and DAVINCI for 3–D simulations).

2.3.3 Summary

There is little doubt that process simulation tools are becoming increasingly more important to assist in the development of new processes. As the models become more accurate and comprehensive, it should be possible to design entire technologies and predict device performance, before a significant amount of silicon is committed to processing.

2.4 Device modeling for circuit simulation

2.4.1 Background

So far this chapter has dealt with a specific BiCMOS technology and some of the process modelling issues. However, at some stage it is necessary to predict circuit performance for circuits with hundreds of thousands of transistors. The manual prediction of such circuits is practically impossible, so there is a need for automated techniques to give a more accurate and repeatable prediction of performance. This is also driven by market needs where larger scales of integration, increasing significance of 'time to market' and narrow profit margins mean that the need for 'right first time' circuit designs is very important.

The most obvious approach might be to take a process–based device simulator such as PISCES (mentioned earlier) which can predict individual transistor performance and subsequently use it to predict the performance of a whole range of transistor sizes and geometries. However, in order to achieve this, these simulators would require far too much computing power and so are not feasible for full circuits.

Therefore, in most cases 'compact' device models are used, which can predict device performance very rapidly and so are suitable for large circuits. These models have, in fact, been available for much longer than their equivalent process–based device models. Compact models aim to relate physical device parameters to device characteristics by using a set of input parameters that are extracted from transistor characteristics – this is opposed to process–based device models which use process parameters (temperature, implant energies, diffusion times etc.) to predict device performance. The former is clearly more attractive to the circuit designer who may have no knowledge of process variables, but will probably be aware of the characteristics of the device. The most popular circuit simulators used today are mainly based on the original SPICE programme [2] from Berkeley University which is briefly described in the next section.

2.4.2 Circuit simulation with SPICE

SPICE (Simulation Programme with Integrated Circuit Emphasis) allows complex circuits to be configured from the individual circuit components and then simulates the full circuit using compact models. SPICE allows the use of most circuit components (e.g. BJTs, MOSFETs, JFETs, diodes, resistors, capacitors) and has a range of built–in models for semiconductor devices.

Three main types of analysis are available in SPICE – D.C. analysis, A.C. small signal analysis and transient analysis.

The D.C. analysis is important because, as well as predicting D.C. operation, it is also used to determine the operating point of a circuit prior to both A.C. analyses and transient analyses. Therefore, this analysis should be as accurate as possible so that the final results, from any type of analysis, are correct.

The A.C. small signal analysis is performed over a frequency range specified by the user. A noise analysis can also be performed, as part of an A.C. analysis, to model mean–square noise currents and flicker noise.

Transient analyses perform time specified sweeps. Unlike the previous types of analyses, time– dependent components can be studied.

However, SPICE is not without its drawbacks – for example, convergence of some analyses, the speed of solution for complex circuits and the temperature model employed. For these reasons there are now several competitors currently on the market that have moved away from SPICE – it should be noted, however, that many of these competitors still use the same compact device models.

2.4.3 Compact device models

There are two aspects to accurately modelling a circuit transistor for compact models: the first is the accuracy and efficiency of the transistor model; the second is the accuracy and ease of extraction of the model parameters. The former aspect determines how the simulated transistor behaves: for example, the model may not account for junction breakdown, in which case the effect cannot be simulated. The latter aspect determines the accuracy of the simulation (within the constraints of the model): that is to say, even with a perfect transistor model, inaccurate or inappropriate parameter values can easily cause poor simulations.

This section deals with the device models, giving information of the basic public domain BJT and MOSFET models; the next section deals with parameter extraction and discusses some of the difficulties involved for current technologies. Only MOSFETs and BJTs are considered because they form the basis of most integrated circuits today and present one of the greatest challenges for accurate device modelling.

Device models can be classed by their 'philosophy'. Some models try to wholly implement the device physics that occur in a given device. This has the advantage that the model can be directly related to real effects. The main disadvantage is that the model can become cumbersome and is no longer 'compact'. Other models are based on curve–fitting to the device characteristics. These models tend to be fast, but can suffer when device geometries reduce and other physical effects become dominant, which cannot be accounted for in the curve fitting algorithms.

Another distinction between models is whether they are designed for scaling. Some models try to predict the performance of transistors of all sizes with one set of input parameters, whereas others require a different set of parameters for each device size. Whilst the former is easier to manage, because it only has one set of input parameters, more and more difficulties are being encountered as device sizes shrink; therefore, many recently published models use the latter approach for both BJTs and MOSFETs.

2.4.3.1 MOSFET Models

MOS technology has dominated the semiconductor market for many years and technological and manufacturing advancements have generally been aimed at this market. As a consequence, MOSFET models have received a great deal of attention in order to keep pace with technology. Many semiconductor companies have their own proprietary models to tackle the problems of small geometry devices – however, these are not generally public domain.

The public domain models available in SPICE are termed level 1, level 2 and level 3. These three models are all designed to be scalable over different device sizes and as such only require one set of input parameters.

Level 1 is the simplest model and is computationally the fastest and so tends to be used if a rapid prediction of circuit performance is required. It is sufficient to produce the basic I–V (current–voltage) characteristics of a MOSFET (see Figure 2.13) and is largely based on physical effects. However, it does not model many second–order effects which means that it is not accurate for most modern–day MOSFETs which utilise short

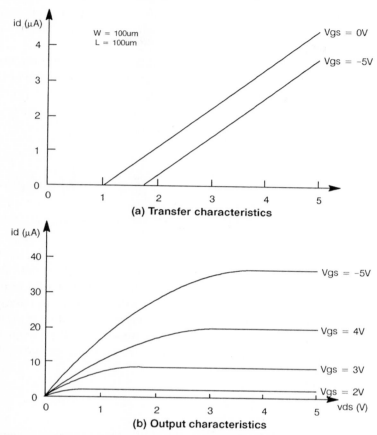

Figure 2.13 Basic I-V characteristisc for MOSFET using level 1 model

channels and very thin gate oxides. For example, the 'rule of thumb' that the point of saturation for a MOSFET is the difference between the gate voltage and the threshold voltage is based on the level 1 model – unfortunately, this rule does not hold for most modern–day MOSFETs. The full model has twenty–five input parameters.

Level 2 is an extension of the level 1 model and was proposed by Meyer [3]. The model attempts to stay with physical effects and accounts for many short channel effects, such as small length/width effects, channel–length modulation, sub–threshold conduction and velocity saturation of carriers. However, it has the major problem (especially for analogue design) that there is a discontinuity between the linear and saturated regions of operation, which can lead to convergence difficulties. The model has an extra eleven parameters over the level 1 model.

Level 3 is also an extension of the level 1 model and accounts for many second–order effects, like the level 2 model above. Whilst it tends to be more accurate than the level 2 model, it can still suffer from convergence difficulties, because the linear and saturation regions of operation are described by different equations. The model was proposed by Dang [4] and relies on a large amount of empirical formulation, particularly for the small size effects on threshold voltage. The model can be used down to about 1.5µm geometries, but will probably not scale over the full range of geometries – the answer is to produce two or three sets of parameters for different sets of channel lengths. For example, a model for 1.5 to 3µm and another for 3µm and above. There are an additional nine parameters over level 1.

Owing to the short–falls of the above models, Berkeley University have recently developed a MOS model called BSIM [5]. This model has a basis in the level 3 model described above, but has device parameters that are extracted from different geometry devices and then merged to provide a scaling model – however, there are still potential difficulties regarding discontinuities between the linear and saturated regions of operation. The model is usable for devices below 1mm.

2.4.3.2 BJT Models

In contrast to the MOSFET case, bipolar modelling has not received a great deal of attention over the past decade or so. However, since bipolar technology has recently become far more important, especially with the emergence of production BiCMOS processes, bipolar modelling is now becoming an important issue.

The dominant bipolar model used today was originally proposed more than twenty years ago by Gummel and Poon [6]. The model is physical and is based on the integral charge in the emitter, base and collector – it explains some of the second order effects in BJTs, such as low current effects and some high level injection effects. There are various short–falls in the model though, when related to a real device – these include the variation of output conductance with bias, some high current effects in the collector and base, bias dependence of the transit times, etc. The full model contains forty parameters.

Another BJT model that is still used is that of Ebers and Moll [7] which was a fore–runner to the Gummel–Poon (G–P) model. It is not as accurate, nor does it have such a strong physical base, as the G–P, but it is sometimes used to produce fast results (compare with level 1 MOSFET model).

2.4.4 Parameter extraction

2.4.4.1 Equipment requirements

There are generally three distinct equipment sets that are required to extract the full set of parameters for BJT or MOSFET models – d.c., A.C. and capacitance test equipment.

The basic d.c test needs are for accurate voltage and current sources and meters – 'SMUs' (source– measure units) are popular because they provide integrated sources and meters for current and voltage (see Figure 2.14).

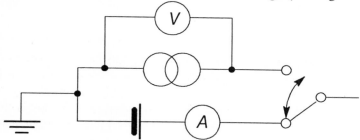

Figure 2.14 A typical SMU

This equipment enables the majority of the D.C. parameters to be extracted. Also, it is normal to automate D.C. parameter extraction where possible to produce statistically valid and repeatable parameters. This requires a computer for control, a wafer prober and a switching matrix to enable any D.C. source/measure to be applied to any pin on the wafer prober.

Capacitance parameters are generally extracted using a capacitance bridge. These tend to be used at frequencies of either 100 kHz or 1 MHz and extract data for the capacitance associated with the various device junctions – these values of capacitance are often less than 0.1pF and so automation is often difficult.

Equipment to extract A.C. parameters is the most specialist and is mainly applied to BJTs in cases where high frequency performance is most important. The actual equipment used will vary, depending on the extraction methods chosen, but will typically consist of a network analyser with an s–parameter test set. Also, since many modern high speed BJTs have cut–off frequencies in excess of 5GHz, specialist wafer probes or packages may be required – this makes complete automation extremely difficult.

In addition to the above equipment, there should also be the capability to extract parameters over a range of temperatures – therefore, there may be a need for thermal chucks and package holders to be used in conjunction with the measurement equipment above.

Various other methods are also used for parameter extractions and can require noise measurement equipment and time–based equipment.

2.4.4.2 Choice of extraction

Even the perfect equipment set and the perfect model are worthless if incorrect parameter extractions are performed. Part of the problem concerns parasitic components in the extraction – as device geometries shrink, greater accuracy tends to be required for measurements and so parasitic resistance, capacitance and inductance can be very important.

Another major problem is that many of the model parameters represent physical properties of a device, so there is a great temptation to perform an extraction based purely on the physical effect. However, as will be demonstrated, this is not necessarily the exact definition of the parameter in the model; therefore, extractions based on a physical effect will give a parameter value that does not necessarily match its function in the model. A simple example of this is shown below for the G–P model.

Figure 2.15 shows a simulated 'Gummel Plot' for a single BJT – key parameters used for the simulation are given in Table 2. In Figure 2.16, a 'standard' parameter extraction has been applied which performs curve fitting to the collector and base current lines and extracts the five parameters (see table 2).

Table 2.2 Simulation parameters and their extracted equivalent

Key simulation parameters		Extracted parameters – 'standard' extractions	
(Gummel–Poon model)		(from 'Gummel plot')	
BF	100	BF	73.4
IS	35E–18A	IS	36E–18A
NF	1.00	NF	1.01
ISE	300E–18A	ISE	510E–18A
NE	1.5	NE	1.56
VAR	3.0V		

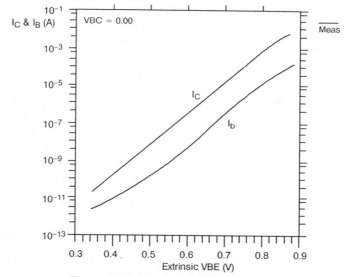

Figure 2.15 Simulated Gummel plot *(vbc =0)*

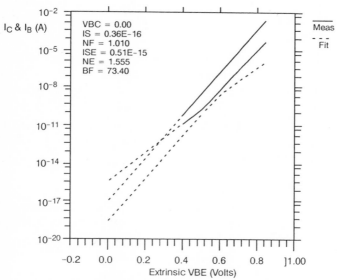

Figure 2.16 Standard extraction form Gummel plot

For this example, consider, the current gain parameter (BF) which is extracted as the ratio of collector to base current – the physical current gain (β_F). By consideration of the G–P model equations, the following relationship is derived for β_F in terms of β_F, neglecting high and low current effects:

$$BF = bF/(1 - vbe/VAR) \qquad \boxed{2.1}$$

where vbe is the internal voltage across the base–emitter junction.

The major problem is the VAR parameter which was much higher for older technologies and so had little effect on the extracted parameters. However, for more recent technologies, VAR can be quite low and so although BF was approximately equal to bF in old technologies, that is no longer the case – for example, with VAR equal to three, there is a thirty percent modulation of BF. Similar analyses can be applied to the other extracted parameters, so clearly there is a need to account for VAR in the extractions if meaningful model parameters are to be obtained.

The most relevant extraction routine for a particular parameter is determined by two criteria: (i) the technology or device under test – for example, a more advanced technology is likely to require more sophisticated extraction techniques than an older one; (ii) the accuracy required in the final simulation – if it is determined that there is only a need for fifteen percent accuracy in the final model, then less accurate extractions can be employed.

If possible, it is convenient to use simulated extractions to assess extractions (as shown in the BF example) as this can save a significant amount of time. Using this, a simple step–by–step approach can be employed for parameter extraction which involves simulating extractions first to check their accuracy. Of course, consideration should be given to the importance of the parameter as it may not influence the characteristics significantly for the particular technology under test.

2.4.4.3 Statistical models

There is need to maximise yield for ASICs as this allows more competitive pricing of the final product; it is, therefore, important for circuit designers to know the process variations with which they are working. For example, if a designer is unaware that a manufacturing process allows a 20% variation in the gain of a transistor, he may design a circuit which is very sensitive to gain – consequently, the circuit yield will be low.

The way to tackle this problem is to produce model parameters that describe the manufacturing variations inherent in a process. In order to do this successfully, parameters must be extracted from many batches of wafers to allow the data–basing of the parameters – this will then allow a statistical analysis of the results to check the variation (standard deviation) of each parameter. From this data, so called "best" and "worstcase" parameter sets can be generated.

However, the generation of "best" and "worstcase" models is not trivial. Firstly, it is not good enough to take all the low parameter values and turn them into a worstcase model this will not represent a real transistor. Secondly, even intelligent choice of the low/high parameter values may still not represent a circuit transistor – very often this results in simulations that are too pessimistic. Thirdly, it is not easy to classify "best" and "worstcase" for many transistor applications; low gain may make one design worstcase and another design bestcase!

For these reasons, there is a move towards parameter correlation using a technique known as factor analysis [8]. This powerful technique allows any group of data to be correlated into sets, each of which has one key parameter. Relationships are formed that describe each parameter in terms of its key parameter so that only this key one need be considered. For example, for MOSFETs, the full set of parameters often correlates into one set which means there is one key parameter. Therefore, studying the variation of this parameter allows the prediction of a nominal model (mode of key parameter) and of extreme models (extremes of the key parameter: best– and worst–case). The situation is often more complicated for BJTs as the model parameters can split into two or more correlated sets; this means that at least two key parameters have to be studied, resulting in at least nine final parameter sets! If the circuit simulates successfully under all nine parameter sets, though, there should be no unnecessary yield loss due to the design.

One potential hazard to the statistical generation of model parameter sets is the use of optimization. As stated earlier, there are problems with most public domain models and with many 'standard' parameter extractions – the net result of these points is that the extracted models often do not accurately match the transistor characteristics. Therefore, parameter optimization is used to 'manipulate' parameters, within given boundaries, until the simulated characteristic matches the measured characteristic. Although this produces good simulated–measured fits, the statistical validity of parameters can be lost, unless the optimization is performed identically in each case. Furthermore, the physical link is lost between the extracted parameter and the simulation, which leads to a general loss of understanding for the model. Therefore, although optimization is often unavoidable due to short–falls in the actual model, it should be minimized as much as possible by optimization of the actual extractions and, where it is used, it should be done in as controllable and reproducible way as possible.

2.4.5 Summary

Two important areas have been covered in this section – (i) the main public domain compact device models used for circuit simulation and (ii) parameter extraction issues for such models.

(i) The public domain models currently available will model most current technologies. Most of the current problems are for BJT models, which are behind the technology due to the reasons discussed above. However, since BiCMOS technology is becoming more important, this area is currently being addressed by many companies and universities.

(ii) Parameter extractions are often not appropriate for the model they represent, as they often do not account for some second order effects which are important for many of today's technologies. This can lead to a heavy dependence on parameter optimization, with the result that extracted parameter sets can lose their physical and statistical significance. Therefore, 'standard' extractions must be improved to accurately represent the model.

2.5 Final conclusions

This chapter has covered various aspects of an advanced BiCMOS process and some of the process and device modelling issues associated with it.

Starting from an overview of the technology itself, we have presented a detailed description of the process architecture and discussed some of the modifications necessary to ensure a successful blend between analogue and digital processing techniques.

Process and device modelling have been shown to represent a means of cutting costs as they encourage the development of more stable processes and the design of higher yielding circuits, both of which are vital for most ASIC technologies. However, it is important that process and device modelling are continually improved to keep pace with technology, because many effects, which are only regarded as second and third order for current processes, will become more critical as device dimensions shrink and new technologies are introduced.

With regard to the latter, the best way forward is probably to complete a 'technology modelling circle' – see Figure 2.17. This 'circle' would allow technologies to be modelled using process–based parameters and these to

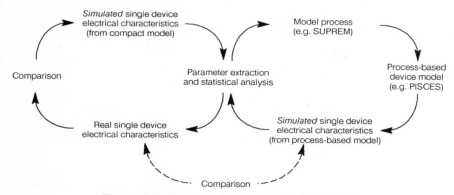

Figure 2.17 The 'Technology modeling circle'

be linked directly to simulated electrical parameters via parameter extraction – this also links directly to real devices and to compact device models. The extension of statistical analysis allows process variations to be predicted as well. If such a system was calibrated against a real technology, this set of modelling packages would provide an invaluable tool with which to develop future generations of the technology and to explore alternative processing scenarios.

Finally, the authors would like to acknowledge Sion Quinlan and Brian North for their assistance and support during the writing of this article.

2.6 References

[1]   D. Antoniadis et al, "SUPREM1 – A Program for IC Process Modeling and Simulation", Stanford Electronics Laboratories Rep. SEL–77–006, Stanford, California, 1977

[2]   L. Nagel and D. Pederson, "Simulation Program with Integrated Circuit Emphasis (SPICE)", Electronics Research Laboratory Rep. ERL–M520, University of California, Berkeley, 1975

[3]   J. Meyer, "MOS Models and Circuit Simulation", RCA Rev., Vol. 32, 1971

[4]   L. Dang, "A Simple Current Model for the Short Channel IGFET", IEEE J. Solid State Circuits, Vol. 14, 1979

[5]   Sheu, B., Scharfetter, D. et al, "BSIM: Berkeley Short–Channel IGFET Model for MOS Transistors", IEEE J. Solid State Circuits, Vol. SC–22, 1987

[6] H. Gummel and H. Poon, "An Integral Charge Control Model of Bipolar Transistors", Bell Syst. Tech. J., Vol. 49, 1970

[7] J. Ebers and J. Moll, "Large Signal Behaviour of Junction Transistors", Proc. IRE 42, 1954

[8] R. Dutton and D. Divekar, "Bipolar Models for Statistical IC Design", published in "Process and Device Modelling for Circuit Design" (F. Van De Wiele et al (eds.)), Noordhoff International Publishing, 1977

High Performance Operational Amplifiers and Comparators

M. Steyaert, W. Sansen
Katholieke Universitiet Leuven

3.1. Introduction

In analogue circuits the most used building block is the operational amplifier. In order to realize high performance circuits opamps with high open loop gains, high gain bandwidths and low power consumption are desired. The most suitable structure for high resistive loaded opamps is the well known transconductance amplifier. Nowadays it can be found in many applications such as switched capacitor filters, medical integrated circuits and telecommunication circuits. The design of mixed mode circuits requires digital and analogue circuits integrated in the same circuit. Because CMOS technology is very suitable for this purpose, high performance CMOS amplifiers and high performance CMOS comparators are discussed. Extra requirements for mixed mode circuits are analyzed in detail and the different presented structures are compared with their performances.

In paragraph 3.2, several CMOS operational transconductance amplifiers are reviewed. First a differential pair in combination with an active load, and a very simple OTA, is shortly discussed. Then the load compensated amplifier is studied and analyzed. The well known Miller compensated amplifier is then described and its performance compared with the load compensated amplifier. Recently an improved Miller compensated amplifier, the Core amplifier, has been introduced. The main advantages of this structure are reviewed and compared with other structures. Because in mixed mode designs the rejection of noise on power supply lines is extremely important, the presented structures are compared for their ability to reject this power supply noise (power supply rejection ratio).

Translation algorithms for converting analogue signals into digital signals usually require a decision building block: the comparator. Because higher A/D performances are required (more bits, higher sampling rates,...) the design of high speed comparators is discussed in paragraph 3.3. Several structures, including continuous time structures and clocked comparators are compared and discussed. Finally a high speed accurate latched CMOS comparator is analyzed.

3.2 High performance amplifiers

In the design of analogue circuits the operational amplifier (opamp) is an indispensable building block. Opamps allow the user to realize with high accuracy special functions such as inverting amplifiers, summing amplifiers, integrators and buffers. All this can be achieved with an opamp and only a few extra integrated passive circuit elements. The combination of these functions can result in very complex circuits such as higher order switched capacitor filters, telecommunication circuits and very sensitive amplifiers for medical purposes. However the first order calculations of those building blocks are based on feedback structures using an ideal opamp: an ideal voltage dependent voltage source with an infinite gain factor. Practical opamps can only reach high gain factors ($> 10 \, E + 5$) at low frequencies (< 10Hz !). Usually the gain decreases for frequencies above a few Hertz. The product of the –3dB point (the open loop bandwidth) and the low frequency open loop gain (Ao) is called the gain bandwidth (GBW). For frequencies (f) above the open loop bandwidth (f–3dB) the open loop gain is given by $|A| = Ao.f$–3dB/f = GBW/f. Because the accuracy of a feedback network structure (settling error, distortion, power supply rejection ratio, gain accuracy, bandwidth, ...) depends on the open loop gain of the opamp, the gain bandwidth is a very important parameter.

However, a practical opamp has a lot of extra non ideal performances, which can further degenerate the total performance of the system. For example noise on power supply rails can be amplified towards the opamp output. This noise on power supply rails is extremely important in switched capacitor and mixed mode designs. In such designs digital switching circuits are integrated in the same circuit as the analogue part. Due to their switching behaviour these digital circuits create large voltage spikes (up to 100mV and higher [1]) on the power supply line. To reduce the problem separate analogue and digital power supply lines are required. But even then, large spikes on the analogue power supply lines can be noticed due to the transient step responses of the opamps [1]. So the analogue circuits must reject this power supply noise in order to achieve high performance in mixed mode analogue circuits. This ability is described by the parameter power supply rejection ratio (PSRR).

An other non ideal parameter is the offset voltage. In an ideal opamp the output voltage is zero when the two inputs are equal. In a practical opamp a voltage (the offset voltage) must be applied in order to have zero output voltage. The offset voltage is quite important in the design of large analogue systems, such as higher order switched–capacitor filters. In such circuits the systematic offset voltages of the opamps can accumulate. For example, in a structure with twenty opamps, each having a systematic offset of 10mV, and a gain of 20dB can easily result in a total output offset voltage of 2V!. Of course such large values can not be tolerated at all.

In this section the performance of different amplifier structures are discussed regarding the GBW, systematic offset voltage and PSRR. Because the PSRR is defined by the ratio of the open loop gain and the transfer function from the power supply line to the output, the PSRR is directly related to the GBW of an opamp (the higher the GBW, the higher the PSRR). Therefore, in order to be able to compare more accurately different amplifier structures, the normalized parameter PSRR.f/(2πGBW) is used for frequencies higher than the bandwidth of the amplifier.

3.2.1 Differential pair with active load

A very simple and well known amplifier structure is a differential pair with an active load (OTA), as presented in Figure 3.1.

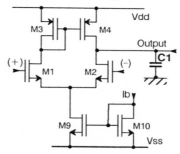

Figure 3.1 A differential pair and active load as transconductance amplifier.

The low frequency open loop gain can easily be calculated and is given by:

$$Ao = \frac{gm1 + gm2}{2(go2 + go4)} \approx \frac{gm}{2go}$$

$$\boxed{3.1}$$

Due to the capacitive load the dominant pole is situated at a frequency:

$$f1 = \frac{go2 + go4}{2\pi C1}$$

$$\boxed{3.2}$$

and hence the GBW is given by (GBW = Ao.f1):

$$GBW = \frac{gm}{2\pi C1}$$

$$\boxed{3.3}$$

Because opamps are used in feedback networks, the stability of the system must be guaranteed. This is usually described by the parameter phase margin. In order to be able to calculate the phase margin, the position of the second pole should be known. The second pole in this system is a result of the active load or the current mirror. This pole is approximately given by (neglecting the overlap and drain bulk capacitances):

$$f2 = \frac{gm3}{2\pi Cgs.2} \approx \frac{3.KPp.(Vgs - Vt)}{8\pi Cox.L^2}$$

$$\boxed{3.4}$$

with KPp the transconductance parameter, L the effective length and Vgs–Vt of the PMOS current mirror transistors. In order to reach a phase margin of 65° the second pole should be located at a frequency twice the GBW [5,6] or:

$$f2 = 2.GBW = \frac{2.gm}{2\pi C1} = \frac{3.KPp.(Vgs - Vt)}{8\pi Cox.L^2}$$

$$\boxed{3.5}$$

Increasing Vgs–Vt increases the GBW. However increasing Vgs–Vt decreases the output voltage swing and also the input common mode range. For practical use this voltage is limited to approximately Vgs–Vt = 500mV. So the maximum GBW (GBWmax) is as a result given by (3μm process):

$$GBWmax \approx \frac{3.(Vgs - Vt).KPp}{16\pi.L^2.Cox}$$

$$\approx \frac{3.(0.5V).10\mu A/V^2}{16\pi(3\mu m)^2.0.6fF/\mu m^2} \approx 50MHz$$

$$\boxed{3.6}$$

Due to parasitic capacitances, such as drain bulk and drain gate overlap capacitances, the maximum value is typically 10% smaller. In order to simplify the further calculations and comparisons, this effect will be neglected in this text.

The main disadvantage of this structure is its limited output voltage swing. As soon as the output voltage is lower than the voltage at the inverting input, one of the input transistors is forced in the triode region. Hence the performance of the amplifier is drastically decreased.

An other disadvantage of this structure is the offset voltage. If the output voltage is zero a voltage drop equal to Vdd appears across the drain source of transistor M4. On the other hand Vds,M1 is also approximately Vdd (if the inputs are grounded). Due to the drain source conductance of these transistors, an extra current is injected in the output node. This extra current can only be compensated by the applied offset voltage or:

$$gm.Voffset \approx Vdd.(go1 + go4)$$

$$Voffset \approx Vdd.\frac{2go}{gm}$$

$$\boxed{3.7}$$

So in this structure the offset voltage is inversely proportional to the low frequency open loop gain of the amplifier. Hence by designing a high open loop gain a low systematic offset is achieved.

Concerning the PSRR, it can be calculated that the PSRR,vdd is to first order given by [2].

$$PSRR,vdd = \frac{gm1}{go1 + go4 + s(Cp1 + Cp4)}$$

$$\approx \frac{gm}{2go + s2Cp}$$

3.8

with Cp the parasitic capacitances between the drain and the power supply line. The normalized PSRR for frequencies higher than the bandwidth of the opamp becomes.

$$\frac{PRSS.f}{2\pi GBW} = \frac{\frac{gm}{s.2Cp}}{\frac{gm}{s.C1}} \approx \frac{C1}{2.Cp}$$

3.9

The PSRR at low frequencies can again be improved by increasing the low frequency open loop gain (PSRR = gm/(2go) = Ao). On the other hand the PSRR at frequencies above the bandwidth of the opamp can be improved by increasing the load capacitance. But increasing the load capacitance influences the GBW or the power consumption of the circuit (see equation 3.3).

3.2.2 The load compensated OTA (LC–OTA)

The main disadvantage of the first OTA structure is the limited output voltage swing. To overcome this problem, two extra current mirrors, as presented in Figure 3.2 are usually added. At the output a rail–to–rail

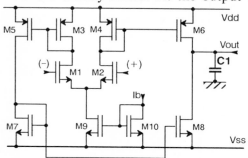

Figure 3.2 The load compensated OTA.

voltage swing can now be obtained. For the upper two current mirrors a current ratio (B) can be used. However this will decrease the higher order poles of the OTA (larger transistor areas = > more capacitance) and will increase the total equivalent input voltage noise. Therefore a ratio $B = 1$ is advised and will be further assumed in this text.

Concerning the open loop gain specifications, this structure has the same relationships as the OTA:

$$Ao = \frac{gm1 + gm2}{2(go6 + go8)} \approx \frac{gm}{2go}$$

and

$$GBW = \frac{gm}{2\pi C1} \qquad \boxed{3.10}$$

Due to those extra current mirrors, this structure has two main high frequency poles: one as a result of the upper two current mirrors and an extra one due to the bottom current mirror. The first pole (fa) is the same as the one of the differential pair with active load:

$$fa = \frac{3.(Vgs - Vt).KPp}{16\pi L^2.Cox} \qquad \boxed{3.11}$$

The other pole (fb) is given by

$$fb = \frac{3.(Vgs - Vt).KPn}{16\pi L^2.Cox} \qquad \boxed{3.12}$$

with KPn the transconductance parameter of the NMOS current mirror transistors. As can be seen from relationship 3.11 and 3.12, the only difference for the same transistor length and (Vgs–Vt) is the mobility (KP $= \mu.Cox$). Hence fb is typically a factor $\mu n/\mu p \approx 3$ higher than fa. As a result the extra phase shift is 10°–15°. Hence this amplifier has a slightly smaller GBW (approximately 20–30%) for the same phase margin.

To overcome the problem of this reduced GBW a folded cascode structure, as presented in Figure 3.3, can be used. The pole associated with the upper folded current mirrors is shifted to higher frequencies (less capacitance at the source of the cascode transistors) and given by:

$$fa = \frac{gm5}{2\pi Cgs} = \frac{3.(Vgs - Vt).KPp}{4\pi L^2.Cox} \qquad \boxed{3.13}$$

which is a factor of two higher than the usual current mirror structure. As a result a gain in phase shift of approximately 10°–15° is obtained. So the phase loss due to the bottom current mirror is compensated by the phase gain in the folded cascode structure. Hence, a folded cascode transconductance amplifier has approximately the same GBW (for a given phase margin) as the differential pair with active load.

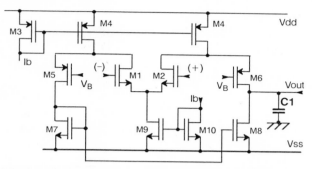

Figure 3.3 An LC–OTA with a folded cascode current mirror structure.

Although the voltage output swing is solved in the load compensated OTA, the structure (Figure 3.2) still suffers from a high systematic offset voltage. If the output voltage is zero, and assuming $|Vgate,M7| \approx |Vss-(Vgs-Vt)M7| \approx |Vss| \approx Vdd$, the extra current that flows into the output must be compensated by the offset voltage or (go6≈go8).

$$gm.\ Voffset = 2.go5.Vdd$$

$$Voffset = \frac{2.go5}{gm}.Vdd$$

$$\boxed{3.14}$$

This drawback can be solved by inserting cascode transistors as presented in Figure 3.4a or using a folded cascode structure as presented in Figure 3.4b.

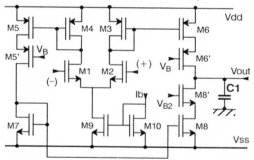

Figure 3.4a An OTA with cascode transistors to Improve the offset specifications and the low frequency open loop gain.

Using the same analyzing technique as above the offset voltage becomes for those structures.

$$Voffset = \frac{2.go5}{gm}.\frac{go5'}{gm5'}.Vdd \approx 2.\left(\frac{go}{gm}\right)^2.Vdd$$

$$\boxed{3.15}$$

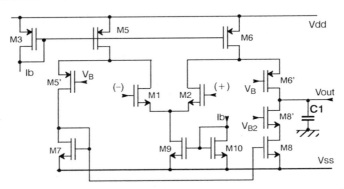

Figure 3.4b The folded cascode OTA with an extra cascode transistor to improve the low frequency open loop gain.

$$Ao = \frac{1}{2} \cdot \left(\frac{gm}{go}\right)^2$$

3.16

As a result the systematic offset voltage becomes so small (Voffset << 1mV) that it can be neglected compared to the statistic offset voltage of the input differential pair (typical 2mV).

An extra advantage of these (folded) cascode transistors is that they decrease the output conductance by a factor go/gm. Hence the low frequency gain is increased (not the GBW) with the same factor gm/go, or:

Let us now analyze the PSRR of the LC-OTA (Figure 3.2). In this structure the effects of both power supply lines must be studied. Concerning the positive power supply line (Vdd), the PSRR is approximately given by [2].

$$PSRR, Vdd \approx \frac{gm1}{(go6 - go5) + (go1 - go2) + s.(Cp6 - Cp5) + s.(Cp1 - Cp2)}$$

$$\approx \frac{gm}{2\Delta go + s.2\Delta Cp}$$

3.17

and the normalized PSRR becomes:

$$\frac{PSRRvdd.f}{2\pi GBW} = \frac{C1}{2\Delta Cp}$$

3.18

The PSSR of the negative power supply line is:

$$PRSS, vss \approx \frac{gm1}{go5 + go8 + s.(Cp5 + Cp8)}$$

$$\approx \frac{gm}{2go + s2Cp}$$

3.19

and the normalized parameter is:

$$\frac{PSRRvss.f}{2\pi GBW} = \frac{C1}{2Cp}$$

3.20

It can be concluded that a high PSRR,vdd can be realized by designing a very symmetrical OTA ($\Delta Cp < <$). On the other hand, the phase of the PSRR,vdd can be 0° or 180° at low frequencies depending on the mismatch in the drain source conductances of the transistors. Also the pole of the PSRR transfer function can be situated in the left or in the right half plane, depending on which parasitic capacitance is the largest. For the PSRR,vss it can be concluded that the drain source conductances and the sum of the drain–gate and the drain– bulk capacitances have to be made very small in order to realize a high PSRR,vss. If these results are compared with the differential pair with active load, it can be concluded that the PSRR,vdd is approximately equal to the PSRR,vss of the load compensated OTA.

If a cascode structure or a folded cascode structure is used, and assuming that the (folded) cascode transistors have the same geometry as the drive transistors, the PSRR becomes (with ' the parameters of the (folded) cascode transistors):

$$PSRR, vdd \approx \frac{gm1}{go1 - go2 + s(Cp1 - Cp2 + Cp6' - Cp5')}$$

$$\approx \frac{gm}{\Delta go + s2\Delta Cp}$$

$$PSRR, vss \approx \frac{gm}{\frac{go5go5'}{gm5} + \frac{go6go6'}{gm} + s(Cp5' + Cp6')}$$

$$\approx \frac{gm}{\frac{2gosop2}{gm} + s.2Cp}$$

3.21

So only the low frequency PSRR will be improved. It has to be remarked that the (folded) cascode transistor must be biased to the correct power supply line. This means for the PMOS transistors to the Vdd and for the NMOS transistors to the Vss. If this is not accomplished the PSRR of the structures are degraded to those of the simple load compensated OTA.

3.2.3 The Miller compensated OTA (M–OTA)

A widely used transconductance amplifier is the two stage amplifier presented in Figure 3.5. The first stage (M1–M4) corresponds to the source–coupled input structure with active load. The second stage (M5–M6) is an inverter structure. Due to its second stage the Miller OTA has also a rail–to–rail output swing. However two high impedance nodes are

High Performance Operational Amplifiers and Comparators

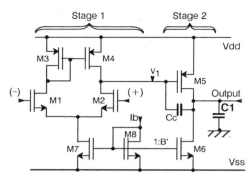

Figure 3.5 A two stage Miller compensated OTA.

present in this structure: one at the output of the first stage (node V1) and the other at the output of the OTA. In order to ensure stability a pole splitting technique is used: the compensation capacitance (Cc) is placed between the two gain nodes. As a result the compensation capacitance benefits from the Miller effect to split the two poles apart [5]. The open loop parameters are as a result given by:

$$Ao = \frac{gm^2}{4.go^2}$$

$$\boxed{3.22}$$

$$GBW = \frac{gm}{2\pi Cc}$$

$$\boxed{3.23}$$

Due to the pole splitting the second pole (f2) is approximately given by:

$$f2 = \frac{gm5}{2\pi C1}$$

$$\boxed{3.24}$$

However it is well known that the open loop transfer function also has a negative zero, located at a frequency:

$$fz = -\frac{gm5}{2\pi Cc}$$

$$\boxed{3.25}$$

In order to decrease the effect of this negative zero, this zero is designed one order of magnitude higher than the GBW (C1 $\approx$ 5 – 10 times Cc). This results in a much higher (by factor of 5–10) total integrated equivalent input noise voltage power compared to the LC–OTA. However this drawback will not be further discussed or analyzed in this text.

If the relationship of the second pole (3.24) is compared with the GBW relationship of the LC–OTA (3.10), it is clear that they are identical for the same drain current and Vgs–Vt (gm = gm5 = 2Idrain/(Vgs–Vt)). From this fact a very important drawback can be derived. In order to realize 65º phase margin the second pole should be twice the GBW. Hence the GBW of a Miller amplifier is half the GBW of the LC–OTA.

An advantage of the Miller OTA is that the systematic offset voltage can be made equal to zero. In order to achieve this, the biasing transistors must be correctly designed. To avoid any offset due to the first stage, the drain source voltage of transistor M3 and M4 must be equal. This can be achieved by designing the Vgs,M5 equal to the one of M3 or:

$$Id5 = \left(\frac{W}{L}\right)_5 \cdot \frac{KP}{2}\,(Vgs - Vt)^2 = B'.Ib$$

$$Id3 = \left(\frac{W}{L}\right)_3 \cdot \frac{KP}{2}\,(Vgs - Vt)^2 = \frac{Ib}{2}$$

$$\left(\frac{W}{L}\right)_5 = 2.B'.\left(\frac{W}{L}\right)_3$$

3.26

Of course the total systematic offset voltage will not be zero due to the offset voltage of the second stage. But this offset voltage is divided by the gain of the first stage (gm/(2go) ≈ 50 – 100). Hence the systematic offset voltage is so small (< < 1mV) that it can be totally neglected compared to the statistic offset voltage of the input stage (≈ 2mV).

The worst drawback of the M–OTA, especially in mixed mode designs, is the bad PSRR. To explain its cause the equivalent circuit presented in Figure 3.6a is used. If the PSRR of the first stage is neglected and remarking that the first stage is a transconductance amplifier, the equivalent circuit can be simplified into Figure 3.6b. C1 can be neglected because the

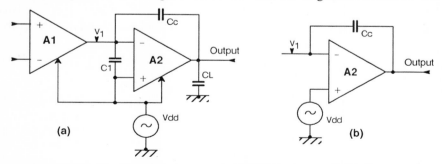

Figure 3.6a Equivalent circuit of the Miller OTA. C1 is the parasitic capacitance between node V1 (Figure 3.5) and Vdd (C1 ≈ Cgs,M5).

Figure 3.6b The simplified equivalent circuit of the Miller OTA.

inverting mode of an opamp follows always the non–inverting input. The capacitance Cc can be replaced for AC signals by a short connection. Now it is clear that Vout/Vdd = Ap ≈ 1, or that the PSRR,vdd is equal to:

and as a result the normalized parameter becomes:

$$PSRR, vdd = \frac{Ao}{Ap} \approx \frac{\frac{gm}{s.Cc}}{1} = \frac{gm}{s.Cc}$$

| 3.27 |

$$\frac{PSRRvdd.f}{2\pi GBW} = 1$$

| 3.28 |

If this relationship is compared with the one of the LC–OTA, it can be concluded that the Miller OTA has at least one order of magnitude lower PSRR !

The PSRR at low frequencies (below the bandwidth of the opamp) is given by:

$$PSRR, vdd \approx \frac{gm.gm5}{(go1 + go4).go5} \approx \frac{gm^2}{2.go^2}$$

| 3.29 |

This relationship is equal to the one of the (folded) cascode LC–OTA.

3.2.4 The core–amplifier (C–OTA)

In the previous section it is shown that a Miller–OTA has a bad power supply rejection ratio performance. However, a two stage amplifier, compared with a single stage OTA, has other advantages such as a higher low frequency open loop gain and a smaller systematic offset. Therefore a two stage amplifier is designed which combines the advantages of the symmetrical input stage of a single stage OTA and the advantages of a two stage amplifier. The total circuit is presented in Figure 3.7 and this structure

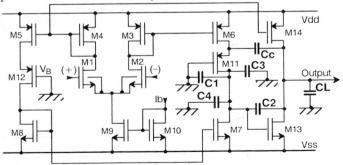

Figure 3.7 An improved Miller OTA : the core amplifier (C–OTA).

is called a core–amplifier [3]. This totally symmetrical input stage structure is an advantage for improving the offset voltage and common mode rejection ratio specifications.

In this structure transistor M11 improves the PSRR specifications, which will be discussed later on. The small signal specifications of this core amplifier are similar to these of a two stage amplifier. The gain bandwidth (GBW), with a current gain factor of one in the current mirrors, is:

$$GBW = \frac{gm1}{2\pi Cc}$$

3.30

However, due to the extra transistor M11, the second and the third pole are not widely spaced. Hence these poles normally form a complex–pole pair and can cause stability problems. Especially the gain margin (see Figure 3.8) can become very critical. The relationships for the circle frequency (ωn) and the damping ratio (ζ) of these complex poles as a function of the transistor parameters are given by [2,3,4]:

$$\omega n = \sqrt{\frac{gm11.gm13}{C1.CL}}$$

3.31

$$\frac{1}{2Q} = \zeta = \frac{CL + Cc}{2Cc} \cdot \sqrt{\frac{gm11.C1}{gm13.CL}}$$

3.32

In order to be able to calculate the transistor specifications as a function of the stability requirements, the gain margin (GMR) is expressed as a function of ωn, the quality factor ($Q = 1/2.\zeta$) and the GBW of the amplifier. Assuming that the maximum peak of a complex–pole pair is situated at the circle frequency ωn (exact $\omega max = \omega n.\sqrt{(1-\zeta^2)}$), we can deduce (see Figure 3.8) that:

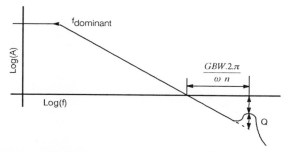

Figure 3.8 **Open loop transfer function of an amplifier with complex poles.**

$$\frac{2\pi.GBW}{\omega n} .Q.GMR = 1$$

3.33

Using relationships 3.31 and 3.32, the relation above leads to:

$$GMR = \frac{gm11}{2\pi.GBW.Cc}$$

$$\boxed{3.34}$$

(with $Cc < CL$). To reach stability requirements the GMR should be less than 2 or:

$$\frac{gm11}{2\pi.Cc} = 2.GBW$$

$$\boxed{3.35}$$

If this relationship is again compared with the GBW of the LC–OTA, we can derive the same conclusion as for the M–OTA, the GBW of the core amplifier is for the same transconductances and stability requirements half the one of the LC–OTA.

To reduce the systematic offset voltage the same technique as for the M–OTA must be applied, Vgs,M13 should be equal to Vgs,M8. This can be achieved if the aspect ratio of M13/M18 is equal to the ratio M14/M3 (if M5,M3,M4 and M6 are identical transistors). Hence the systematic offset voltage of the first stage becomes zero. The total systematic offset voltage, including the one of the second stage, becomes so small that it can be neglected compared to the statistic offset voltage.

The main improvement in the PSRR is due to transistor M11. To understand this effect, the equivalent circuit presented in Figure 3.9 is

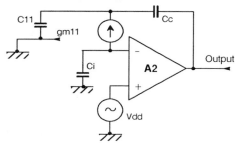

Figure 3.9 The simplified equivalent circuit of the C–OTA (Ci= C3+C4 of Figure 3.7).

analyzed. The only difference with Figure 3.6b (the one of the M–OTA) is transistor M11 (gm11 and C11). If the ratio Vout/Vdd is calculated it becomes:

$$\frac{Vout}{Vss} = \frac{sCi(gm11 + s(Cc + C11))}{gm11 + sCc} \approx \frac{Ci}{Cc}$$

$$\boxed{3.36}$$

which is at least one order of magnitude smaller than for the M–OTA (Vout/Vdd ≈ 1). The PSRR,vss becomes:

$$PSRR, vss = \frac{gm1}{go6\left(\frac{gm1-gm2}{gm1+gm2}\right) + sCi}$$

(3.37)

The normalized parameter is:

$$\frac{PSRRvss.f}{2\pi GBW} = \frac{Cc}{Ci}$$

(3.38)

The PSRR,vdd is similar to the one of a LC–OTA which is:

$$PSRR, vdd = \frac{gm1}{2\Delta go + s2\Delta Cp}$$

(3.39)

It can be concluded that the PSRR of a C–OTA is a factor Cc/Ci (which is typically 2pF/ 200 fF ≈ 20dB) better than the M–OTA. However it has to be remarked that the bulks of the transistors M11 and M10 must be connected to their sources (separated wells). Otherwise the PSRR,vdd will decrease drastically so that it becomes even worse than the M–OTA.

3.2.5 Conclusion

In this section different transconductance amplifiers are studied. Because the PSRR is very important in mixed–mode designs, the PSRR of the different presented structures are calculated. Also the systematic offset voltage of the different structures are analyzed. Table 1 shows the performance obtained for the different OTA designs presented.

It can be concluded that for low systematic offset voltage the M–OTA and the C– OTA are preferred. However the drawback of the M–OTA is its bad PSRR. As a result the C–OTA must be preferred above the M–OTA. Concerning the normalized maximum GBW (= GBWmax/GBW,ota) it can be concluded that the OTA and the folded cascode LC–OTA are preferred. For the folded cascode LC–OTA the PSRR is of the same order of magnitude as the C–OTA and even the systematic offset voltage is so small (< < 1mV) that it can be neglected compared to the statistic offset voltage.

So it can be concluded that overall the folded cascode LC–OTA has to be preferred in the design of complex mixed mode integrated circuits, because it has a high GBW, a high open loop gain, low systematic offset voltage and an excellent power supply rejection ratio.

Table 3.1 Summary of the different analysed parameters of the different presented operational transconductance amplifiers.

	OTA	LC-OTA	Folded cascode LC-OTA	M-OTA	C-OTA
Low Frequency Gain	$\dfrac{gm}{2go}$	$\dfrac{gm}{2go}$	$\dfrac{gm^2}{2go^2}$	$\dfrac{gm^2}{4go^2}$	$\dfrac{gm^2}{2go^2}$
GBW	$\dfrac{gm}{2\pi C1}$	$\dfrac{gm}{2\pi C1}$	$\dfrac{gm}{2\pi C1}$	$\dfrac{gm}{2\pi Cc}$	$\dfrac{gm}{2\pi Cc}$
$\dfrac{GBWmax}{GBWota}$	1	0.7	1	0.5	0.5
$\dfrac{Voffset}{Vdd}$	$\dfrac{2go}{gm}$	$\dfrac{2go}{gm}$	$\dfrac{2go^2}{gm^2}$	negligible	negligible
Worst PSRR	Vdd	Vss	Vss	Vdd	Vss
Worst $\dfrac{PSRR.f}{2\pi GBW}$	$\dfrac{C1}{2Cp}$	$\dfrac{C1}{2Cp}$	$\dfrac{C1}{2Cp}$	1	$\dfrac{C1}{Cp}$

3.3 High performance comparators

In the design of mixed mode circuits analogue circuits are integrated in combination with digital circuits. The information interchange requires high accuracy A/D and D/A convertors. For the realization of these circuits analogue decision circuits such as comparators are required. In order to reach high performances (e.g. with over sampling or flash A/D) high speed high accuracy comparators are indispensable. In this second part, several CMOS comparators are analyzed and compared. First a continuous time comparator is discussed. It is shown that the main drawback is its speed limitation. To overcome this problem latched or clocked comparators are very often used.

Because the speed capability, or the so called response time, is so important for comparators, the different structures are compared towards this specification. The response time (Δt) is defined as the time interval between the application of a predetermined input step and the time when the output crosses the corresponding logic state level. The input step drives the comparator from some initial saturated input condition to the output switch state. The step input voltage is called the overdrive. The comparator response speeds up as the overdrive is increased. For comparison purposes, response times are normally stated with a 5mV overdrive. If the response time is unsymmetrical for positive and negative transitions, the longer time is normally specified.

3.3.1 The OTA as comparator

A very simple comparator is a transconductance amplifier as presented in Figure 3.10. The inverting input is connected to a reference voltage and the non–inverting input is the comparator input. As soon as the input voltage is

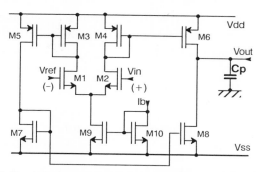

Figure 3.10   A transconductance amplifier as comparator.

higher (lower) than the reference voltage, the output voltage switches high (low) and clips against the power supply line. So the analogue input voltage is converted into a one bit high (low) digital level. The relationship between the step input voltage (Vin) and the output voltage (Vout) as function of time ($\Delta t = t - t_{start}$) is given by (assuming a constant gm and neglecting the effect of higher order poles):

$$Vout(\Delta t) = Vin.\frac{gm}{go6 + go8}.\left[1 - e^{\frac{\Delta t.(go6 + go8)}{Cp}}\right]$$

3.40

which can be simplified for small (go6 + go8) values into:

$$Vout(\Delta t) = Vin.\frac{gm}{Cp}.\Delta t$$

3.41

The disadvantage of the OTA as a comparator is its limited speed: for a small overdrive, let us use Vin = 5mV, the output voltage will reach its high level value (90%.(Vdd + |Vss|)) after a time Δt given by:

$$\Delta t = \frac{0.9(Vdd + |Vss|).Cp.(Vgs - Vt)}{Vin.Ib}$$

$$= \frac{0.9(5V).0.2pF.0.2V}{5mV.100\mu A} \approx 400nSec$$

3.42

with Ib the OTA biasing current, Cp the load capacitance (the input capacitance of one inverter) and (Vgs–Vt) the one of the input transistors. However this is an optimistic value. As a result of the higher order poles and due to the clipping effect the response time further increases. It can be concluded that a simple OTA is only suitable for use at low frequencies (,1MHz). Because an OTA can have remarkable systematic offset voltage (which reduces the comparator accuracy) a cascode or folded cascode load compensated OTA is advised. With such structures, accuracies better than a few millivolts can be obtained, but the response time is not improved compared to the simple OTA structure.

3.3.2 Latched comparators

To overcome the limited response time of an OTA–comparator, latched comparators are very often used. A simple transistor circuit of a latched comparator is presented in Figure 3.11 [7,8,9]. The comparator consists of a

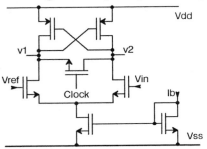

Figure 3.11 A latched comparator circuit

differential stage with a latch as load. Latched comparators have two phases: the reset mode and the compare mode. In the reset mode the nodes of the latch are shorted to set it to an unstable high gain mode. In the compare mode these nodes are released. Depending on the input voltage the latch will switch very fast to the high (low) state due to the positive feed back in the latch. The draw backs of this simple structure are:

- At the end of the compare mode one of the input transistors is forced into the triode region. As a result the total speed of the comparator (including the reset phase) is increased. This is due to the required extra settling time in order to achieve the necessary accuracy in the reset phase.

- The switch which controls the two modes is connected to the input transistors. As a result clock feed-through can be fed to the input nodes. This can limit the comparator accuracy especially when a voltage with a high source impedance is being compared.

- The accuracy of this comparator is mainly limited by the unsymmetrical latch structure (only NMOS devices) and by the latch transistor mismatches. These effects are analyzed more in detail in the next section.

To overcome the two first problems the input stage of the circuit is usually modified using an extra current mirror[7] as is presented in Figure 3.12. With such a structure a response time (reset phase included) of 30–40nSec can be achieved.

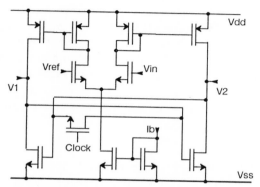

Figure 3.12 An improved latched comparator

3.3.3 A high speed accurate comparator

In the realization of high speed CMOS flash A/D converters and high accuracy CMOS over sampling A/D converters, high speed CMOS comparators are indispensable. The high speed structure is presented in Figure 3.13 [10]. It consists of two inverters (M13, M14 and M11, M12),

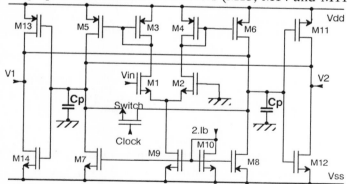

Figure 3.13 The high speed latched comparator circuit.

assembled as a latch, in combination with a transconductance amplifier (M1–M10). By closing the switch, the latch is set in its unstable state. This structure has several important advantages compared with the other structures using the latch principle. First, the input stage does not contain any switches. As a result no clock feed-through is injected to the input nodes. This is very important for an over sampling A/D where an integrator drives the input of the comparator. Secondly, due to M7 and M8, no current from the transconductance amplifier flows into the latch when the switch is closed. As a result the latch voltage is not influenced by the biasing current when the switch is closed. Hence the biasing current can be increased which results in a higher resolution. Thirdly, only one switch is used in a totally symmetrical structure. So the effect of clock feed-through is minimized. Finally a two inverter latch structure is employed. This is important because the connected circuit is usually a digital building block (an inverter). As a result, if identical inverters are used, the output voltage in the reset mode is preset at the threshold voltage of the connected digital inverter. This results in almost equal response times for the high (low) transitions. This fact remains true even with process variations, as long as the inverter latch transistors are equal to the one of the connected digital inverter.

First the comparator output voltage as function of time is discussed. Then the comparator accuracy due to transistor mismatches and clock feed-through are studied. Finally the comparator speed is discussed.

The comparator : Assuming a constant transconductance (gm) model for the inverters, and if the transconductance of the amplifier is Gm, the output voltages of the comparator as a function of time are (the switch opens at $t=0$):

$$v_1 = \left(V_D - \frac{Gm}{2.gm}.Vin\right).e^{a.t} + \left(V_C\right).e^{-a.t} + \frac{gm}{2.gm}.Vin + \frac{\Delta IB}{gm} \qquad \boxed{3.43}$$

and

$$v_2 = \left(V_D - \frac{Gm}{2.gm}.Vin\right).e^{a.t} + \left(V_C\right).e^{-a.t} + \frac{gm}{2.gm}.Vin + \frac{\Delta IB}{gm} \qquad \boxed{3.44}$$

with $\alpha = gm/Cp$. ΔIb is the current that flows into the latches due to mismatches in M7 and M8, V_C is the common mode signal and $2.V_D$ is the differential signal across the switch at $t=0$ (ideal case $V_C = V_D = 0$, $\Delta Ib = 0$). V_C and $2.V_D$ are generated due to clock feed-through, the finite switch resistance and transistor mismatches in the latch. If M7 and M8 are not used, ΔIb becomes equal to Ib. As can be seen from the equations, this would result in an extra output voltage (Ib/gm) which slows down the speed of one comparator output and increases the speed of the other comparator output. However, the comparator speed is determined by the slowest output. Any differential signal V_D can result in a comparator error, so the maximum resolution is $Vin > (2.gm/Gm).V_D$.

Limited accuracy due to Transistor mismatches in the latch: Suppose that one of the latch transistors has a certain mismatch. As a result an offset voltage is generated at the input of the inverter. If the threshold voltages (Vt) of PMOS and NMOS transistors are approximately equal, a symmetrical power supply is used (Vdd = −Vss) and the transistors are designed so that KP(W/L)nmos = KP(W/L)pmos, then the voltages at t = 0 without mismatch are $v_1 = v_2 = 0$. However, due to mismatches, let say a ΔW and ΔVt between PMOS transistors M11 and M13, the maximum error voltage at t = 0 becomes $v_1 \approx -v_2 \Delta(Vt/4).(\Delta W/2.W + \Delta Vt/Vt)$. As a result an extra V_D is generated which limits the accuracy of the comparator. The comparator reaches a resolution Vr given by:

$$Vr \approx \frac{gm.Vt}{2.Gm} \cdot \left(\frac{\Delta W}{2.W} + \frac{\Delta Vt}{Vt} \right)$$

3.45

For example, with $(\Delta W/2.W + \Delta Vt/Vt) \approx 2\%$, gm = 300μA/V and Gm = 500μA/V, Vt = 0.8V, Vr ≈ 5mV. In order to decrease this error, the transconductance of the latch must be decreased or Gm must be increased.

Clock feed-through: When the switch opens, a charge is injected into the capacitances Cp. This results in a common mode signal V_C and a differential signal $2.V_D$. As can be seen from equation 3.43 – 3.44, V_C will not effect the accuracy. However, the resolution of the comparator is limited by V_D, which is: [11]

$$V_D = \frac{Cox.(W.L)s.Vstep}{2.Cp} \cdot \frac{\Delta Cp}{Cp + Cox.(W.L)s}$$

$$\approx \frac{\Delta Cp}{2.Cp} \cdot \frac{Ws}{2.Wi + Ws}.Vstep$$

3.46

where Vstep = ((Vclock,high−Vclock,low)/2−Vt). (W.L)s are the switch dimensions, ΔCp is the mismatch between the capacitances Cp, (W.L)i are the dimensions of M12 and M14. The aspect ratio of M11,M13 are equal to twice the value of M12,M14 and as a result Cp is approximately equal to 2.Cox(W.L)i (it can be shown that capacitances other than the inverter transistors have a positive effect on the accuracy). Using equations 3.43 – 3.44 a resolution Vr = 5mV is obtained when (2.Wi < Ws).

$$(W/L)i \approx \frac{Gm.Vr}{\frac{\Delta Cp}{Cp}.KP.(Vgs - Vt)i.Vstep}$$

3.47

where KP.(Vgs−Vt)i are the parameters of the NMOS transistors M12 and M14. With $\Delta Cp/Cp = 1\%$, Vstep = 1.7V, Gm = 500μA/V, KP.(Vgs−Vt)i = 50μA/V (W/L)i is approximately 5μm/2μm. It is interesting to remark that decreasing (W/L)i will result in a higher accuracy. However as it is discussed later, this will also decrease the comparator speed. The way to increase the accuracy is to increase Gm.

The comparator speed: Neglecting the second order effects, equation 3.43 – 3.44 can be simplified into ($v_1 \approx -v_2$).

$$t \approx \frac{1}{a}.\ln\left(\frac{-V1.2.gm}{Vin.Gm} + 1\right)$$

$$\boxed{3.48}$$

With $1/\alpha$ = Cp/gm $\approx$ 0.7nSec, 2.gm/Gm $\approx$ 1.2, v1 = –2 V and Vin = 5mV, the comparator speed is t$\approx$4.3nSec. In Figure 3.14 the Spice output

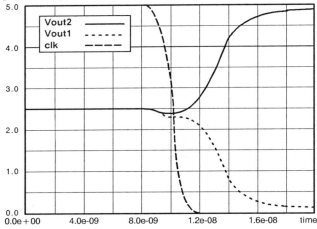

Figure 3.14 Output response of the high speed comparator (Vin = 5mV).

response is represented for an input voltage Vin = 5mV. As can be seen the speed (4.2nSec) is very close to the calculated one. From equation 3.48 it can be concluded that the speed mainly depends on Cp/gm ($=1/\alpha$). Cp is approximately the capacitance of the latch inverter plus the capacitance of the connected digital building block. As a result, the speed is approximately equal to the speed of a loaded inverter ($\approx$2nSec) or is equal to the speed of digital circuits. However, if the W/L's of the latch transistors are further decreased (to increase the accuracy), the parasitic capacitances of the transconductance amplifier and the switch will become dominant, and as a result the comparator speed is going to decrease. To further increase the accuracy it is best to design the input stage with a higher transconductance (more GM but also more power drain). The total response time of this configuration (including the reset phase) for an accuracy of 5mV becomes less than 10nSec.[12]

3.3.4 Conclusion

CMOS comparator structures have been discussed. It is shown that an OTA can only be used for low frequency applications. If a fast response time is required, latched comparators must be employed. It is shown that the resolution of a latched comparator is a function of the clock feed-through and the transistor mismatches. To overcome those problems the advantages of the OTA–latch structure are analyzed. The effect of clock feed-through and transistor mismatch can now be reduced by increasing Gm. This can best be achieved by designing the input stage at the boundary of strong inversion (Vgs–Vt ≈ 200mV). As a result the accuracy of the comparator is reduced approximately to the offset voltage of a differential pair. The speed of this comparator is approximately equal to a loaded inverter. Hence accurate high speed comparators can be obtained.

3.4. References

[1]   K.Halonen, W.Sansen, Effect of current spikes in power supply rails on PSRR performances of SC–filters, Procs.IEEE–ISCAS'87, May 87, pp.64–67.

[2]   M.Steyaert, W.Sansen, Power supply rejection ratio in operational transconductance amplifiers, IEEE–Trans. on Circuits ans Systems, Vol.CAS-37, Sept 1990, pp.1077–1084.

[3]   J.Fisher, A high performance CMOS power amplifier, IEEE–JSSC,vol.SC–20, pp.1200–1205, Dec.85.

[4]   D.Ribner, M.Copeland, Design technique for a cascoded CMOS opamp with improved PSRR and common–mode input range, IEEE–JSSC, vol.SC–19, pp.919–925, Dec.84.

[5]   P.Gray, M.Meyer, Analysis and design of analogue integrated circuits, Wiley 84.

[6]   W.Sansen, M.Steyaert, P.Vandeloo, Measurements of operational amplifier characteristics in the frequency domain, IEEE Trans. Inst. & Meas., Vol.IM–34, March'85, pp.59–64.

[7]   D.Sallaerts et al., A single chip U–interface transceiver for ISDN, IEEE–JSSC Vol.SC–22, No.6 December 1987.

[8]   R.Koch et al., A 12–bit sigma–delta analogue–to–digital converter with a 15Mhz clock rate, IEEE JSSC Vol.SC–21, No.6 December 1983.

[9] B.McCarrol et al., A high speed CMOS comparator for use in an ADC, IEEE–JSSC, Vol.SC–23, Feb.88, pp.159–165.

[10] M.Steyaert, V.Comino, High–speed accurate CMOS comparator, Electronics Letters, Vol.24, No.16, Aug.88, pp.1027–1029.

[11] P.Van Peteghem, Accuracy and resolution of switched–capacitor circuits in MOS technology, Ph.D. dissertation, K.U.Leuven, June 1983.

[12] V. Comino, M. Steyaert, and G. Temes, "A first order current steering sigma delta modulator", IEEE Journal of Solid-State Circuits, Vol.26, No.3, March 1991, pp.176-183.

| Switched Capacitor Filters | **Chapter 4** |

D. G. Haigh, C. Toumazou,
J. E. Franca, B. Singh

4.1 Introduction

Before 1970, communication systems were primarily analogue and relied heavily on analogue circuits including high-precision filters. About 1970, these systems began to be replaced by digital communication systems. One might have thought that this would have eliminated the need for high-precision analogue filters. However, the harmful effects of aliasing and imaging when the digital system interfaced with analogue signals, which is necessary at the transducers, made it necessary to introduce band limiting filters at the digital-analogue interface. For a time, these filters were generally realised as active RC networks realised as micro-electronic hybrid modules combining integrated circuit operational amplifiers with high-precision thin-film or thick-film RC networks [1]. As digital systems developed, there was a move toward systems comprising fewer numbers of ever more complex digital integrated circuits, and it was recognized that there could be significant advantages in terms of cost and reliability if the analogue interface circuits could be realised on the same chips along with the digital circuits. This was not possible for active RC filters because they rely on high precision RC networks which can not be realised in integrated circuit technology and because, except at very high frequencies, the resistor values required would imply excessive chip areas. A breakthrough occurred in 1977. Researchers at the University of California at Berkeley [2] and at the University of Ottawa [3] independently proposed a new kind of analogue filter - the switched capacitor filter, whose transfer function is a function of ratios of capacitor values which can be realised with high precision using an MOS integrated circuit process. Before discussing switched capacitor filters in some detail, it is appropriate to compare their performance capabilities with the performance of other filter types.

The performance capabilities of some filter types are illustrated in Figure 4.1. The vertical scale shows the approximate transfer function Q-factor (or loaded pole Q-factor) which can be achieved and the horizontal scale the approximate operating frequency range. Inductor-Capacitor–Resistor (LCR) filters can achieve Q-factors of the order of 100 up to a frequency of up to about 100MHz but below 1kHz LC filters become very bulky and are not very attractive. For active RC filters, the upper frequency range is limited to around 10MHz but the Q factors realisable are somewhat

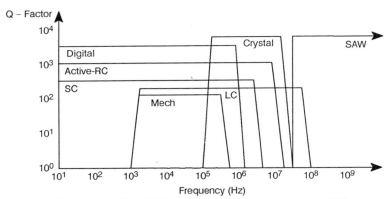

Figure 4.1 Filter Q/frequency performance capabilities

higher, say up to 1000, because it is possible to compensate to some extent for loss effects [4,5]. Digital filters can operate up to about 1MHz, for a reasonable number of bits such as 12, and achieve very high Q factors. Mechanical filters are similar to digital filters as far as Q-factor is concerned but the upper frequency range is limited to about 500kHz. For switched capacitor filters, we see that the upper frequency limit so far achieved using CMOS integrated circuit technology is about 5MHz. Q-factors are limited to about 100 due to a limit on maximum capacitor ratios (maximum capacitance is limited by chip area, minimum capacitance by lithographic tolerance) and are thus similar or even slightly higher than those achievable with LC filters. Crystal and SAW filters achieve the highest Q-factors of any filter of around 10,000. Crystal filters operate from 100kHz to about 50MHz. SAW filters start at about 50MHz and can go up to about 5GHz. Another type of filter, referred to as the integrated continuous-time filter, overcomes the problems of integrating active RC filters by realising resistive elements using active devices such that chip areas are small and tolerance effects can be accommodated by automatic tuning of the resistance values. This type of filter has been recently discussed in [6]. Another type of integrated filter which is at a relatively early stage is the switched current filter which avoids the need for precision capacitor ratios and can be therefore realised using a basic digital integrated circuit process. Switched current filters are covered in detail in Chapter 5 of this volume.

Switched capacitor filters can be integrated on a chip along with substantial amounts of digital circuitry to realise a system or a substantial sub-system of the mixed-mode type. One such application we have already referred to is where the switched capacitor filters are realising the interface circuits between the digital system and analogue input and output signals, as schematically shown in Figure 4.2. An example of this analogue-interface type application would be the PCM CODEC [7]. Another application which falls into the area of mixed-mode is not necessarily of the interface type. Here we have a system which is predominantly digital but where there

Figure 4.2 The analogue interface in a mixed-mode system

is a requirement for high precision filtering within the system which can best be performed by switched capacitor techniques from the points of view of chip area and power consumption. An example here would be an FSK modem for data communications [8]. Another example would be the DTMF or Dual Tone Multi Frequency receiver [9] for telecommunications and data transmission applications where banks of band pass filters are needed within the chip to select various tones which are used for signalling.

Switched capacitor filters have now been around for a decade and many powerful and sophisticated design techniques have been developed [10, 11, 12]. In view of the excellent literature on this subject to which the reader may refer, the present paper has the rather specific aim of presenting some advances developed by the authors which have made a contribution towards extending the limits for switched capacitor filters beyond the generally accepted performance boundaries shown in Figure 4.1. The areas of performance improvement considered include performance precision, maximum achievable Q-factor and maximum operating frequency.

4.2 Building blocks for switched capacitor systems

4.2.1 Sensitivity to grounded parasitic capacitances

The basic elements of switched capacitor filter design are shown in Figure 4.3. At the ideal level these are the ideal capacitor, the ideal operational amplifier and the ideal periodically actuated switch which presents a short-circuit or an open circuit between the switch terminals x

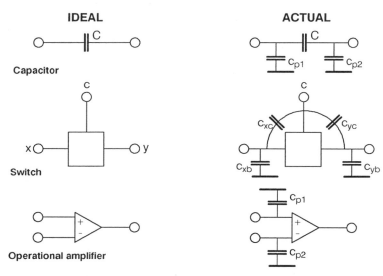

Figure 4.3 Basic elements of SC filters

and y depending on the value of the signal on control terminal c. When these components are realised in integrated circuit form we find that parasitic capacitances are associated with them [11]. In the case of the capacitor there are parasitic capacitances, C_{P1} and C_{P2} associated with the bottom plate and the top plate of the capacitance. The bottom plate capacitance can be as high as 10% of the wanted capacitor value C. The operational amplifier has parasitic capacitances from each terminal to ground. The switch has grounded capacitances C_{XB} and C_{YB} and also non-grounded capacitances C_{XC} and C_{YC} associated with its control terminal.

In view of the high values of the parasitic capacitances and the serious effect of them on filter response, it is appropriate to consider them from the outset of the design process. We therefore classify switched capacitor circuits according to the extent to which they are affected by grounded parasitic capacitances. One such category is called 'parasitic sensitive' where grounded parasitic capacitances affect the circuit transfer function. An example of a circuit in this category, the toggle-switched integrator, is shown in Figure 4.4a [2]. It is sometimes possible to pre-distort the original circuit capacitor values in order to allow for the parasitics which will be introduced as, for example, Ca in Figure 4.4a may be reduced to allow for the parasitic capacitance C'.

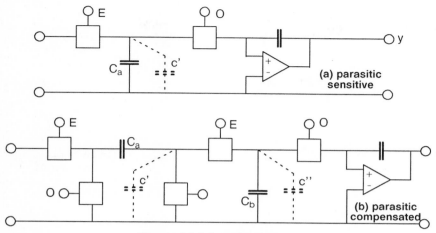

Figure 4.4 Integrator types

A second category is called 'parasitic compensated'. Here the parasitic capacitances do not affect the transfer function provided the ratios of parasitic capacitances match the ratios of certain circuit capacitor values [13]. An example of a circuit in this category is the parasitic compensated integrator of Figure 4.4b, for which the effect of parasitic capacitances is zero provided C'/C" = Ca/Cb.

The final category is called 'parasitic insensitive' [14]. Here the transfer function is completely independent of grounded parasitic capacitances under the assumption that the operational amplifiers are ideal. In practice, with real operational amplifiers, the parasitic insensitive type of circuit is much less sensitive to grounded parasitic capacitances than the other categories.

Fortunately, most of the applications of switched capacitor filters can be met entirely by using parasitic insensitive circuits. Under certain circumstances, for example, where it is desired to multiplex operational amplifiers, or where exceptionally long circuit delays are necessary, parasitic compensated circuits have a special place [15]. It is usually possible to completely avoid the use of parasitic sensitive circuits.

4.2.2 Parasitic Insensitive Integrator

The concept of parasitic insensitivity was originally applied to integrators [14], and Figure 4.5a shows the differential input parasitic insensitive integrator consisting of one operational amplifier, four periodically actuated switches, and two capacitors. Switched capacitor circuits generally operate with a two-phase switching waveform, as shown in Figure 4.5b. It is

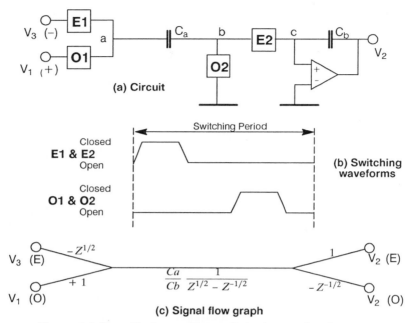

Figure 4.5 Parasitic insensitive switched capacitor integrator

important that the two phases, E (even phase) and O (odd phase) do not overlap and that there is a guard interval between them when they are both open. The insensitivity of this circuit to grounded parasitic capacitances may be simply demonstrated. Consider a parasitic capacitance at node 'a'. This is being switched in both switch phases to an integrator input terminal which would normally be fed from the output of another integrator which can be regarded as a voltage source. Thus parasitic capacitance at node 'a' is simply charged up by the input voltage sources and does not affect the response of the circuit. Node 'b' is being switched in the odd phase to ground and then in the even phase to amplifier virtual ground, which is at the same potential. Therefore parasitic capacitance of node 'b' has a constant voltage across it. Node 'c' is the input of the operational amplifier is also a constant potential and therefore parasitic capacitance here cannot affect the transfer function. And finally the output node is a voltage source. So parasitic capacitance here too does not affect the transfer function.

The transfer function of the integrator in Figure 4.5a may be analysed using the principle of charge conservation [16]. It is convenient to describe the transfer function in terms of a signal flow-graph as shown in Figure 4.5c.

The transfer function is a function of the parameter z, which is the discrete time complex frequency variable, and of the capacitors C_a and C_b. The central branch transmittance:

$$t = C_a/C_b \ (z^{1/2} - z^{-1/2})^{-1}$$

4.1

corresponds to discrete-time integration of the LDI (Lossless Discrete Integrator) type [17]. The transfer function depends on the ratio of the capacitor values. In integrated circuits, tolerances on the absolute values of capacitances are high, but it is possible to realise a ratio of capacitances to a very high tolerance of the order of 0.1% as required for high precision filters. In order to achieve this, the capacitors will be placed close together on the chip and for non- unity ratios, the larger capacitor will be realised as an interconnection of unit capacitors equal to the smaller capacitor value. This allows absolute capacitor values to be small leading to small chip area while the ratios remain accurately defined giving high precision filter response [18].

4.2.3 Optimum switching schemes

The parasitic insensitive integrator circuit has a performance which is very much better than that of previous types of switched capacitor integrators but performance is still not good enough for some high quality applications. For example, total harmonic distortion levels can be at around -40dB and gain errors of the order of 1dB can be obtained [19]. The parasitic insensitive type of circuit, as we have said, provides insensitivity only to grounded parasitic capacitances. Research showed that the non-grounded parasitic capacitances associated with the control terminals of the switches, C_{XC} and C_{YC}, in Figure 4.3, can be responsible for significant performance degradation [19]. However, further research indicated that by slightly delaying the times at which the E1 and O1 switches, in Figure 4.5a open relative to the E2 and O2 switches, the residual performance degradation could in theory be eliminated, and in practice, dramatically reduced. Switching waveforms which satisfy these conditions are referred to as optimum switching waveforms and an example of such switching waveforms is shown in Figure 4.6.

Figure 4.7 shows the gain frequency curve at the peak of a second order switched capacitor bandpass filter which is operated with both conventional and optimum switching waveforms. It can be seen that using optimum switching waveforms the mid-band gain and the mid-band frequency are very much closer to the ideal values. The use of optimum switching waveforms provides reduced sensitivity to power supply voltage variations

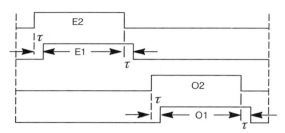

Figure 4.6 Optimum switch phasing

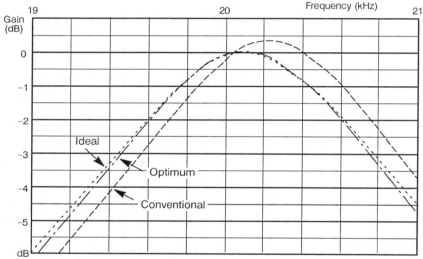

**Figure 4.7 Response precision improvement using
optimum switching waveforms**

and also, as has been shown by practical measurements, a reduced sensitivity to the wave shape of the switch control signals.

Figure 4.8 shows the measured harmonic distortion level for a first order switched capacitor filter operated with different kinds of switching waveforms. With conventional switching waveforms the 2nd. harmonic is at a level of about -40dB. When switching waveforms are changed to optimum, second harmonic distortion falls to -65dB and higher order harmonics reduce accordingly. A third type of switching waveform which we do not have time to discuss here, is referred to as ultra-low distortion and assures that all harmonic components are less than or equal to -80dB [20].

Thus we see that the basic parasitic insensitive integrator, when operated with suitable switching waveforms, which ensure insensitivity to non-grounded parasitic capacitances, provides a building block for high

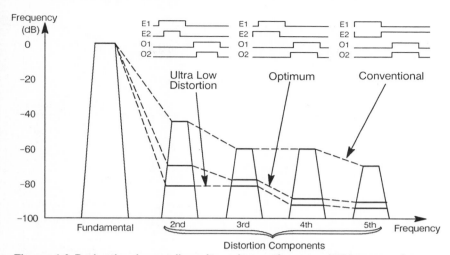

Figure 4.8 Reduction in non-linearity using optimum switching waveforms

order switched capacitor systems which has superb gain accuracy or transfer function accuracy and a very high degree of linearity exemplified by extremely low distortion levels.

4.3 High Order Filter Design

The design of high order switched capacitor filters is a mature subject and many methods have evolved [10,11,12]. The LCR filter has the property that the sensitivity of its response to changes in its component values is very small and therefore many of the methods for switched capacitor filter design begin from a prototype LCR filter. There are, in fact, many ways of simulating LCR filters using switched capacitor circuits and a popular method is based on a signal flow-graph approach. As an example, Figure 4.9a shows a 5th. order polynomial low-pass LCR filter. The relationships between the shunt-branch voltages (Vi, V1, V3 and V5) and the series-branch currents (I2 and I4) in the prototype filter can be represented by the signal flow-graph shown in Figure 4.9b. Since switched capacitor filters are discrete time systems, they are described as we have already seen by the discrete time complex frequency variable z and therefore it is necessary to carry out a transformation from the s-domain to the z-domain.

The transformations commonly used are the bi-linear transformation [11]

$$s \rightarrow (2/T) \ (1 - z^1)/(1 + z^{-1})$$
4.2

and the LDI transformation [11]

$$s \rightarrow (1/T) \ (z^{1/2} - z^{-1/2})$$
4.3

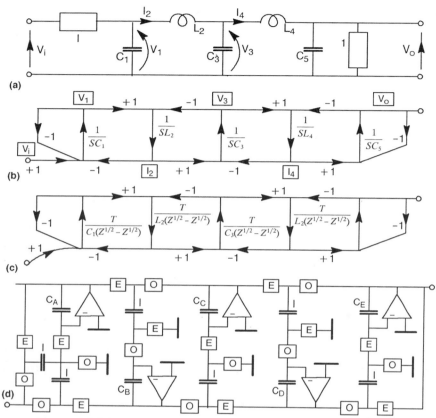

Figure 4.9 5th order switched capacitor filter design illustration

In our example, we apply the LDI transformation to obtain the discrete time signal flow-graph shown in Figure 4.9c. Discrete time integrator branches with transmittance inversely proportional to $(z^{1/2}-z^{-1/2})$ conveniently correspond with the signal flow graph that we obtained by analysis of the parasitic insensitive integrator in Figure 4.5. Thus these branches may be simply replaced by differential-input parasitic insensitive integrators to obtain the circuit in Figure 4.9d. Such methods of design can be classified either as exact or approximate [10]. Exact methods tend to be more involved than approximate methods [21, 22]. Approximate methods introduce deviations in the transfer functions of the switched capacitor filter compared with that of the LCR filter, but usually these errors are small when the switching frequency is much higher than the cut-off frequency, which is usually the case on account of the need to avoid aliasing and imaging which we will discuss presently. A problem which can occur in high order filters is that the amplifier output voltages can peak to high levels at certain frequencies. However, the freedom in many cases exists to exploit degrees of freedom in the choice of capacitor values to scale the nodal voltages for maximum input signal level or optimum dynamic range [11], [Chapter 4 in reference 6].

Switched capacitor filters designed to simulate LCR prototype filters have a response which is relatively insensitive to changes in capacitor ratios. An alternative method of switched capacitor filter design, not based on simulation of an LCR filter, is to factorize the required transfer function into a product of second-order factors and one first order factor in the case of odd order filters [23]. These factors are then realised by first and second order switched capacitor circuits, or sections, which are connected in cascade to realise the required overall transfer function. This design method is in some ways more flexible than the LCR simulation method , but the method does not lead to the property of a low sensitivity of filter response to capacitor ratio variations and therefore the cascade filter design approach tends to be used for less demanding applications.

4.4 Aliasing and imaging

Switched capacitor filters are discrete time systems since the switches effectively sample a continuous-time signal at the input. Thus, a switched capacitor filter may be equivalently represented by a system with a sampling switch at its input operated at the switching frequency as shown in Figure 4.10a [24]. It is well known that the process of sampling a continuous

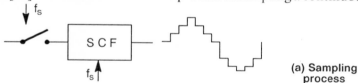

(a) Sampling process

Figure 4.10 Aliassing and imaging

time signal gives rise to the phenomena of aliasing and imaging. In aliasing, a band of frequencies f1 being processed by the discrete time system can be derived by any one of a multiplicity of bands nfs ±f1 where fs is the sampling frequency and n = 1, 2, 3 .. . In imaging, a band being processed by the discrete time system appears at the output in the same frequency location but is augmented by a multiplicity of bands nfs ±f1 where n = 1, 2, 3 ... . The effects of aliasing and imaging can be illustrated by a frequency band diagram as in Figure 4.10b.Since aliasing effectively introduces frequency ambiguity at the input, it corresponds to distortion and is particularly undesirable.

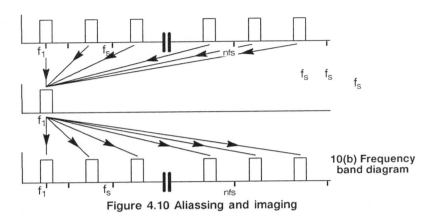

10(b) Frequency band diagram

Figure 4.10 Aliassing and imaging

Because of the problems of aliasing and imaging, switched capacitor filters are generally thought of in the context of a system as shown in Figure 4.10c,

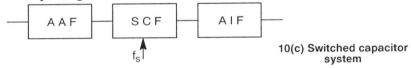

10(c) Switched capacitor system

Figure 4.10 Aliassing and imaging

where the switched capacitor filter AAF is preceded by a low-pass anti-aliasing filter and followed by a low-pass anti-imaging filter AIF [11]. These filters effectively remove the unwanted higher frequency components at the input and the output, as illustrated in Figure 4.10d,.

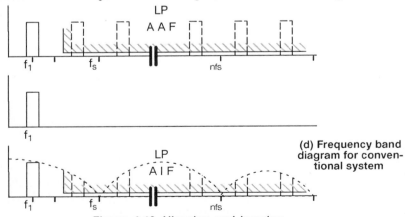

(d) Frequency band diagram for conventional system

Figure 4.10 Aliassing and imaging

Since the output signal of the switched capacitor filter is in sample and hold form, the output higher frequency components are attenuated also by the sample and hold transfer function as also shown in Figure 4.10d. The anti-aliasing and anti-imaging filters in Figure 4.10c are generally realised by low-order Sallen and Key active RC filters realised on chip. The manufacturing tolerance on the RC products is very high for these filters but this is acceptable because these filters do not shape the filter response but only remove spurious frequency components at frequencies very much higher than the filtering frequency. Although aliasing and imaging are generally regarded as detrimental effects, they can in fact be used to considerable advantage as we will show in section 4.6.

4.5 Case Study 1 - Audio frequency filter using CMOS technology

Having derived a switched capacitor circuit, as discussed in section 4.3, it is necessary to implement the required components, namely operational amplifiers, switches and capacitors using the proposed technology, in this case CMOS.

Operational amplifiers for switched capacitor systems generally require gains of greater than or equal to 60dB in order to obtain acceptable precision of filter response. Such amplifiers are designed using a traditional two stage architecture or by using a single stage cascode approach [11]. Figure 4.11a shows a typical two-stage amplifier. The input stage provides a differential input. Two stage amplifiers are unstable with the feedback encountered in a switched capacitor filter and therefore a compensation capacitor C_1 is needed. For higher frequency applications closer to the maximum frequency limit of the technology, single stage architectures are generally used and cascode techniques provide the required high value of gain [11].

The switches required in switched capacitor filters can be realised in CMOS technology using the traditional complementary CMOS switch shown in Figure 4.11b, which consists of an n-channel and p-channel MOSFET pair with their gates controlled by complementary clock lines. This complementary switch has the feature that the analogue input and output voltages may vary over the whole power supply range [11].

As far as capacitors are concerned, several methods of realisation are available [11]. The best method is to realise the capacitance between two layers of poly-silicon but this requires a double poly-silicon CMOS process which is a process that is specifically geared towards high quality analogue applications and may not be generally available in many mixed-mode situations.

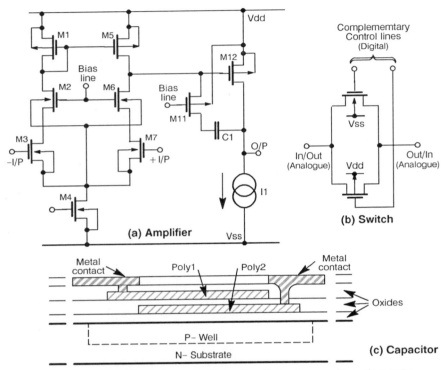

Figure 4.11 CMOS implementation of switched capacitor elements

Figure 4.12 shows a layout plot for a fifth order switched capacitor low-pass filter [25]. The cells containing the five operational amplifiers are realised in a row down the center. Immediately above and below them are the CMOS complimentary switches and immediately above and below them are the switch control lines. The optimum switching waveforms referred to previously are used and therefore four separate switching waveforms are required; in view of the use of complementary switches, a total of eight switch control lines are needed. Immediately above and below the switch control lines are the arrays of capacitors.

Continuous-time anti-alias and anti-imaging filters where required may be added at this stage. Typically, the area occupied by the continuous time filters actually exceeds that of the switched capacitor filters, illustrating the fact that switched capacitor filters provide high quality filtering in a very small chip area. A typical measured gain frequency curve for such a system is shown in Figure 4.13. The pass-band is also shown on an expanded amplitude and frequency scale and it can be seen that the response of this

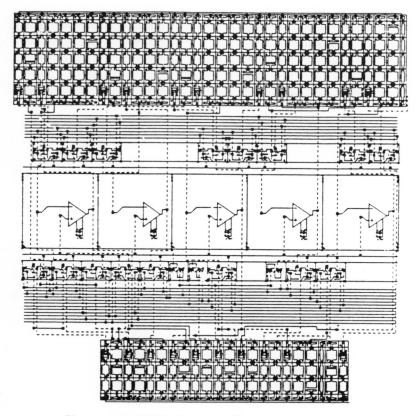

Figure 4.12 CMOS 5th order filter chip layout plot

filter is very close to the ideal response. This serves to demonstrate the high precision of response obtained due to the dependence of the transfer function on capacitor ratios which can be made accurate, the use of parasitic-insensitive circuit structures and the use of optimum switching waveforms within the integrators for low sensitivity to non- grounded parasitic capacitances. Signals on the power supply can feed through to the output terminal and power supply rejection ratio figures as low as -20dB or even -10dB are possible for switched capacitor filters. Thus, power supply voltages need to be well regulated, or alternatively, switched capacitor filters can be realised as fully differential systems which provide immunity to signals in the power supply lines. In this way power supply rejection ratios of the order of 50dB can be obtained. This example illustrates the ability of the switch capacitor filtering technique to provide very high quality filters using CMOS technology.

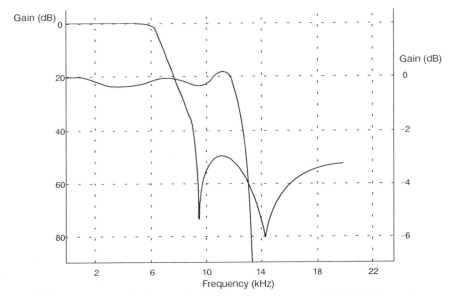

Figure 4.13 Typical measured amplitude response of CMOS filter chip

4.6 Case Study 2 - Narrow-band Filters using Multi-rate SC Techniques

In this section we make use of the multi-band response at the input and the output of a switched capacitor filter due to its discrete-time nature in order to realise Q-enhanced systems with narrow bandwidths below 1%. Such requirements cannot be realised by conventional switched capacitor systems as described in the previous section because capacitor spread increases with reducing bandwidth and spreads above the technological limit of about 100 are obtained. It has been shown that the conditions for avoiding aliasing distortion and imaging may be met by anti-alias and anti-imaging filters which, instead of being low-pass, as in the conventional system, illustrated in Figure 4.10, have band-pass responses which select one of the high frequency bands at the input and the output, as illustrated in Figure 4.14 [26]. The frequency translation from a high input frequency to a lower processing frequency and back to a high output frequency has the property that the absolute bandwidth is preserved and therefore the relative bandwidth of the switched capacitor bandpass filter is reduced (or its Q-factor increased) by the ratio by which the frequency bands are translated [27]. Such a system requires bandpass anti-alias and anti-imaging filters with considerable response precision which cannot be realised by active RC circuits. However, a range of switched capacitor decimators and

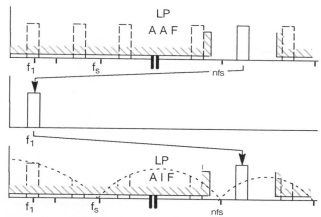

Figure 4.14 Frequency band diagram for single-path frequency-translated system

interpolators has been developed which efficiently realise the required anti-alias and anti-imaging filters [28]. They make full use of the dependence of the transfer function of the switched capacitor circuit on the ratio of capacitor values and are based on poly-phase techniques which can minimize the settling-time requirements for the operational amplifiers [29].

Figure 4.15 shows an example of a narrow-band system based on such a single-path frequency-translated approach [26]. At the centre is the switched capacitor bandpass filter which has a mid-band frequency of 4kHz and a relative bandwidth of 2% which is close to the minimum that can be realised due to the technological restriction on the ratio of capacitor values achievable. The switched capacitor bandpass filter is preceded by a switched capacitor decimator and followed by a switched capacitor interpolator. These circuits select, at the input and the output, high frequency bands centred at 20kHz and thus the relative bandwidth of the complete system is effectively reduced from 2% to 0.4%. The response of this system is shown in Figure 4.16. This single-path frequency-translated approach could be used to realise relative bandwidths as low as 0.1%. Extensions of the approach have been proposed which could allow the realisation of relative bandwidths as low as 0.01% [30] and these approaches are now being evaluated.

Having illustrated the implementation of an audio frequency switched capacitor filter using CMOS technology, and seen how more complex switched capacitor systems may be used to realise specific system requirements, such as very-narrow-band filters, it is appropriate to ask the

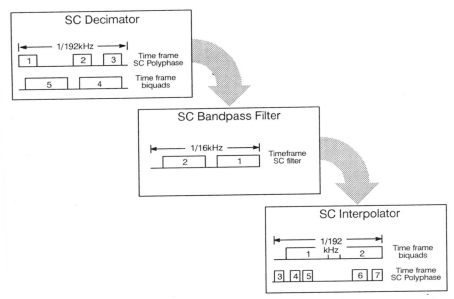

Figure 4.15 Single-path frequency-translated narrow band system example

question - up to what frequency CMOS switched capacitor filters may be used? It may be shown that the limitation on maximum frequency of operation depends primarily on the dynamic performance of the operational amplifier. An analysis has indicated that the maximum switching frequency is given by:

$$f_{max} = 0.25 \; d \; \mu \; 1^{-2}(V_{gs} - V_T) \; (\ln \; \epsilon^{-1})^{-1} \qquad \boxed{4.4}$$

when d is the duty cycle of the switching waveforms, μ is the mobility of the electron in the semiconductor material, 1 is MOSFET gate-length, Vgs is the MOSFET gate-source voltage, VT is the threshold voltage and ϵ is equal to the maximum permissible relative error in the amplifier output voltage [30]. Using typical parameters of d = 0.25 (25% duty cycle), μ = 0.06 m^{2}v^{-1}s^{-1} (n type silicon), Vgs-VT = 1 V and ϵ = 0.001 (0.1% accuracy), we obtain:

$$f_{max} = 0.0005429 \; 1^{-2} \qquad \boxed{4.5}$$

For a typical CMOS process, having a gate-length of 2 microns we obtain an fmax of 136MHz. When we take into account the problems of aliasing and imaging, the maximum filtering frequency will be in the range 5 to 10MHz. However for radar and communications systems, there is a need for integrated filters at higher frequencies and it is tempting to look at equation (4.5) and ask whether the maximum switching frequency can be increased in any way. One possible approach is to use semiconductor material with a mobility μ which is higher than that of Silicon. One such material is Gallium Arsenide (GaAs) and we now explore the use of GaAs technology for the realisation of high frequency switched capacitor filters.

4.7 Case study 3 - High Frequency Filter using GaAs Technology

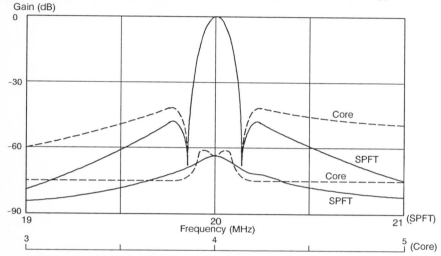

Figure 4.16 Typical measured gain and noise responses for narrow-band SPFT system and core band-pass filter

The mobility of the electron in GaAs is about six times higher than in silicon and therefore GaAs technology is eminently suitable for high frequency circuits [31, 32]. In addition the GaAs substrate material is semi-insulating, as compared with the semiconducting silicon substrate and this means that parasitic capacitances of components to ground are reduced and there is excellent isolation between components. The principle active component of GaAs technology is the metal semi-conductor FET or MESFET. It is most generally manufactured in depletion-mode or normally-on form. N-channel devices are generally the only type available because p-channel devices are very slow due to the low mobility of the hole in GaAs. Device gate-lengths are typically of the order of 1 micron. The open circuit voltage gain of a single device gm/go, where gm is the transconductance and go is output conductance, has a value of only about 20 which is very much lower than for MOS devices. Other components available in GaAs technology are Schottky barrier diodes, which have a forward voltage drop of about 0.7 volts and capacitors which are generally of the metal-insulator-metal type where the insulator is generally silicon nitride. GaAs technology is suitable for realising high-speed digital circuits [33] and the successful realisation of analogue circuits will permit the realisation of very high speed mixed-mode systems.

The key component of switched capacitor filter design is the operational amplifier and in order to achieve a gain of 1000 or above using MESFET devices with open circuit gains of only of the order of 20 has led to special circuit design techniques [34]. A typical example of a suitable operational amplifier design is shown in Figure 4.17a. The high gain is achieved by the

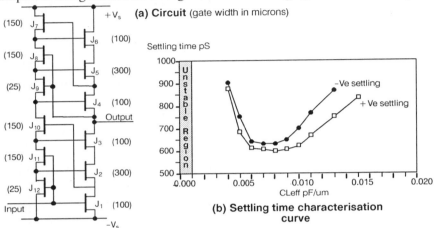

(a) Circuit (gate width in microns)

(b) Settling time characterisation curve

Figure 4.17 GaAs operational amplifier and performance

use of double cascode techniques and a special biasing approach has been developed which maximises the bandwidth [35]. The circuit achieves maximum performance for a given chip area and power consumption by using using a push-pull architecture based on the use of an inverting current mirror [36]. Simulation of amplifier performance yields a gain of about 60dB, gain bandwidth product of 3.5GHz and a phase margin of 60 degrees with a 1pF capacitive load. The amplifier parameter which most affects maximum switching frequency in switched capacitor applicators is settling time, and Figure 4.17b shows simulated settling time for the amplifier in Figure 4.17a as a function of load capacitance [37]. We can see that with a suitable load capacitance, settling times of the order of 600 picoseconds are possible. The feasibility of GaAs technology for switched capacitor applications was evaluated by the design and fabrication of a second order bandpass filter, having a design switching frequency to mid-band frequency ratio of 25, a mid-band gain of unity and a Q-factor of 16 [38]. The circuit diagram is shown in Figure 4.18. We have already seen that amplifier settling time is critically dependent on load capacitance and the capacitors Ca, Cb and Cc have been introduced in order to optimise the loading on the

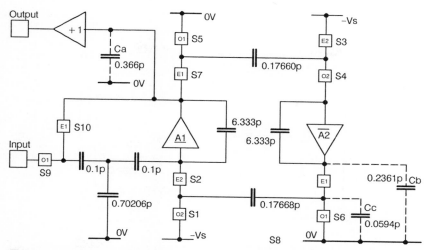

Figure 4.18 2nd order bandpass filter circuit for implementation in GaAs

amplifiers for optimum settling performance. The output is fed via a unity-gain buffer amplifier. The switches are implemented by using MESFET's as switching devices, fed by suitable switch control circuits [39].

Figure 4.19 shows a layout plot for this filter, which was fabricated by Anadigics Inc. The size of the chip is 3.3mm by 2.7mm and total power consumption is 400mW. About half of the chip area and half of the power consumption are involved in the buffer amplifier for driving off-chip loads. Figure 4.20a shows a typical measured gain frequency response for this filter, operated with a switching frequency of 500MHz. It can be seen that the mid-band frequency is very close to the ideal value of 20MHz and that mid-band gain is also close to 0dB. Figure 4.20b shows the pass-band response using an expanded scale, further confirming the accuracy of the response. The peak measured noise of this filter using a bandwidth of 10kHz is about 70dB below the nominal output signal level. This figure is about 20dB worse than would be obtained for a similar test conducted at a lower frequency on a CMOS filter but is acceptable for many applications. This filter used an operational amplifier with minimum settling time of 600 pico-seconds. Further research has succeeded in developing a family of fast settling operational amplifiers which achieve settling times as low as 200 pico-seconds. It is expected that these amplifiers will allow the realisation of switched capacitor filters with switching frequencies as high as 1GHz and above, making possible filtering frequencies of 100MHz and above [40].

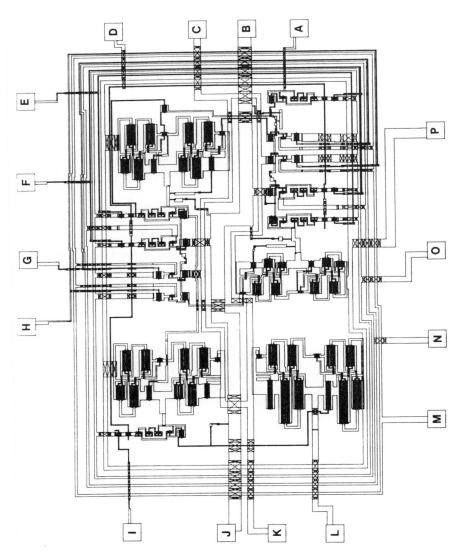

Figure 4.19 Layout plot of 2nd order GaAs filter chip

4.8 Conclusions

Figure 4.21 shows the Q-frequency performance capability chart of Figure 4.1, but now the Q and frequency capabilities of switched capacitor filters have been extended, according to the developments which have just been described. On the vertical axis the Q capabilities have already been

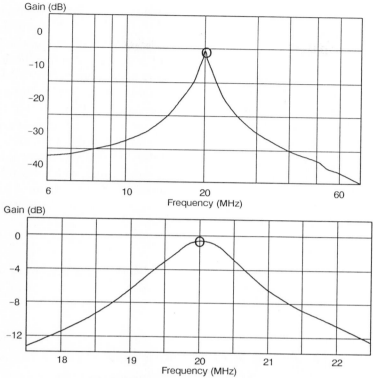

**Figure 4.20 Measured GaAs' filter overall response
and expanded pass-band response**

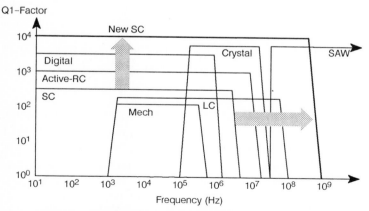

**Figure 4.21 Projected operating frequency/Q-factor
potential for switched capacitor filters**

increased from about 200 to 1000 using the single path frequency translated approach and further increases to Q values of 10,000 are anticipated by further development of this technique, at present under investigation. On the frequency scale, the maximum frequency filtering frequency limit of about 5MHz of CMOS technology has already been increased to 20MHz using GaAs technology and further developments making use of fast-settling operational amplifiers are expected to lead to filtering frequencies up to above 100MHz. Thus the switched capacitor filtering technique can provide Q values which compare with the high Q technologies of crystal and SAW filters and they cover a frequency range from sub-audio medical frequencies to hundreds of megahertz. Switched capacitor filtering is an approach which, compared to other technologies, can be described as universal and it will find very demanding applications over its full range of Q values and frequencies.

4.9 Acknowledgements

The authors acknowledge support for the work described from the Admiralty Research Establishment, from British Telecom and from the Science and Engineering Research Council.

4.10 References

[1]   Modern active filter design, IEEE Press (Selected reprint series), eds R. Schaumann et al, 1981

[2]   B J Hostika, R W Brodersen and P R Gray, MOS sampled data recursive filters using switched capacitor integrators, IEEE J Solid State Circuits, vol SC-12, pp 600-608, Dec 1977

[3]   J T Caves et al, Sampled analog filtering using switched capacitors as resistor equivalents, , IEEE J Solid State Circuits, vol SC-12, pp 592-599, Dec 1977

[4]   W Saraga, D G Haigh and R G Barker, Microelectronic active-RC channel bandpass filters in the frequency range 60-108kHz for FDM SSB telephone systems, IEEE Trans on Circuits and Systems, Vol CAS-25, no 12, pp 1022-1031, December 1978 (included in 'Modern Active Filter Design', IEEE Press, 1981).

[5]   W Saraga, D G Haigh and R G Barker, A design philosophy for microelectronic active-RC filters, Proc IEEE, Vol 67, no 1, pp 24-33, January 1979.

[6] R Schaumann and M A Ali, Chapter 9 (Continuous-time filters) in Analogue IC Design: the current-mode approach, eds C Toumazou, F J Lidgey and D G Haigh. Peter Peregrinus, London, 1990

[7] D G Marsh, B K Ahija, T Misawa, M R Dwarkanath, P E Fleischer and V R Saari, A single- chip CMOS PCM CODEC with filters, IEEE J Solid State Circuits, vol SC-16, no 4, pp 308- 315, Aug 1981

[8] L T Lin, H F Tseng and L Querry, Monolithic filters for 1200 baud modems, Digest of technical papers, International Solid State Circuits Conference, pp 148-149, 1982

[9] B J White, G M Jacobs and G F Landsburg, A monolithic dual-tone multifrequency receiver, IEEE J Solid State Circuits, vol SC-14, no 6, pp 412-418, Dec 1979

[10] MOS switched capacitor filters: analysis and design, ed G S Moschytz, IEEE Press, 1984

[11] Analog MOS integrated circuits for signal processing, R Gregorian and G C Temes, John Wiley, 1986

[12] Design of analog filters: passive active-RC and switched capacitor, R Schaumann, M S Ghausi and K R Laker, Prentice Hall, 1990

[13] P E Fleischer, A Ganesan and K R Laker, Parasitic compensated switched capacitor circuits, Electronics Letters, vol 17, no 24, pp 929-931, 26th Nov 1981

[14] K Martin, Improved circuits for the realisation of switched capacitor filters, IEEE Trans Circuits Systems, vol CAS-27, no 4, pp 237-244, April 1980

[15] K R Laker, P E Fleischer, and A Ganesan, Parasitic insensitive biphase switched capacitor filters realised with one operational amplifier per pole pair, Bell System Tech J, vol 61, pp 685-707, May-June 1982

[16] Y Tsividis, Principles of operation and analysis of switched capacitor circuits, Proc IEEE, vol 71, pp 926-940, Aug 1983

[17] L T Bruton, Low sensitivity digital ladder filters, IEEE Trans Circuits Systems, vol CAS- 22, pp 168-176, Mar 1975

[18] C W Soloman, Switched capacitor filters: precise compact inexpensive, IEEE Spectrum, June 1988

[19] D G Haigh and B Singh, A switching scheme for switched capacitor filters which reduces the effect of parasitic capacitances associated with switch control terminals, Procs 1983 IEEE International Symposium on Circuits and Systems (Newport Beach), pp 586-589, May 1983.

[20] D G Haigh and J T Taylor, On switch-induced distortion in switched capacitor circuits, Procs 1988 IEEE International Symposium on Circuits and Systems (Espoo, Finland), June 1988, pp 1987 - 1990

[21] M S Lee and C Chang, Switched capacitor filters using the LDI and bilinear transformations, IEEE Trans Circuits Systems, vol CAS-28, pp 265-270, April 1981

[22] J T Taylor, Exact design of elliptic switched capacitor filters by synthesis, Electronics Letters, vol 18, no 19, pp 807-809, Sept 1982

[23] U W Brugger, D C von Grunigen and G S Moschytz, A comprehensive procedure for the design of cascaded switched capacitor filters, IEEE Trans Circuits Systems, vol CAS-28, pp 803-810, Aug 1981

[24] D A Linden, Discussion of sampling theorems, Proc IRE, vol 47, pp 1219-1226, July 1959

[25] D Wilcox et al, Analogue CMOS standard cells and circuits, Procs 1984 European Solid State Circuits Conference, Edinburgh, Sept 1984

[26] J E Franca, A single-path frequency-translated switched capacitor bandpass filter system, IEEE Transa on Circuits and Systems, vol 32, no 9, pp 938-943, Sept 1985

[27] J E Franca and D G Haigh, Design and applications of single-path frequency-translated switched capacitor systems, IEEE Transactions on Circuits and Systems, vol 35, no 4, pp 394 - 408, April 1988

[28] J E Franca, Nonrecursive polyphase switched capacitor decimators and interpolators, IEEE Transactions on Circuits and Systems, vol 32, no 9, pp 877-887, Sept 1985

[29] J E Franca and D G Haigh, Optimum implementation of IIR switched capacitor decimators, IEEE Int Symp Circuits and Systems, pp 76-79, 1987

[30] D G Haigh and J T Taylor, High frequency switched capacitor filters for CMOS technology, Procs 1988 IEEE International Symposium on Circuits and Systems (Espoo, Finland), pp 1469 - 1472, June 1988

[31] GaAs integrated circuits:design and technology, ed J Mun, BSP Books, 1988

[32] GaAs techonology and devices and their impact on circuits and systems, eds D G Haigh and J K A Everard, Peter Peregrinus, September 1989

[33] R Cates, Gallium Arsenide finds a new niche, IEEE Spectrum, April 1990

[34] C Toumazou and D G Haigh, Design of operational amplifiers for sampled data applications using GaAs technology, IEEE Transactions on Circuits and Systems, vol 37, no 7, pp 922- 935, July 1990

[35] C Toumazou and D G Haigh, Design of a high-gain, single-stage operational amplifier for GaAs switched-capacitor filters, Electronics Letters, vol 23, no 14, 2nd July 1987, pp 752- 754.

[36] C Toumazou and D G Haigh, Design and application of GaAs MESFET current mirror circuits, IEE Proceedings Part G Special Issue 'Current-mode analogue circuits', vol 137, no 2, April 1990, pp 101 - 108

[37] C Toumazou and D G Haigh, Some designs and a characterisation method for GaAs operational amplifiers for switched capacitor applications, Electronics Letters, vol 24, no 18, 1st September 1988, pp 1170 - 1172

[38] D G Haigh, C Toumazou, S J Harrold, K Steptoe, J I Sewell and R Bayruns, Design optimisation and testing of a GaAs switched capacitor filter, IEEE Trans on Circs and Systems, vol CAS-38, no 8, Aug 1991

[39] D G Haigh, A K Betts, K Steptoe and J T Taylor, The design of switched capacitor filter circuits for GaAs MSI technology, Procs 1989 European Conference on Circuit Theory and Design, Brighton (UK), 5th - 8th September 1989, pp 32 - 36

[40] D G Haigh and C Toumazou, High performance GaAs analog sampled data circuits and systems, Procs IEEE Int Symp Circuits Systems, Singapore, June 1991

Author Afiliations D G Haigh is with the Department of Electronic and Electrical Engineering, University College London, C Toumazou is with the Department of Electrical Engineering, Imperial College of Science Technology and Medicine, London, J E Franca is with the Instituto Superior Tecnico, Lisbon and B Singh is with Phoenix VLSI Consultants, Towcester, Northants

Chapter 5

Switched-Current Techniques for Analogue Sampled-Data Signals

N. C. Battersby & C. Toumazou – Imperial College University of London
J. B. Hughes – Philips Research Labs. Redhill, Surrey. England

5.1 Introduction

Since its introduction in 1972 [1] the switched–capacitor (SC) technique has come to dominate analogue sampled–data signal processing. Initially, switched–capacitors replaced active–RC filters, offering higher precision and compactness, and as the technique matured they were used to realise more and more complex signal processing systems [2].

From the mid–1980s complete signal processing systems were increasingly being realised digitally and switched–capacitor circuits were only being used in the A/D and D/A interfaces. To implement complete single chip subsystems switched–capacitors had now to co–exist with digital circuitry. Inevitably, with the dominance of digital circuitry, processing technology is being tuned for optimum digital performance; unfortunately the switched–capacitor technique is not fully compatible with this processing.

The switched–capacitor technique requires the availability of a linear floating capacitor, leading to special processing options, such as double polysilicon, not required by digital circuitry. Now that switched–capacitors generally only occupy a small proportion of the total chip area, the cost of providing the extra processing options required is becoming an increasingly high overhead. The ever increasing density of digital circuits has lead to pressure for the standard supply voltage to be reduced from 5V to 3.3V [3]. Reducing the supply voltage directly reduces the maximum voltage swing and so also the dynamic range achievable by switched–capacitor circuits. Additionally, the threshold voltages used in 3.3V processes will not be lowered sufficiently for optimum switched–capacitor performance and so without the provision of extra doping options switched–capacitor performance will degrade. So, whilst performing only peripheral interface functions switched–capacitor circuits are becoming an increasingly costly part of the total system.

With the rapid increase in single chip mixed–mode systems, and their use in high volume consumer products, an analogue sampled–data signal processing technique which is fully compatible with digital processing is needed. The switched–current (SI) technique is an analogue sampled–data technique, which by overcoming the aforementioned drawbacks of switched–capacitor circuits, is compatible with digital processing technology.

In this chapter we introduce the switched–current technique and show how it can be used to implement various analogue sampled–data signal processing functions. The sources of error in the basic cells are identified and a number of enhanced cells, which overcome some of these errors, are described. Finally some future potential areas of application are explored.

5.2 First generation memory cells

The principle on which all switched–current (SI) memory cells operate is that an MOS transistor requires no gate current in order to maintain a constant drain–source current. The application of this principle to the design of filters was initially based around the switched–current mirror [4], now referred to as the first–generation switched–current technique [5].

The fundamental first–generation SI building block is the track and hold (T/H) memory cell and this is shown in Figure 5.1.

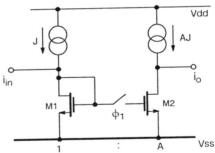

Figure 5.1 First Generation switched-current T/H memory cell

The memory cell shown in Figure 5.1 is a simple current mirror, with a switch S connected between the gates of the input and output transistors. Assuming both mirror transistors are operated in saturation, when switch S is closed (ie ø is high) the circuit operates as a simple current mirror and the oxide capacitances of M_1 and M_2 are both charged to V_{gs} where

$$V_{gs} = V_T + \sqrt{\frac{2(J + i_{in})}{K'(W/L)}}$$

5.1

Where V_T is the device threshold voltage, K' is the device transconductance parameter and W/L is the device aspect ratio.

The output current (i_o) is simply $-Ai_{in}$, where A is the aspect ratio of M_2 relative to M_1. When switch S opens the gate of M_2 is isolated and a voltage of near to V_{gs} is held on its oxide capacitance, thus maintaining an output current close to $-Ai_{in}$. In this way the circuit implements the T/H function.

The basic T/H memory cell can thus perform inversion, discrete time delay and scaling. It can also sum multiple input currents at its low impedance input node and generate multiple output currents by the addition of further mirroring transistors. A combination of these basic functions is all that is required to allow the realisation of sampled–data signal processing systems.

The integrator has proved to be an extremely versatile and useful building block for the synthesis of switched–capacitor filters [2] and it is possible to realise an equivalent integrator using first–generation SI techniques. The generalised form of such an integrator is shown in Figure 5.2.

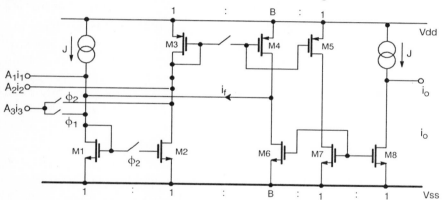

Figure 5.2 Generalised first generation switched–current integrator.

The integrator consists of a two stage delay with two outputs, one of which is fed back to the input summing node. To understand the operation of the circuit consider the A_1i_1 input. During phase ϕ_2 of period $(n-1)$ the total current $I(n-1)$ flowing in the diode–connected transistor M_1 is

$$I(n-1) = A_1i_1(n-1) + i_f(n-1) + J \qquad \boxed{5.2}$$

where $i_f(n-1)$ is simply a copy of the output current $i_o(n-1)$.

By scaling the output current by a factor B, $I(n-1)$ can be written as

$$I(n-1) = A_1i_1(n-1) + Bi_o(n-1) + J \qquad \boxed{5.3}$$

During phase $\phi1$ of period (n), $I(n-1)$ is being stored in the p–channel memory cell consisting of M_3, M_4 and M_5. Thus during phase $\phi2$ of period

(n) the current stored in M_5 is $I(n-1)$ and in M_4 is $BI(n-1)$. However, a current J is sunk by M_7 and BJ by M_6 and so the output current is

$$i_o(n) = \frac{i_f(n)}{B} = A_1 i_1(n-1) + B i_o(n-1)$$

5.4

In the z–domain this becomes

$$\Rightarrow i_o(z) = A_1 z^{-1} i_1(z) + B z^{-1} i_o(z)$$

5.5

$$i_o(z) = A_1 \frac{z^{-1}}{1 - Bz^{-1}} i_1(z)$$

5.6

This corresponds to the forward Euler z–transform of a damped non–inverting integrator. Completing the analysis for the $A_2 i_2$ and $A_3 i_3$ inputs, the overall transfer function is found to be

$$I_o(z) = A_1 \frac{z^{-1}}{1 - Bz^{-1}} i_1(z) - A_2 \frac{1}{1 - Bz^{-1}} i_2(z) - A_3 \frac{1 - z^{-1}}{1 - Bz^{-1}} I_3(z)$$

5.7

The generalised first generation SI integrator (Figure 5.2) can therefore perform both forward and backward Euler integration as well as inverting feedforward (if $B = 1$), and so, in conjunction with the basic T/H cell, it can be used to synthesize a wide range of filters, using the techniques originally developed for switched–capacitors. Furthermore, it has recently been shown [6] that by transposing the SFG (signal flow graph) of a switched–capacitor circuit a switched–current circuit with the same properties can be realised.

Experimental verification of this technique for filter design was first reported in 1990 [7] and subsequently in a number of further papers [9–14]. These papers verified that the switched–current technique can be successfully used to realise a range of different filter structures; however, they also highlighted the major source of error in the first–generation switched–current cell, that being the matching accuracy of the mirror transistors.

Although matching is by no means the only source of error present (other sources are discussed in section 4) the achievable matching accuracy is a serious limitation to circuit performance, and for this reason a second generation switched–current memory cell has been developed.

5.3 Second generation memory cells

The second generation switched–current memory cell is similar to a circuit known as a current copier [15–17], used as a current calibration system [16] and as a method of realising very accurate dynamic current mirrors [17]. The basic cell is shown in Figure 5.3.

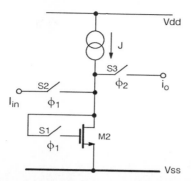

Figure 5.3 Second–generation switched–current memory cell.

The operation of the cell is much the same as that of the first–generation cell, except that this time the same transistor is used in both the sampling and retrieval phases and thus the very notion of mismatch is eliminated. During ø1 the memory transistor is diode–connected, with a current of $J + i_{in}$, and thus it's gate–source capacitance is charged. During ø2 the diode–connecting switch is opened and the memory transistor maintains a drain–source current of near to $J + i_{in}$, thus the output current is $-i_{in}$. Since, in practice, it is very difficult to ensure that S_1 and S_2 open at precisely the same time, a modified clock is used for S_1 so that it opens slightly before S_2, thus ensuring proper operation. By eliminating the problem of mismatch, the second–generation SI memory cell offers a considerable accuracy improvement over the first–generation cell.

Experimental verification of the second generation SI memory cell was obtained by fabricating a two stage delay cell (ie a cascade of two simple memories) using a 1.6μm CMOS process. To obtain high accuracy the single transistor memories (shown in Figure 5.3) were replaced by regulated cascode transistors (see section 5(i) for details) and a microphotograph of part of the chip is shown in Figure 5.4.

The results confirmed that the cell functions correctly, with an accuracy of better than 0.5%, and Figure 5.5 shows a photograph, illustrating the cell's input (upper trace) and output (lower trace) for a clock frequency of 10kHz.

The second–generation memory cell can be used to realise filter structures in the same way as the first–generation cell and, as an example, a realisation of a generalised integrator [18] is shown in Figure 5.6.

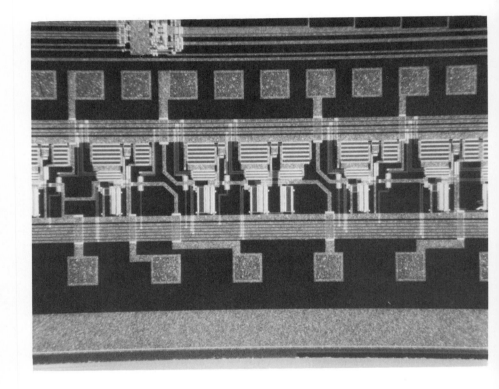

Figure 5.4 Microphotograph of part of the chip.

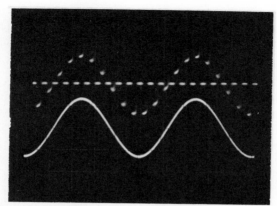

Figure 5.5 Measured current waveforms of a two stage regulated cascode SI delay cell.

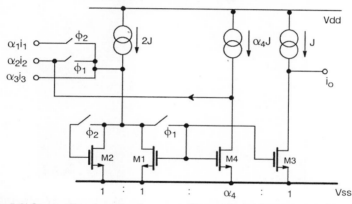

Figure 5.6 Generalised second–generation switched–current integrator.

The transfer function of this integrator may be derived using the same approach as was used for the first–generation SI integrator, and if this is done its transfer function is found to be:

$$i_o(z) = \frac{A_1 z^{-1}}{1 - Bz^{-1}} i_1(z) - \frac{A_2}{1 - Bz^{-1}} i_2(z) - \frac{A_3(1 - z_{-1})}{1 - Bz_{-1}} i_3(z)$$

$$where \quad A_1 = \frac{a_1}{1 = a_4}, \quad A_2 = \frac{a_2}{1 + a_4}, \quad A_3 = \frac{a_3}{1 + a4}, \quad B = \frac{1}{1 + a_4} \qquad \boxed{5.8}$$

Comparing equations (7) and (8) it is clear that the transfer functions of both the first and second–generation SI integrators are of the same form. However, a sensitivity analysis of the two structures [5] shows that, for lightly damped integrators, the low frequency gain and cut–off frequency of the first–generation integrator is highly sensitive to the feedback factor B whereas the second–generation structure is not so sensitive to its feedback factor $\alpha4$. Therefore the second–generation structure is much better suited to filters with a high Q–factor or having high clock frequency to cut–off frequency ratios.

The operation of the second generation SI integrator was verified by the fabrication of an ideal integrator using a 1.6μm CMOS process. As for the two stage delay cell, regulated cascode memory transistors (described in section 5(i)) were used to obtain higher precision. The ideal integrator was based on the generalised integrator (shown in Figure 5.6), but with $\alpha1 = 1$, $\alpha2 = \alpha3 = \alpha4 = 0$ and only the i_1 input used. Figure 5.7 shows a

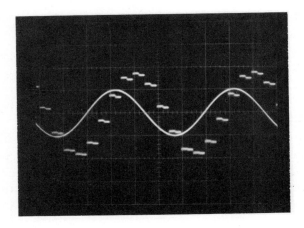

Figure 5.7 Measured current waveforms of an ideal SI integrator.

photograph of the integrator's input and output current waveforms, for a clock frequency of 10kHz and a signal frequency of 1kHz, and Figure 5.8 shows the measured and calculated gain of the integrator for a 100kHz clock.

The results confirm that the fabricated integrator functions correctly and from Figure 5.8 it can be seen that there is good agreement between the measured and calculated gain of the integrator, the low frequency gain roll–off observed being a result of the external feedback system employed to ensure that the integrator remained stable. The maximum clock frequency of the fabricated integrator was found to be around 1MHz, but it is envisaged that significant speed enhancements will be achieved in forthcoming designs.

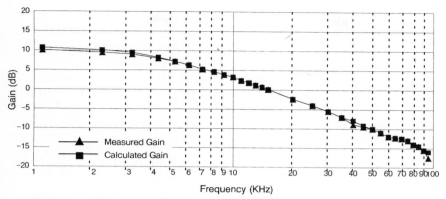

Figure 5.8 Measured and calculated gain of an ideal SI integrator.

5.4 Limitations of the basic SI memory cell

There are five main factors which limit the performance of the basic second generation switched–current memory cell (Figure 5.3) and in this section each of these limitations is briefly discussed.

5.4.1 Channel length modulation

Considering the basic SI memory cell (Figure 5.3), the drain of M_1 is connected to its gate during ø1 (storage phase) and to the load during ø2 (retrieval phase). Thus during ø1 the drain of M_1 will be at the same potential as its gate whereas during ø2 its potential will be determined by the load and will in general be different. The change in the V_{ds1} between the storage and retrieval phases causes a change in the effective channel length and thus results in an output current error. This error can be reduced by conventional cascoding, using a servo–amplifier [19,20], or more recently using a regulated cascode [21]. The use of a regulated cascode transconductor is discussed in section 5(i).

5.4.2 Charge injection

When the diode–connecting switch S_1 (Figure 5.3) is opened, the channel charge of the MOS transistors, used to implement the switch, flows in their respective source, drain and substrate ports [22]. The charge dumped onto the gate of M_1 will cause an error in the voltage sample of ΔV_{gs}, which results in an output current error of $-_{gm}\Delta V_{gs}$. A number of techniques have been proposed to minimise the effect of charge injection, these include the use of dummy switches, special layout, minimising the gate voltage of the switch [23] and an algorithmic cancellation scheme [24]. The algorithmic cancellation scheme is discussed in section 5(ii).

5.4.3 Junction leakage

Even when the MOS transistor(s) implementing the diode–connecting switch S1 (Figure 5.3) is off, a very small leakage current will flow through its reverse biased source–substrate junction causing a leakage of charge from the gate of M_1. This junction leakage current will only be of consequence in systems with very low clock frequencies.

5.4.4 Settling error

Incomplete charging of the memory transistor's gate capacitance (C_{gs}), when it is diode–connected (sampling), will lead to errors in the stored current. Provided that the memory transistor's output conductance is sufficiently low, switch resistance has little effect and the charging time–constant is simply C_{gs}/g_m. Thus for high accuracy, C_{gs}/g_m should be made sufficiently low compared to the clock period.

5.4.5 Noise

The MOS transistors which are used in the implementation of switched–current circuits introduce both thermal and $1/f$ (or flicker) noise and, as in other sampled–data circuits [25, 26], both sampled and direct (or nonsampled) noise are generated. An analysis of the noise generated in the basic second generation switched–current cell (Figure 5.3) has been performed previously [27] and in this section the main results are summarised.

Although all the transistors used in the implementation of the SI memory cell contribute noise, the noise generated by the memory transistor M_1 can be shown to dominate [27], and therefore to simplify the analysis it is assumed that all the noise originates from M1. Further, the noise generated by M_1 is taken to be a gate refered voltage with a noise PSD of S_{M1}.

During the sampling phase, when M_1 is diode–connected, the noise causes fluctuation of the gate voltage. At the end of the sampling phase, the gate of M_1 is isolated and the instantaneous value of the noise is 'frozen' on the gate capacitance. Then during the retrieval phase this 'frozen' gate voltage error causes a corresponding output current error and it has been shown [27] that the PSD of this noise current is given by

$$S_I^{S/H}(f) = g_m^2 \left(\frac{\tau}{T_c}\right)^2 \left(\frac{sin(\omega\tau)}{\omega\tau}\right)^2 \left(\frac{1}{1 + (f/f_2)^2}\right) \cdot \sum_{k=-\infty}^{\infty} S_{m1} (f - kf_c)$$

5.9

where τ = the hold period, T_c = clock period and f_o = loop bandwidth of M_1 diode–connected.

From equation 9 it can be seen that the sampling process alters the shape of the noise by folding high frequency noise down into the baseband and by imposing the sinc function characteristic on the noise.

During the retrieval phase noise is still being generated by the memory transistor and this directly gives rise to a noise component at the output, whose PSD can be shown [27] to be

$$S_I^D(f) = g_m^2 \left(\frac{\tau}{T_c} \right)^2 S_{M1}(f)$$

5.10

The total noise in the SI memory is therefore the sum of the sampled (equation 9) and direct (equation 10) noise PSDs.

To illustrate more clearly the noise characteristics of switched–current circuits, the noise generated in the fabricated two–stage delay cell and ideal integrator were measured. The noise generated by the two–stage delay cell, for a clock frequency of 100kHz, is shown in Figure 5.9.

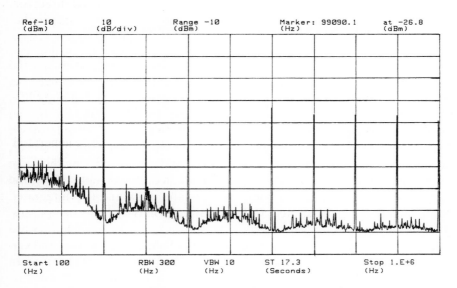

Figure 5.9 Measured noise in the two–stage SI delay cell.

It can clearly be seen that the noise is shaped by a sinc function with nulls at multiples of twice the clock frequency and is consistent with the theoretical expectation.

The noise generated by the ideal integrator, for a clock frequency of 10kHz, is shown in Figure 5.10.

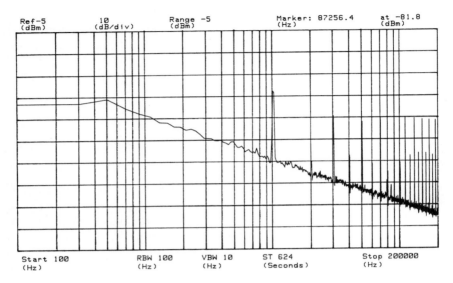

Figure 5.10 Measured noise in an ideal SI integrator.

The main observation from Figure 5.10 is that the integrator has the property that it integrates the unwanted noise as well as the signal. An intuitive insight into this noise shaping effect can be obtained without a rigourous noise analysis, by considering the transfer function of the generalised integrator shown in equation 5.8. In equation 5.8 it is shown that if the input current is sampled on $\phi 1$ or $\phi 2$ it will be integrated, however, a continuous input will be fedforward (if $B = 1$). Now considering the noise of M_1 and M_2 (Figure 5.6) to be represented as drain currents. The noise currents will be sampled on both phases of the clock and two integrated components will be generated. If the signal was correlated between phases, the two integrated components could be subtracted to leave a feedforward component, however, because noise is uncorrelated the powers of the two components must be added and this results in overall integration. In addition to this there will be noise folding effects and the noise generated by the output transistor.

The noise characteristic of the switched–current integrator creates difficulties for the design of low noise filters which are integrator based and consequently the minimisation of noise remains a challenge to the designer.

5.4.6 Limited signal swing

Both the first and second–generation switched–current memory cells (figures 5.1 and 5.3) described so far are class A circuits and so it is not possible for the signal swing to exceed the bias current level. Thus for a large signal swing it is necessary to provide a large bias current, which leads to high power consumption. In section 5(iii) a class AB switched–current memory cell [28], which is a way of overcoming this limitation, is described.

5.5 Improved SI memory cells

5.5.1 The regulated cascode

In section 5.4.1 it was seen that channel length modulation is a serious performance limitation of the basic switched–current memory cell (Figure 5.3). By cascoding the memory transistor the output resistance of the cell is increased by approximately $g_m r_{ds}$, typically 100. This is usually sufficient, however, large signal currents can lead to loss of saturated operation when similar cells are cascaded.

By using the regulated cascode switched–current cell [21] the output resistance during saturated operation is boosted by approximately 10,000 times when compared with the basic cell (Figure 5.3) and remains very high even when the memory transistor is non–saturated. The regulated cascode switched–current memory cell is shown in Figure 5.11.

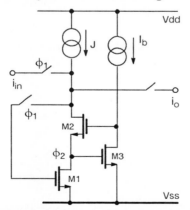

Figure 5.11 Regulated cascode switched–current memory cell.

The operation of the cell is as follows. During ø1, M_1 is diode–connected and its gate capacitance charges up until V_{gs1} is sufficient to maintain the drain current. The gate of M_3 is connected to the drain of M_1 and so senses the drain–source voltage of M_1 (V_{ds1}). As M_3 carries a constant current I, any difference between V_{ds1} and the V needed to maintain I is amplified by the loop formed by M_2 and M_3. Thus Vds1 is maintained constant at V_{gs3}.

During phase ø2, M_1 maintains the sampled–drain current, and the effect of any output voltage variations on V_{ds1} is reduced by the gain of the loop ($g_{m2}r_{ds2}g_{m3}r_{ds3}$), which when all the transistors are saturated is approximately 10,000.

The very high gain of the loop allows the memory transistor M_1 to be operated in non–saturation, whilst still maintaining an output resistance similar to that of a saturated cascode. By operating the memory transistor non–saturated, its g_m is made independent of the signal magnitude and so gives the opportunity to improve charge injection performance.

Simulated performance results [21] show that the regulated cascode cell settles in approximately the same time as the simple memory cell and that with M_1 in saturation a simulated output resistance of $10G\Omega$ could be achieved and with M_1 unsaturated $80M\Omega$.

5.5.2 The algorithmic cell

The charge injection error that occurs in the basic SI cell was discussed in section 5.4.2 and a number of schemes, aimed at reducing the size of the error, were mentioned. The algorithmic cell [24] uses an algorithmic means to cancel the charge injection error whilst not relying on component matching and is thus able to achieve an improvement in charge injection error of about two orders of magnitude.

The algorithmic memory cell is shown in Figure 5.12a [24] and its timing

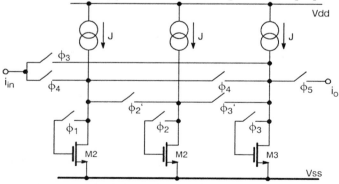

Figure 5.12a Algorithmic switched–current memory cell.

diagram in Figure 5.12b. The cell comprises of 3 basic memory cells and requires 5 clock phases; it should be noted that phases ϕ2' and ϕ3' (not

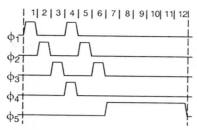

Figure 5.12b Algorithmic switched–current memory cell.

shown) are the same as $\phi2$ and $\phi3$ with a delayed trailing edge to avoid additional charge injection.

The concept of the circuit is to access the charge injection error of a memory cell, invert it and then pass it through the same cell again, thus cancelling the charge injection error without the need for matching.

The operation of the cell is as follows. During periods 1 and 2 the charge injection error of the first two cells is accumulated. During period 3 the charge injection error of the third cell is added; however, this is done in the presence of the input current (i_{in}), to ensure that the g_m of M_3 is the same as it will be in the output phase ($\phi5$). In period 4 the output of the third cell is summed with the input (i_{in}), leaving just the accumulated charge injection errors to be sampled by the first cell. The uneven number of signal inversions around the loop ensures that the charge injection error of the first cell in period 1 cancels with its charge injection in period 4. The same cancellation occurs in the second cell and then again in the third cell, where the input signal is reintroduced. Finally, during periods 7–12, the input current (i_{in}) is outputted together with any residual error. Thus using the proposed cell, charge injection errors have been cancelled, according to the following relationship

$$i_{out} = -i_{in} - (\delta_{14} - \delta_{11}) + (\delta_{25} - \delta_{22}) - (\delta_{36} - \delta_{33}) \qquad \boxed{5.11}$$

where δ_{nj} is the charge injection in the nth cell during period j.

To assess the performance of the algorithmic cell it was simulated using HSPICE and the parameter set of a typical 2μm N–well CMOS process. All the switches used were 2/2μm (W/L); the memory transistors were 100/20μm for the saturated regulated cascode and 25/20μm for the unsaturated regulated cascode. The bias current J used in all cases was 100μA and the clock frequency was 2MHz.

The charge injection error of the algorithmic cell (Figure 5.12a), using regulated cascode memory transistors, and the basic regulated cascode memory cell (Figure 5.11) was compared for both saturated and unsaturated memory transistor operation. The simulated results showed that the improvement in charge injection error, obtained by using the algorithmic cell, was approximately one order of magnitude improvement for the saturated case and two orders of magnitude improvement for the unsaturated case.

The use of this more complex clocking scheme reduces the speed of the algorithmic cell to be one sixth of the speed of a single cell using the same size transistors. However, because such a large improvement in charge injection performance has been achieved the speed/accuracy ratio of the cell has still significantly improved and this is the more important factor [23].

This improved performance has also been achieved at the cost of increased area and power consumption; however, in typical system applications where most of the chip area will be occupied by digital logic the price paid for the improvement in performance is not prohibitive. Furthermore class AB techniques can be used to reduce power consumption.

5.5.3 The class AB cell

The class A nature of the first and second–generation switched–current memory cells (figures 5.1 and 5.3) limits their maximum signal swing to no more than the supply bias current J. By employing class AB techniques it is possible to extend the maximum signal swing well beyond the bias current level.

The class–AB switched–current memory cell [28] is based on the operational amplifier supply current sensing techniques used to realise current–conveyors [29], and the basic cell is shown in Figure 5.13.

The class AB memory cell (Figure 5.13) is a CCII + current–conveyor [29] in which the two current mirrors have been replaced by switched–current memory cells of the type shown in Figure 5.3. The operation of the cell is as follows. During $\phi 1$, M_5 and M_6 are diode connected (sampling mode), and during $\phi 2$, M_5 and M_6 are output transistors (hold mode). The switches connecting the drains of M_2 and M_4 to the power rails ensure that M2 and M4 maintain current during the hold period $\phi 2$, thus reducing the recovery time during sample period $\phi 1$.

The input voltage following action ensures that the iin node remains at approximately ground potential. The sharing of i_{in}, assuming matched devices and equal n– and p–channel quiescent transconductance, can be described by:

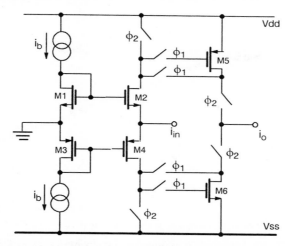

Figure 5.13 Basic class AB switched–current memory cell.

$$i_{D4} = \frac{(4I_b + I_{in})^2)}{16I_b} \qquad \frac{(4I_b + i_{in})^2}{16I_b} \qquad \text{for } |i_{in}| \leq 4I_b$$

5.12

where i_{D4} and i_{D2} are the drain currents of M_4 and M_2 respectively.

So from KCL at the output

$$i_{D6} = i_{D8} = i_{in} = i_{out}$$

5.13

When the input current exceeds 4Ib the distribution will become non–linear, but equation (13) must still be satisfied.

To verify the operation of the class AB cell it was simulated using HSPICE and the level 2 parameters of a typical 5V 2μm digital CMOS process. A comparison between the transfer characteristics of the class A (100μA bias current) and class AB (10μA bias current) regulated cascode memory cells is shown in Figure 5.14.

It can be seen from Figure 5.14 that the class AB memory cell achieves the same signal swing as the class A cell, whilst requiring only a tenth of the bias current. Further simulations of the class AB cell showed that it was possible for the signal current to exceed the bias current by 100 times, although performance was somewhat degraded, and that the voltage headroom required by the current memories ultimately limits the maximum current [28]. Further simulations also demonstrated no ???? in output signal distortion of the class AB cell compared with the class A cell for the same current swing.

The class AB cell can be used for the realisation of filter structures using techniques similar to those employed for class A structures [5]. The use of

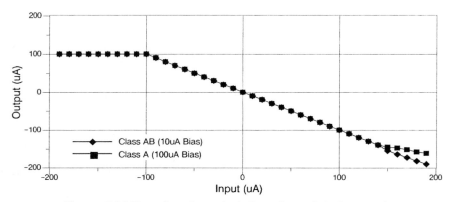

Figure 5.14 Transfer characteristics of regulated cascode class A and class AB switched–current memory cells.

the class AB cell to realise a generalised switched–current integrator, with exactly the same transfer function (equation 8) as the second–generation class A integrator (Figure 5.6) is shown in Figure 5.15.

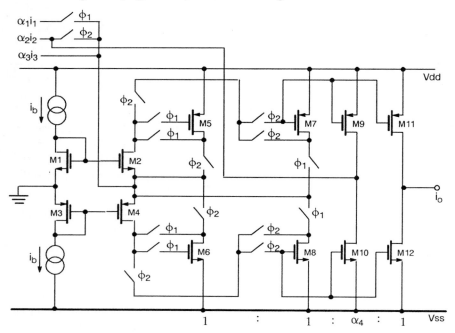

Figure 5.15 Generalised class AB switched–current integrator.

From the integrator example (Figure 5.15) it can be seen that the bias chain, consisting of the bias current sources and transistors M1 and M3, does not need to be replicated when cells are cascaded, since the gate voltages of M1 and M3 can be used to bias more than one cell; this results in a hardware and power consumption saving. It can also be seen that the integrator is built from three basic blocks; the input/bias block, the current memory block and a current mirror block; these blocks can easily be assembled into a variety of different structures.

5.6 Applications

The switched–current technique can be applied to the whole range of analogue sampled–data signal processing applications and in this section we point out suitable applications and in particular show how FIR and biquadratic filters can be realised.

Switched–current techniques can be applied to the synthesis of general filtering functions and the design of data converters. Two data converter architectures to which the technique is well suited are the sigma–delta [30] and algorithmic [20] converters. Another area of possible application is the design of analogue neural networks.

The application of switched–current techniques to the realisation of filters has been demonstrated using the first–generation cell, which has been used to realise biquads [7,10], ladder filters [7,13,14], FIR filters [8,13] and digitally programmable filters [11]. The major sources of error were found [14] to be nonunity current gain, due to device mismatch, and clock feedthrough effects, sources of error which can be considerably reduced using the enhanced cells described in the previous section.

5.6.1 Integrator based biquad

We now show, in block diagram form, how the second–generation of SI cell can be applied to the design of an integrator based biquad [5]. The block diagram of the biquad is shown in Figure 5.16.

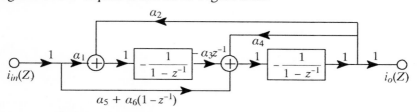

.Figure 5.16 Block diagram of an integrator based biquad.

The z–domain transfer function of this biquad is

$$H(z) = \frac{i_o(z)}{i_{in}(z)} = -\frac{a_5 + a_6 z^2 + (a_1 a_3 - a_5 - 2a_6)z + a_6}{(1 + a_4)z^2 + (a_2 a_3 - a_4 - 2)z + 1} \qquad \boxed{5.14}$$

The s–domain biquadratic transfer function is

$$H(s) = -\frac{k_2 s^2 = k_1 s + k_0}{s^2 + (\omega_0/Q)s + \omega_0^2}$$

5.15

Applying the bilinear–z transform, with a clock period T, to equation 15 and equating coefficients the filter coefficients show in Table 1 were calculated.

Table 5.1 Coefficients for use with the integrator based biquad.

Coefficient	Value
$\alpha_1\alpha_3$	$4k_0 T^2/D$
$\alpha_2\alpha_3$	$4\alpha_0 T^2/D$
α_4	$4\alpha_0 T^2/QD$
α_5	$4k_1 T/D$
α_6	$4k_2 - 2k_1 + k_0 T^2/D$
D	$\alpha_0^2 T^2 - (\alpha_0/Q)T + 4$

Using the coefficients in table 1 an integrator based biquad can be realised, and using the same approach a range of different biquads can be synthesized, including ones based on differentiators [5].

5.6.2 FIR filters

Traditionally FIR filters have been implemented using charge–coupled device (CCD) technology, but for high performance CCD filters non–standard processing options are required [31]. FIR filtering can readily be achieved using the switched–current technique [9,14] and is shown in block diagram form in Figure 5.17.

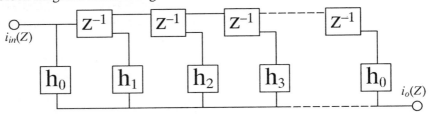

Figure 5.17 Block diagram of a switched–current FIR filter.

Each z^{-1} delay is produced by a cascade of two SI memory cells realising a full period delay. The main output of each delay cell is fed to the next cell and an extra mirrored output is also provided. The mirrored output is scaled by a coefficient hk (k = 0,1,2 ... n) and summed with the other mirrored output currents to form the overall output. The z–domain transfer function of the filter is given by:

$$H(z) = \sum_{K=0}^{n} h(k)z^{-k}$$

5.16

By choosing the coefficient values hk a variety of FIR filters can be realised.

The reported FIR filter [9] was 11 stages long and used first generation class A techniques; its accuracy was therefore limited by matching and charge injection errors. However, by using enhanced cells [21, 24, 28], a whole range of different FIR filters featuring higher speed and accuracy will be possible.

5.6.3 Sigma–Delta modulators

The application of switched–current techniques to the realisation of sigma–delta modulators is very promising and allows a complete sigma–delta A/D converter to be fabricated on a single digital chip.

A second–order sigma–delta modulator, based on switched–current techniques, has recently been reported [30]. Operating at a clock frequency of 1.024MHz and with a decimation factor of 64, a resolution of 13 bits, THD of better than −85dB and a signal to noise ratio of 72dB was achieved.

With this level of performance the sigma–delta modulator [30] can be applied to telecommunications and, with further development, performance levels suitable for, such applications as, digital hi–fi audio should be achieved. This is a very exciting and promising area of future research.

5.7 Conclusions

The switched–current technique is an analogue sampled data signal processing technique which by operating in the current domain offers several advantages over the switched–capacitor technique. Switched–current circuits do not require a linear capacitor and are able to function from reduced supply voltages, hence they are fully compatible with digital VLSI technology.

The switched–current technique can be applied to any analogue sampled–data signal processing task, but will find most application in mixed–mode environments where its compatibility with digital processing gives it a significant advantage over the conventional switched–capacitor technique.

In this chapter we have presented a review of the current state of switched–current techniques, pointed out the shortcomings of existing techniques, shown some enhanced memory cells and finally shown some promising areas of application.

At the present time there is considerable interest in the switched–current technique and future work will concentrate on improving the performance of the building blocks, developing applications and exploring the use of design automation tools.

5.8 Acknowledgements

The authors gratefully acknowledge D. M. Pattullo of Philips Research Labs (Redhill) for his contribution to this work and the S.E.R.C and Philips Research Labs (Redhill) for their financial support.

5.9 References

[1] D. L. Fried, "Analog sample–data filters," IEEE J. Solid–State Circuits, vol. SC–7, pp. 302–304, Aug. 1972.

[2] R. Gregorian, K. W. Martin and G. C. Temes, "Switched–capacitor circuit design," Proc. IEEE 1984, vol. 71, pp. 671–678, No. 88.

[3] Solid State Products Engineering Council, "JEDEC standard No. 8, Standard for reduced operating voltages and interface levels for integrated circuits," Electronic Industries Association, Washington D. C., Dec. 1984.

[4] J. B. Hughes, N. C. Bird and I. C. Macbeth, "Switched–currents – A new technique for sampled–data signal processing," Proc. IEEE International Symposium on Circuits and Systems, pp. 1584–1587, May 1989.

[5] J. B. Hughes, "Switched–current filters," in C. Toumazou, F. J. Lidgey and D. G. Haigh (Eds) 'Analogue IC design: the current–mode approach,' (Peter Peregrinus Ltd, 1990).

[6] G. W. Roberts and A. S. Sedra,"Sythesizing switched–current filters by transposing the SFG of switched–capacitor filter circuits," IEEE Trans. Circuits and Systems, vol. CAS– 38, no. 3, pp. 337–340, March 1991.

[7] T. S. Fiez and D. J. Allstot, "A CMOS switched–current filter technique," International Solid State Circuits Conference Digest of Technical Papers, pp. 206,207,297, Feb. 1990.

[8] D. J. Allstot, T. S. Fiez and G. Liang, "Design considerations for CMOS switched–current filters," Proc. IEEE Custom Integrated Circuits Conf., pp.8.1.1–8.1.4., May 1990.

[9] G. Liang and D. J. Allstot, "FIR filtering using CMOS switched–current techniques," Proc. IEEE International Symposium on Circuits and Systems, pp. 2291–2294, May 1990.

[10] T. S. Fiez, B. Lee and D. J. Allstot, "CMOS switched–current biquadratic filters," Proc. IEEE International Symposium on Circuits and Systems, pp. 2300–2304, May 1990.

[11] A. Begisi, T. S. Fiez and D. J. Allstot, "Digitally–programmable switched–current filters," Proc. IEEE International Symposium on Circuits and Systems, pp. 3178–3181, May 1990.

[12] H. C. Yang, T. S. Fiez and D. J. Allstot, "Current–feedthrough effects and cancellation techniques in switched–current circuits," Proc. IEEE International Symposium on Circuits and Systems, pp. 3186–3188, May 1990.

[13] T. S. Fiez and D. J. Allstot, "CMOS switched–current ladder filters," IEEE J. Solid–State Circuits, vol. 25, pp.1360–1367, no. 6, Dec. 1990.

[14] T. S. Fiez, G. Liang and D. J. Allstot, "Switched–current circuit design issues," IEEE J. Solid–State Circuits, vol. 26, pp.192–201, no. 3, March 1991.

[15] S. J. Daubert, D. Vallancourt and Y. P. Tsividis, "Current copier cells," IEE Electronics Letters, vol. 24, pp. 1560–1562, 8 Dec. 1988.

[16] W. Groeneveld, H. Schouwenaars and H. Termeer, "A self calibration technique for monolithic high–resolution D/A converters," International Solid State Circuits Conference Digest of Technical Papers, pp. 22–23, Feb. 1989.

[17] G. Wegmann and E. A. Vittoz, "Very accurate dynamic current mirrors," IEE Electronics Letters, vol. 25, pp. 644–646, 11 May 1989.

[18] J. B. Hughes, I. C. Macbeth and D. M. Pattullo, "Second–generation switched–current signal processing," Proc. IEEE International Symposium on Circuits and Systems, pp. 2805–2808, May 1990.

[19] D. Vallancourt, Y. P. Tsividis and S. Daubert, "Sampled–current circuits," Proc. IEEE International Symposium on Circuits and Systems, pp.1592–1595, 1989.

[20] D. G. Nairn and C. A. T. Salama, "Ratio–independent current–mode algorithmic analog–to– digital converters," Proc. IEEE International Symposium on Circuits and Systems, pp.250– 253, 1989.

[21] C. Toumazou, J. B. Hughes and D. M. Pattullo, "Regulated cascode switched–current memory cell," IEE Electronics Letters, vol. 26, pp. 303–305, 1 March 1990.

[22] W. B. Wilson, H. Z. Massoud, E. J. Swanson, R. T. George and B. R. Fair, "Measurement and modeling of charge feedthrough in n–channel MOS analog switches," IEEE J. Solid–State Circuits, vol. SC–20,no. 6, pp. 1206–1213, Dec. 1985.

[23] G. Wegmann and E. A. Vittoz, "Basic principles of accurate dynamic current mirrors," Proc. IEE Pt. G, vol. 137, no. 2, pp. 95–100, April 1990.

[24] C. Toumazou, N. C. Battersby and C. Maglaras, "High–performance switched–current memory cell," IEE Electronics Letters, vol. 26, pp.1593–1595, 13 Sept. 1990.

[25] J. H. Fischer, "Noise sources and calculation techniques for switched capacitor filters," IEEE J. Solid–State Circuits, vol. SC–17, no. 4, pp. 742–752.

[26] C.–A. Gobet and A. Knob, "Noise analysis of switched capacitor networks," IEEE Trans. Circuits and Systems, vol. CAS–30, no. 1, pp 37–43, Jan. 1983.

[27] S. J. Daubert and D. Vallancourt, "Operation and analysis of current copier circuits," Proc. IEE Pt. G, vol. 137, no. 2, pp. 109–115, April 1990.

[28] N. C. Battersby and C. Toumazou, "Class AB switched–current memory for analogue sampled data systems," IEE Electronics Letters, vol. 27, pp. 873–875, 9 May 1991.

[29] A. S. Sedra and K. C. Smith, "A second–generation current conveyor and its applications," IEEE Trans. Circuit Theory, vol. CT–17, pp. 132–134, Feb. 1970.

[30] S. J. Daubert and D. Vallancourt,"A transistor–only current–mode SD modulator," Proc. IEEE Custom Integrated Circuits Conf., pp. 24.3.1–24.3.4, May 1991.

[31] D. D. Buss, D. R. Collins, W. H. Bailey and C. R. Reeves, "Transversal filtering using charge–coupled devices," IEEE J. Solid–State Circuits, vol. SC–8, pp.138–146, April 1973.

Data Converters

Franco Maloberti
University of Pavia, Italy

6.1 Introduction

Data converters are becoming increasingly important in application specific integrated circuits (ASIC's). Today's technologies allow integration of a very large number of transistors in a single chip, making possible the realization of mixed analogue and digital functions in integrated form. Modern trends are towards the processing of signals partially in the analogue domain (usually pre–processing and post–processing sections) and partially in the digital domain with the boundary between the analogue and the digital placed in the most appropriate position in the processing chain, according to the complexity of the processing functions and the features of the technology used.

At the analogue–digital interfaces the transition between the analogue and the digital world requires data converters. The capacity of mixed analogue–digital chips is rapidly increasing (Figure 6.1 [1]); allowing very complex systems to be designed. The number of analogue inputs and outputs is likewise increasing, requiring a large number of data converters in a single chip.

In the past, data converters were designed and fabricated as stand alone components, so that, the most appropriate processing technology could be used for these functions. As bipolar and JFET technologies display better analogue performances in terms of accuracy and speed compared to their CMOS counterpart, design techniques based on bipolar and JFET were preferred. Nowadays, data converters are integrated in an entire signal processing system where the digital section is the dominant part of the system. As a results the choice of technology employed is driven by global considerations such as area consumption, power dissipation, circuit compactness rather than accuracy and speed. For this reason the decision on technology is often restricted to CMOS. Thus, data converters must be realized using a technology that is not optimum for the application and specific design approaches for overcoming the associated limitations must be used. In this chapter, after an initial discussion of the basic static and

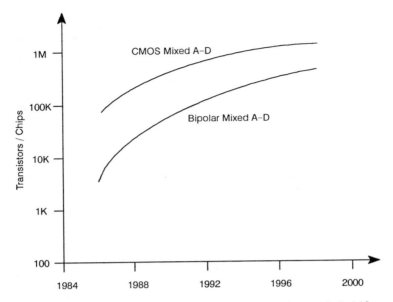

Figure 6.1 Complexity trend of mixed analogue-digital IC

dynamic parameters which characterise data converters, conventional conversion approaches and specific techniques used in CMOS technology are described. Non–conventional and more advanced solutions are considered in two successive chapters.

6.2 Parameters for data converter characterisation

It is known that an A/D converter generates a digital code as a response to an analogue input signal (usually as in the form of voltage or current) and by contrast, a D/A converter generates an analogue signal (voltage or current) to represent an input digital code [2]. The number of bits of the digital code is referred to as the resolution of the converter. This parameter is also represented as percentage or in the form of parts per millions (ppm) of the full scale range. For example, a 16 bit converter has a resolution of 16 bits or of 0.00152% (15.2 ppm) of the full scale. The parameters qualifying a data converter can be divided into static and dynamic ones. The static parameters are deduced from an analysis of the input–output transfer characteristic. This curve displays the relationship between the analogue signal and the digital code. For an ideal N bit converter it is an ideal staircase with 2^N steps, as shown in Figure 6.2 for a 4 bit converter. It should be noted that, because of the finite number of bits, one digital code does not

Data Converters

represent only one analogue signal value but a given interval of the analogue dynamic range. This interval is usually referred to as the quantization step, (QS). Its amplitude depends on the full scale range (FS) and the number of bits (N) of the converter:

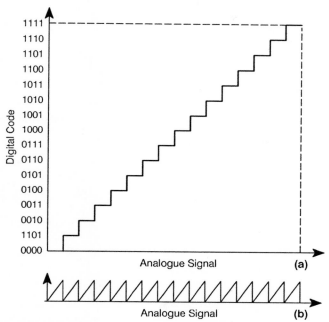

Figure 6.2a Input-output transfer characteristics of an ideal 4 bit converter. b. Quantization error

$$QS = \frac{FS}{2^N}$$

6.1

The coding of different analogue signal values with the same digital word, depending on the intrinsic finite resolution of a converter implies an error, the so called quantization error. It is defined as the difference between the transfer characteristic of a converter with a finite resolution and that of a converter with an infinite resolution. In the absence of other errors, the graphic representation of the quantisation error is a sawtooth shape (Figure 6.2b).

The static transfer characteristic of a real converter always display differences with respect to an ideal staircase. These differences come from one or more of the following errors (Figure 6.3).

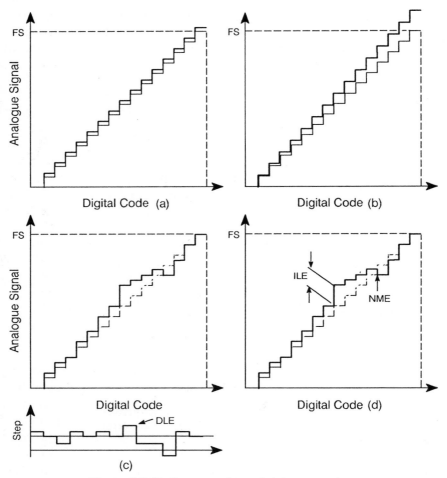

Figure 6.3 Static errors in real data converters

- Offset error,

- Gain error,

- Differential non-linearity Error (DLE),

- Integral non-linearity Error (ILE),

- Non-monotonicity Error (NME),

The offset error is a constant shift of the entire transfer characteristic (Figure 6.3a). The gain error describes an incorrect full scale value (Figure 6.3b). For a negative gain error a digital code made of all "1s" corresponds to an analogue voltage smaller than the full scale while for a positive gain error the analogue full scale is obtained with a digital code smaller than all "1s". When the steps of the transfer characteristic are not all equal the converter is affected by a differential non–linearity error (DNL) and by an integral non–linearity error (INL). The differential non linearity defines the maximum deviation of the step amplitudes from the nominal value, QS (Figure 6.3c). The integral nonlinearity gives the maximum deviation of the transfer characteristic from the ideal curve (Figure 6.3d). The non monotonicity error (NME) specifies that the transfer characteristic is not always a rising curve (Figure 6.3d). The dynamic performance of a converter is described by the following specifications:

- maximum sampling rate,

- slew rate,

- settling time,

- glitch power.

Any conversion operation takes a finite amount of time because the circuits do not have an infinite bandwidth and because in some cases the algorithm employed goes through a sequence of operations. For this, it is necessary to sample the analogue input signal or to enter the digital data at a given rate, the maximum sampling rate. This parameter is very important since it determines, according to Shannon's theorem, the maximum bandwidth of the processed signal. Moreover, it indirectly indicates the slew rate and settling time of the active analogue components employed. Often, slew rate and settling time are also explicitly specified for a more precise qualification of the data converter. An important dynamic qualifier is the glitch power. When the digital input changes from one code value to its neighbouring one, the analogue output is expected to display a steep transition of one QS. In reality the transition is not steep and is often accompanied by a glitch (Figure 6.4). In general this effect is due to different speeds in the analogue component used to transform each bit into an analogue signal. The largest glitch normally occurs in the middle of the range at the transition 10000..., 01111.... (or vice–versa). The parameter used to quantify this effect is the glitch power which is defined as the time integral of the largest glitch, expressed in V–Sec or W–Sec (Joule).

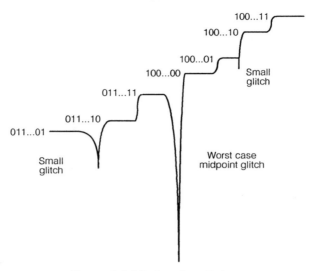

Figure 6.4 Glitches in a D/A converter

6.3 Data converters: Basic design considerations

Implementation of any conversion algorithm demands the use of active elements and passive components. The active elements are operational amplifiers and comparators; their characteristics and design techniques are extensively considered in chapter 3 of this book. The passive components are integrated resistors and capacitors. These elements, together with current mirrors used in data converters to get, in some sense, a "measurement" of analogue variables. In D/A converters a reference signal (voltage or current), is generated internal to the chip or is provided externally. This is then used to determine the analogue full scale value and, by suitable active or passive "attenuation", the analogue output is generated. In an A/D converter, in contrast, the basic reference is used to "measure" the amplitude of the analogue input signal. The "attenuation" and "measure" elements in data converters are very important and together with accuracy and stability of the basic reference, mainly determine the overall performance of the converter. The stability (in temperature and time) and absolute value of the basic reference principally affects the gain error. In contrast accuracy and stability of "attenuation" and "measure" elements have a more important impact on the overall performance. Mismatch is the major cause of linearity errors (differential and integral) while circuit speed determines the dynamic characteristics.

Current mirrors are always based on the well known Widlar architecture, shown in Figure 6.5. Modifications with added transistors, like the cascode or the Wilson structure, aimed to increase output resistance, are also used. However, in all these cases, accuracy of the mirror factor is principally

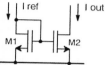

Figure 6.5 Simple current mirror (Widlar architecture)

determined by the basic architecture and, in the final analysis, by the voltage–current characteristic of an MOS transistor in the saturation region. It is approximately expressed by:

$$I_D = k' \frac{W}{L} \ (V_{GS} - V_{Th})^2$$

6.2

A mismatch in the physical parameters defining k' (surface mobility and gate oxide capacitance) as well as the error in gate length or width determines inaccuracy. However, the inaccuracy of the threshold voltage, either random or associated with gradients, produces greater inaccuracies. Even the gate–to–source voltage can contribute to errors. The gate connection is usually firmly defined but the voltage of the source can not be well established because of voltage drops across the resistance of the metal or poly lines of the VDD or ground connection.

As a consequence of the above the matching of current mirrors (and in particular in MOS technology) is not very accurate, and so current–mirror–based techniques are basically used for low and medium resolution (up to 8 bits).

The matching accuracy of integrated resistors is of the order of 0.1–0.4% making it possible to obtain a resolution of up to 9 bits. However, this relatively good resolution can be achieved only with careful design of the circuit layout. In particular inter–digitized structures with dummy strips around the structure and minimisation and control of resistance–metal contacts must be used. Specific guide–lines for this are given in the Chapter 11 of this book which discusses practical aspects of design.

A matching better than that possible with integrated resistors can be achieved with integrated capacitors (0.05–0.2%). Moreover, the temperature and voltage coefficients of capacitors are quite good. These superior characteristics make capacitor–based conversion algorithms very attractive. In general, an array of capacitors is pre-charged at given voltages and the various capacitors of the array are connected according to the specific algorithm used.

It follows that in capacitor–based data converters, analogue switches are necessary . These components are quite simple to implement in CMOS technology. An MOS transistor (or a pair of complementary transistors), with a suitable voltage driving the gate, works very well. After a transient, the current in the switch goes to zero and the transistor goes into the triode region with zero voltage across the drain–source terminals. In this condition the on–resistance is inversely proportional to the W/L aspect ratio and the applied overdrive voltage (V_{GS}–V_{Th}). For a minimum area transistor (W/L = 1) with overdrive of the order of 1V an on–resistance of few KΩ results. A value that is satisfactory for most applications. In the off condition (V_{GS} is set well below the threshold voltage) the switch is almost an open circuit. The only limitation to static ideal operation can come from the reverse biased diodes between the substrate and source or drain, but for the switching intervals normally used, the problem of leakage current does not exist. However, the opening and closure of switches is always accompanied by the so called clock–feedthrough effect. A charge (positive or negative), coming from the channel of the transistor and/or the capacitive couplings of the gate, is injected on both sides of the switch; thus, a parasitic coupling between the analogue circuit and the logic controlling the switches is established. The injected charge, integrated on the capacitors, causes a significant error even for medium–resolution data–converters. For this, and in particular for high resolution converters, the clock–feedthrough is one of the major design problems. It is mandatory to attenuate this effect by use of suitable cancellation techniques. Many solutions have been proposed which suggest the use of dummy elements, or, more generally the use of various compensation schemes. One of the most successful technique is based on the use of a fully differential scheme.

6.4 Simple data conversion techniques

Data conversion techniques can be classified according to several parameters. The two possibilities, used here, are to consider the complexity and the speed. Data conversion may be performed in only one clock phase or, sequentially in a number of clock phases. In general, for sequential algorithms, the number of clock cycles required for performing the conversion increases as the resolution increases. Depending on the specific algorithm the associated speed is reduced linearly or exponentially. The complexity of the architecture of a data converter is generally related to the required accuracy and speed. Simple algorithms, and simple circuit implementations, are sufficient to achieve a low resolution. When the resolution approaches the limit of the basic component accuracy, complex architectures, with specific compensation tricks, or dedicated circuits for error correction, or trimming of analogue elements, are used. Examples of conversion algorithms belonging to the "simple technique" category are considered in this section.

6.4.1 Serial A/D converters

One of the simplest (and least expensive) techniques for A/D conversion is the serial approach. Its principle of operation (single–slope) is shown in Figure 6.6. after a reset command, a ramp generator starts to integrate a reference voltage and its output is compared with the input signal. A digital counter evaluates the number of clock cycles required by the ramp to cross the input signal. The binary content of the counter, eventually converted to the desired digital format, represents the digital conversion of the input signal [3].

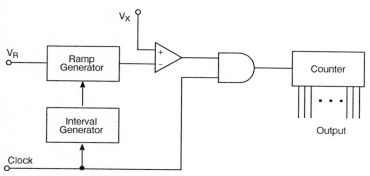

Figure 6.6 Block diagram of a ramp converter

The ramp generator can be implemented conventionally by integrating a fixed current onto an initially discharged capacitor C (Figure 6.7a). The major disadvantage of this simple solution comes from the significant nonlinearity associated with the parasitic affecting integrated capacitors which leads to an integral nonlinearity error. A better solution is shown in Figure 6.7b, where the integration of the current is in active form. Moreover the operation of the operational amplifier is made insensitive to its offset. During the reset phase instead of being discharged, the capacitor C is pre-charged to the offset of the operational amplifier, made available by the unity buffer connection. The main advantages of the serial approach are its simplicity and inherent monotonically (the ramp is certainly always rising). Its disadvantage, however, is its slow conversion speed.

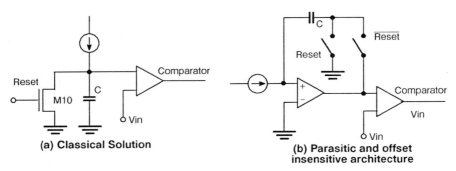

(a) Classical Solution

(b) Parasitic and offset
insensitive architecture

Figure 6.7 Serial D/A converter

6.4.2 Resistive division D/A converts

Resistors in MOS technology are fabricated using resistive layers that are
made by diffusion (n+, p+ or well) or poly-silicon. The typical specific
resistance is 200–600Ω/□ for diffused resistors, and 20–40Ω/□ for
poly-silicon resistors. The accuracy of integrated resistors is usually quite
poor (~40%); however, their matching (~0.1–0.4%) [4] can be satisfactory
enough for low and medium resolutions. Thus, the attenuation of a
reference voltage can be simply obtained by a resistive divider. Figure 6.8
shows a three–bit resistive–divider converter. The selection of the required
voltage is made by a tree of switches. The most significant bit (b3) and its
complement ($\bar{b}_3$) control two switches. The upper one permits the selection
of the upper half of the resistive string, the other one the lower half. The
remaining switches in the tree, driven by the other bits (b2, b1) and their
complements, allow a finer and finer selection of the attainable voltage
interval. The selected voltage is applied to the input of a buffer whose
impedance must be significantly larger than that of the resistive divider.
This condition is certainly satisfied by using a CMOS gain stage. The output
voltage, however, is affected by the offset of the output buffer, which for a
CMOS op–amp, is of the order of 10mV.

The area of the resistor string, as well as the area of the switches used for the
selection, increases exponentially with the number of bits. Thus, the
resistive divider technique is only suitable for low resolution, usually lower
than 8–9 bits. For higher resolution many problems must be addressed.
Among them the process parameters gradient, the additional resistance of
the contacts between metal lines and the resistive string (of the order of
20–100Ω) and the temperature and voltage dependence of the resistive
layers used.

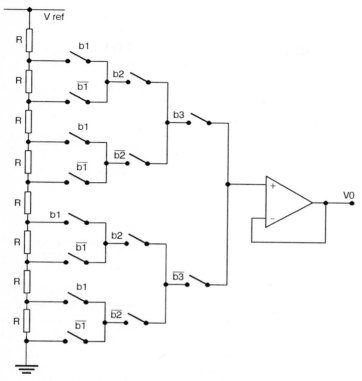

Figure 6.8 Three bits resistive division DAC

Interesting advantages of the technique are the inherent monotonicity and the speed. However, for medium resolution (around eight bits) the tree of switches imposes a large resistance in the path to the buffer, which, in turn, together with the input capacitance of the buffer slows down the conversion rate. The number of switches in series with the signal is reduced by the solution shown in Figure 6.9 where an x–y decoding scheme is used.

6.4.3 Capacitive division D/A converter

An alternative technique for generation of a defined fraction of the reference voltage is given by the capacitive divider [5]. The principle of operation is shown in Figure 6.10. An array of binary–weighted capacitors is first discharged (during the reset phase FR) and then, under the control of the input digital code, a fraction of it is connected to the reference voltage,

Data Converters

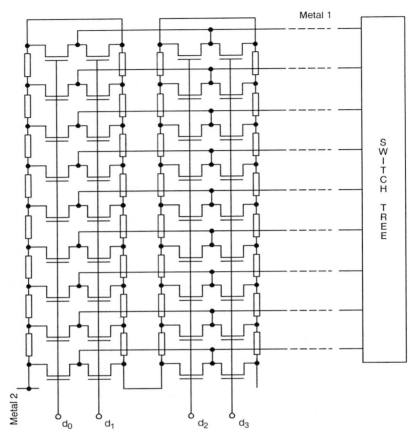

Figure 6.9: X–Y decoding scheme for resistive divider DAC

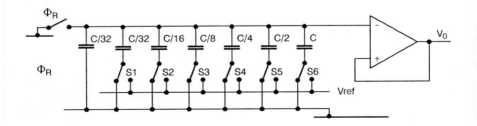

Figure 6.10 Capacitive divider DAC

while the rest of the array remains connected to ground. The voltage Vx that is established at the input of the buffer is given by:

$$V_x = V_{REF} \sum_{i=1}^{n} \frac{C\, b_i 2^i}{C\, 2^N}$$

6.3

Here V_{REF} is the reference voltage, N is the number of bits, and bi are the logic value of the bits. The sum in equation 6.3 corresponds to the fraction of the total capacitance connected to V_{REF} and 2C is the total capacitance of the array. The voltage Vx is then replicated at the output by a buffer. The main advantage of the capacitive division approach, compared to one where resistive division is used, is its simpler selection structure. It is not necessary to use a tree of switches but it is necessary to use two switches per bit–element. A disadvantage of the solution derives from the large and nonlinear parasitic capacitances that always affect CMOS integrated capacitors. The capacitance connected to ground is not only given by the selected elements but also by the top–plate parasitic of the entire array. Since its contribution cannot be kept lower than a fraction of a percent, the scheme in Figure 6.10 is suitable for no more than 6–7 bits of accuracy. For higher resolution a parasitic insensitive scheme must be used. A possible realization which is insensitive to the offset of the operational amplifier used, is shown in Figure 6.11. Strictly speaking, the circuit is not a capacitive

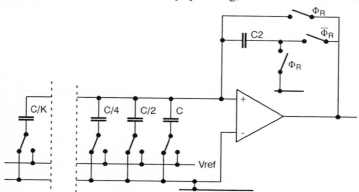

Figure 6.11 Parasitic and offset insensitive capacitor-based MDAC

divider, but it is an amplifier with digital controlled gain (multiplier DAC or MDAC). The operation of the circuit is as follow. During the reset phase the capacitor array and the capacitor 2C around the op–amp are discharged through the virtual ground, thus if an offset affects the op–amp, the capacitors are not completely discharged but are pre-charged to the offset. During the conversion phase, part of the array is connected to the reference voltage, thus injecting a charge into the virtual ground which is integrated into the capacitor 2C in the feedback connection around the op–amp.

In view of the good matching achievable with integrated capacitors, an accuracy of 10 bits can be obtained with the scheme in Figure 6.11. However, for N bit of accuracy 2^{N+1} unity capacitances must be used and a significant area consumption results. The large spread of the capacitors value can be attenuated using solutions with capacitor dividers like the one shown in Figure 6.12.

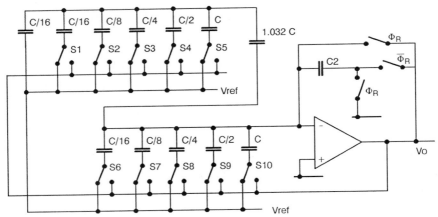

Figure 6.12: A 10 bit MDAC with capacitor divider

6.4.4 Successive approximation A/D converters

Very often A/D data conversion is realized by use of the successive approximation algorithm. This is based on the method shown in Figure 6.13. A D/A converter generates suitable analogue voltages that are successively compared with the input analogue signal. The result of each comparison is used to generate a new analogue voltage which is a better and better approximation the input signal. The digital logic used to control the D/A converter is usually referred to as SAR (Successive Approximation Register) whose operation allows the following sequence of steps. The first generated voltage is set in the middle of the dynamic range ($V_R/2$). The output of the comparator determines if the input voltage is larger or smaller than V_R and consequently sets the value of the MSB. On the basis of the MSB the SAR drives the D/A converter in order to generate $V_R/4$ or $3V_R/4$. The output of the comparator determines the second bit, and so on. It should be noted that the described method requires one clock cycle per bit, thus N clock cycles are needed for an N bit conversion.

Data Converters

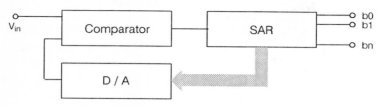

Figure 6.13 Principle of operation of the
successive approximation algorithm

A popular CMOS implementation of the successive approximation technique utilises a capacitive divider as D/A converter [6]. The schematic, for a 6 bit conversion scheme, is shown in Figure 6.14. During the sampling

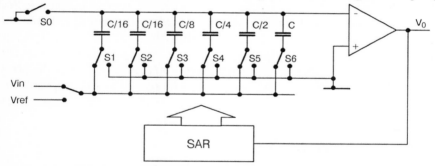

Figure 6.14 Schematic of a charge distribution ADC

phase the entire capacitor array is charged to the input voltage with the upper plate of the capacitors grounded. This is achieved by closing the switch S0 and connecting all the switches S1,..., S6 toward the input voltage. This preliminary operation is important for two reasons. Firstly, an inherent sample and hold is performed. A charge proportional to the input voltage is stored on the entire array when the switch S0 is opened. A second important point is that the terminal connected to the input of the comparator is the common top–plate of the array. Since during the sampling phase, it is connected to the analogue ground, the voltage to be used as reference term in the comparator is the analogue ground. This gives rise to the important advantage that the conversion operation is insensitive to the parasitic of the capacitive array. In fact the successive connections of the bottom plates are such that the common top plate is as close as possible to the analogue ground.

For bipolar input it is also possible to determine the sign of the input signal. If after the switch S0 is opened all the bottom plates are grounded, the voltage of the top plate is pushed to −Vin and the sign is determined by the comparator. This results allows selection of a proper sign for the reference voltage. The other bits are then determined starting from the MSB. The following procedure is used. The largest capacitor of the array is connected to the reference voltage (assumed positive for now) testing the value "1" for this bit The voltage of the top plate is pulled up and if it becomes positive this means that the amount of pre–charge put on the array during the sample and hold phase was not sufficient. The comparator does not confirm the bit and the bottom plate connection of the capacitor is reset to ground. Conversely if the comparator confirms the bit, the larger capacitor remains connected to the reference voltage. The process is then reiterated for every bit. i.e every capacitor.

At the end of the conversion cycle some of the capacitors remain connected to the reference voltage and the rest of the array is grounded. Since the voltage of the top plate is very close to ground, the grounded capacitors will be almost discharged while the charge that was put on the array during the sample–and–hold phase is completely stored on the capacitors of the array which are connected to the reference voltage. For this reason, the architecture described is usually referred to as the charge redistribution technique.

6.5 High speed data conversion techniques

Applications requiring processing of signals in the video band are increasing greatly (Chapter12). The level of complexity of the signal processing required is such that, very often, despite the very high speed logic circuits required a digital approach is suitable. Thus, techniques for high speed A/D and D/A conversion and their implementation with the same technology used for digital logic are becoming more and more important. This section considers typical conventional techniques for high speed data conversion in CMOS technology. Namely, the current switched approach for D/A conversion and the flash and half flash for A/D conversion will be discussed.

6.5.1 Current switched D/A converters

For very high speed applications (several MHz) the output of a D/A converter is usually connected to a coaxial cable with low characteristic impedance (typically 50Ω or 75Ω) [7]. Since it is difficult to design an output stage capable of driving such a low impedance, it is quite common to use the current as the output variable instead of the voltage. The output signal is generated by the use of binary–weighted current sources. If a given bit is one the corresponding current source is activated; all the currents are then summed up at the output node. The simplest way to activate a current source is to switch it from off to on. However, the switching on and off of current sources is accompanied by transient delays and current glitches and this makes the solution impractical. Alternatively, the current sources can always be kept on with their current switched toward the output node or toward a dummy load.

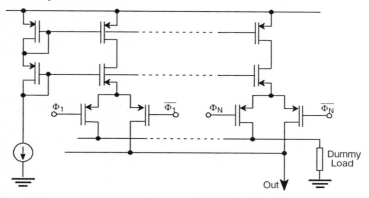

Figure 6.15 Current switch DAC

The use of cascode current mirrors as shown in Figure 6.15, may ensure a good output resistance which, in turn, gives rise to a lower dependence of the output current on the output node voltage. From equation 6.2 it results that binary weighted current sources can be easily realized by mirroring a master current using transistors whose W/L aspect ratio increases as the power of 2, $2^i (i = 1,..., N)$. Another possibility is to use a total of 2^N equal current sources, 2j of which are selected to decode the j–th bit. The latter solution results in greater flexibility. In fact, it allows random selection of the current sources to moderate errors or to follow given strategies to compensate errors. In particular it is possible to take into account mismatches due to technological gradients or due to the voltage drops across the ground or the VDD connections [8]. However, the control logic is more complex.

The key point for fast operation of converters concerns the correct driving of differential switches. If both the switches are left opened, even for a small time interval, the output node of the controlled current source is rapidly charged to VDD (or discharged to ground, for n–channel current mirrors) and a glitch and a delay in the output current results. Conversely, if the switches are both kept closed, an uncontrolled connection between dummy load and output results. From the above, it is not advisable to use just the control bit and its logical inversion to drive the differential switches.

Carefully designed drivers should be adopted which should generate symmetrical waveforms with controlled overlap. With careful design of the driver very fast D/A converters, even with conventional technologies, can be obtained [9].

Generally speaking the resolution of the current–switched converters is limited by the required speed and by the current source matching. The upper limit is the speed of the converter, the lower limit is the time that is permitted to settle within 0.5 LSB. Thus, assuming an exponential settling, an increase in the speed by a factor 2 will correspond to a reduction of the resolution by the same factor. With a 2–μm technology it is possible to achieve, for 8 bits of accuracy, a maximum speed of 50–70MHz.

As already mentioned current source matching is critical in CMOS. This depends on the achievable accuracy of the W/L ratio, on technological gradients, responsible for variations of the threshold voltage, and on the overdrive voltage [$V_{GS} - V_{Th}$] used. In order to get the best result in the current source matching, non minimum length, common centroid structures and high overdrive voltage should be used. However, the achievable resolution is usually limited to 8–10 bits.

6.5.2 Flash and two–step flash converters

The principle of operation of flash converters is very simple. The input voltage is compared with all the possible thresholds that define the transition between two successive codes. Since for N bits there are 2^N quantization steps, 2^{N-1} comparators are necessary. The comparison operations are performed simultaneously and only one clock cycle is required to perform the entire conversion. A typical architecture of a flash converter is shown in Figure 6.16. The necessary 2^{N-1} reference voltages are

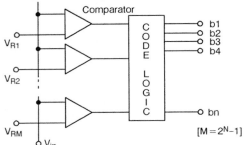

Figure 6.16 Block diagram of a flash converter

obtained with a resistive divider. The outputs of the comparators are the input of a logic circuit encoding the result into its digital code [10]. The speed of a flash converter is determined by the speed of the comparators and by the encoding logic. In general logic is very fast and the comparator speed is the major concern. Moreover, even for medium resolution, the number of comparators is very large and, in order to have a reasonable chip area, their schematic must be simple. Simplicity and speed are generally achieved by using a simple chain of inverters for the comparator scheme. The input signal is successively amplified along the chain to obtain a logic level. The response of the system is very fast, since the capacitances driven are relatively small and the current available in an inverter in the transition from one to zero is fairly large. The number of inverters in the chain should be such as to allow a gain much larger than 2^N. The main drawback of the solution is its large and uncontrolled input referred offset that is mainly caused by the first two or three inverters of the chain. Very often the offset is cancelled by the use of the auto-zero technique (Figure 6.17). During the sample and hold phase, Φs, the comparators are connected in the buffer configuration making their offset available at the inputs. The voltage is stored on the so called auto-zero capacitors, Caz. During the successive phase, the auto-zero switches are opened and the auto-zero capacitors act as level shifters, hence cancelling the offset.

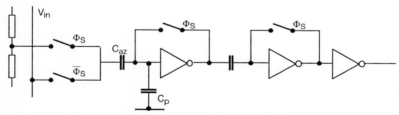

Figure 6.17 CMOS comparator with autozero commonly used in flash converters

However, switching transistors ON and OFF introduces two additional drawbacks namely clock feedthrough and charge pumping. The former consists of the injection of a fraction of the charge stored under the gate in the "on" state into the auto-zero capacitor and, in addition, a capacitive coupling between the logic command and auto-zero capacitor. The clock feedthrough effect introduces an error that is described in terms of a residual offset. The latter effect is a consequence of the alternative charging and discharging of the input parasitic capacitances of the comparators from the input signal to the reference voltages. Charges must be provided (or pumped) from the resistive divider with a consequent transient limitation.

Its relevance depends on the values of the parasitic capacitances and on the value of the resistors of the string. The above effects pose a practical limit to the flash technique to 7–8 bits of accuracy.

Another evident disadvantage of the flash approach comes from the exponential increase of the number of comparator when the resolution increases linearly. Thus, the silicon area and the power dissipation rapidly reaches unacceptable values. Even more critical is the increase of the capacitive load of the comparators. These capacitance affects the input buffer; so, if for a single comparator, it is only 0.2 pF, a load as large as 50 pF results for an 8 bit flash converter. The limitation on speed resulting from having to drive such a large capacitor overrides the advantage of the flash approach.

For eight or more bits a more convenient technique is the so called two–step or sub-ranging flash. It is schematically shown in Figure 6.18 and consists of a sample and hold, a coarse and a fine flash ADC converter, a DAC and a sub-tractor. The conversion is performed in two successive phases. In the

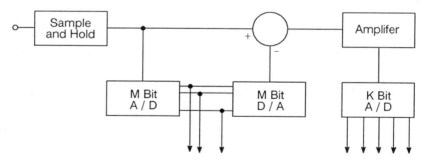

Figure 6.18 Principle of operation of two-step flash ADC

first phase the most significant bits are determined. The DAC converts the coarse digital code and the result is subtracted from the held input to give a fine remainder which is then amplified to a more convenient level. The second flash ADC converts the remainder and determines the LSB's. The speed of the two flash architecture is limited by the speed of the two ADCs, the DAC, the sub-tractor and the amplifier. The architecture of the two–step flash allows many variants [11]. One possibility is to reduce the complexity of the system by using only one DAC multiplexed for the two phases of operation. Another possibility is to increase the throughput by pipe-lining the two stages.

6.6 Limits to speed and resolution in data converters

Two of the key performance parameters in data converters are speed and resolution. This section considers the fundamental and practical limits to these two parameters using present day technologies.

All the practical architectures of D/A converters need operational amplifiers, while in fast A/D converters comparators and sample and hold circuits are necessary. The speed and the accuracy of data converters is strongly dependent on speed and accuracy of the active elements used, such as operational amplifiers, comparators and sample and hold circuits.

Let us first consider the limits imposed by operational amplifiers [12]. The main effects are characterized by a finite gain, bandwidth and slew–rate and a given input referred noise. The finite DC gain A0 determines, in general, an error that is proportional to $1/A_0$. In order to avoid this having an impact on the accuracy, this error should be smaller than 0.5 LSB; hence, for N bits, the gain of the op–amp must be at least 2^{N+1}. The input referred noise of a CMOS op–amp is characterized by a spectrum with a 1/f dominant component spanning up to a given frequency (the corner frequency) and by a white component in the rest of the frequency spectrum. For a well designed op–amp the corner frequency is about 1kHz. The white noise component is dominated by the input pair and it is given by 4kT/gm, where k is the Boltzmann constant and gm is the transconductance of the input pair. The spectrum of the input referred noise source is filtered by the band–limiting elements of the circuit and, because of sampling operation intrinsic to the data conversion, the result is folded in the band of interest to give rise to a noise voltage. This noise voltage should be smaller than 0.5 LSB. If the band–limiting action derives from the the op–amp itself, then taking into account only the white term, the resulting square of the noise voltage can be approximated by the relationship: $v_n^2 = kT/C_c$, where C_c is the compensation capacitance. This result also gives a figure of the practical lower limit of capacitances to be used in high resolution data converters.

The gain–bandwidth product of an op–amp is determined by the non dominant poles whose position must be well beyond the gain–bandwidth. For this, the pole of the input stage is made dominant by a suitable design. The gain–bandwidth product f_T can be expressed as:

$$f_T = \frac{1}{2\pi} \frac{g_m}{C_c}$$

6.4

where, again, gm is the transconductance of the input differential pair and Cc is the compensation capacitance. Even the slew rate of an op–amp is dependent in a similar way on the same parameter. Thus in order to increase the speed, it is necessary to maximize the transconductance and to minimize the compensation capacitance. In practical cases Cc is as large as a few pF since its value is defined by the rule of thumb that it should be from three to ten times larger than the parasitic capacitance of non dominant nodes. Thus, in order to increase the speed, large transconductances must be used. The transconductance of a transistor can be improved by enlarging the W/L ratio or by using a greater bias current. Unfortunately, an increase of the area of transistors affects the non–dominat poles and an increase of the bias current determines a reduction in the DC gain. Hence, for a defined resolution there is a practical limit on the gain–bandwidth product. Figure 6.19 shows the accuracy limit as a function of the frequency for op–amp based and for current switched based D/A converters. The curves, based on current technologies, are limited to 18 bits because of the fundamental white noise limit in CMOS integrated circuits.

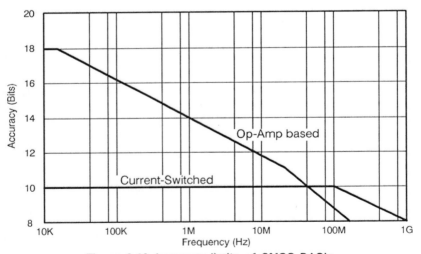

Figure 6.19 Accuracy limits of CMOS DAC's

Let us now consider the factors limiting the speed in A/D converters. As mentioned before the elements to be taken into account are the sample and hold circuit and the comparator. A sample and hold circuit can be modelled as a capacitor that is charged to the input voltage through a switch. Since the voltage stored on the capacitor must be accurate within 1/2 LSB, then assuming an exponential charge with time constant τ, the period of the

Data Converters

sampling phase must at least equal to $(N+1)\pi\log2$. For a typical switch resistance of 5kΩ and for a typical sampling capacitor of 1pF, in a 12 bit CMOS converter, the minimum duration of the sampling phase is 42nSec ($f_{ck} \sim 10$MHz).

As mentioned before the operation of a sample and hold can be modelled with a capacitor C charged to the input voltage through a switch (Figure 6.20). In the on condition the switch can be modelled with its

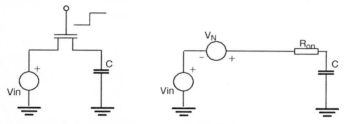

Figure 6.20 Schematic representation of a sample and hold and its noise description

on–resistor, Ron. A white noise source with a spectrum $S_N = 4kTR_{on}\Delta f$ is also associated with the resistor; thus the spectrum at the capacitor becomes:

$$S_{NC}(\omega) = \frac{4kTR_{ON}\ \Delta f}{1 + (\omega\ R_{ON}C)^2}$$

$\boxed{6.5}$

When the switch is turned off not only the signal but also the noise, folded in the band–base, is sampled on C. Its power is given by the integral of the spectrum (Figure 6.21) which equals: kT/c. For discrete applications the value of kT/c is negligible while for monolithic applications it can become significant. For C = 1nF, kT/c is only $(0.2\mu V)^2$, but for C = 0.1pF it rises to $(0.2mV)^2$. The latter value is comparable to 1/2 LSB of a 12 bit converter with 1V as full scale.

Another important limitation to accuracy comes from the coupling with the digital section of the chip. Its effect and the techniques for its limitation are described in chapter 10 of this book. The offset voltage and the matching of passive components are also important. Their effect can be reduced through careful layout, symmetrical structures, common centroid topologies, interconnections with the same length, precise elements designed with non minimum dimension, and matching of the parasitic capacitances . The finite gain of the op–amps can also become a serious

Data Converters

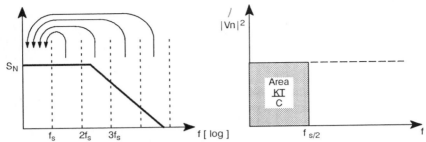

Figure 6.21 Folding of the white noise after it is passed through a low pass filter

problem when very high gain is required. The use of multiple–stage architectures can cause severe problems of compensation. In this case it is better to use techniques which reduce the dependence on high gain [13][14].

The above considerations are summarised in Figure 6.22 which shows limits of accuracy as a function of the frequency for an A/D converter. It should be noted that, in general there are upper limits to the speed and resolution associated with a given technique and pushing speed can compromise resolution.

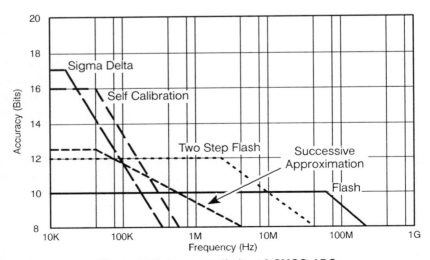

Figure 6.22 Accuracy limits of CMOS ADC

6.7 References

[1] Solomon I. E., "Design Automation for analogue and Mixed analogue/Digital ICs",1989, IEEE Custom Integrated Circuit Conference.

[2] Allen P. E.,Sanchez–Sinencio E., 1984,"Switched Capacitor Circuits, Van Nostrand Reinhold,NY

[3] Smarandoiu G., Fukahori K., Gray P.R., Hodges D.A., "An all MOS analogue–to–digital converter using a constant slope approach" IEEE J. Sol. St. Circ., 1976, SC–11, pp. 408–410

[4] Allstot D. J., Black W. C,"Technological design considerations for monolithic MOS switched–capacitor filtering systems" IEEE Proceedings, 1983, 71, pp. 967–986

[5] Albarran J. F., Hodges D. A., "A charge–transfer multiplying digital–to–analogue converter" IEEE J. Sol. St. Circ., 1976, SC–11 pp. 772–779.

[6] McCreary J. L., Gray P. R., "All–MOS charge redistribution analogue to digital conversion techniques" IEEE J. Sol. St. Circ., 1975, SC–10 pp. 371–379.

[7] Chi K. H., Geisenhainer C. S., Riley M., Rose R. C., Sturges P. J., Sullivan B. M., Watson R. B., Wooside R. H., Wu M. W., "A CMOS triple 100–Mbit/s video DAC with shift register and color map" IEEE J. Sol. St. Circ., 1986, SC–21, pp. 989–996.

[8] Miki T., Nakamura Y., Noyaka M., Asal S., Asakasi Y., Horiba Y., "An 80–MHz 8–bit CMOS D/A converter" IEEE J. Sol. St. Circ., 1986, SC–21, pp. 983–988.

[9] Cremonesi A., Maloberti F., Polito G., "A 100–MHz CMOS DAC for video–graphic systems" IEEE J. Sol. St. Circ., 1989, SC–24, pp.635–639.

[10] Kumamoto T. et al., "An 8–bit high speed CMOS A/D Converter", IEEE J.Sol. St. Circ., 1986, SC–21, pp. 976–982. [11] Cremonesi A., Maloberti F., Torelli G., Vacchi C., "An 8–bit two–step flash A/D converter for video applications" IEEE CICC Proc., 1989, pp.631–634

[12] Gregorian R., Temes G.C.,1986, "analogue MOS Integrated Circuits",J. Wiley & Sons

[13] Nagaraj K., Singhal K., Viswanathan T. R., Vlach J., "Reduction of finite gain effect in switched capacitor filters" Electronic Letters, 1985, 21, pp. 644–645

[14] Haug K., Maloberti F., Temes G. C., "Switched capacitor integrators with low finite gain sensitivity", Electronic Letters, 1985, 21, pp. 1156–1157

Over Sampling Converters

Chapter 7

Franco Maloberti, University of Pavia, Italy
Paul O'Leary Jonnaeum Research Graz, Austria

7.1 Introduction

Conventional high accuracy data converters require the use of accurate analogue components. In general, precise components are obtained by the use of specific technological tricks that led to what is commonly referred to as "analogue technology". However, for high–performances digital applications it is necessary to optimize the technology in directions that are often opposite from the ones indicated by analogue requirements, and, since the digital section of a complex circuit is dominant, a compromise that is very favourable to digital needs must be adopted. As a specific example it is worthwhile remembering that in order to get good capacitors it is necessary to use a double poly layer. However, since the number of interconnection layers for digital circuits is extremely important, it is preferable to make the additional technological effort (and masks) to get one additional metal layer.

The increasing requirement for accuracy in recent years has stimulated intensive research to find suitable analogue design techniques that are capable of achieving good analogue performances while using "digital technologies". For data conversion, the oversampling technique, described in this chapter, seems to be a very profitable solution for low (instrumentation and audio–band) and medium frequency applications. The speed of operation used is much higher than the band of interest and this feature is exploited to improve accuracy [1]. In conventional A/D converters (the so–called Nyquist rate category) the input signal is sampled at a rate that is only twice that of the band of the input signal itself. The digital output, generated at the same rate, following to the sampling theorem, retains all the informative contents of the input signal which represents it. However, in order to avoid aliasing, the input signal must be band–limited by an anti–aliasing filter before sampling. In oversampling converters, by contrast, the input analogue signal is sampled and processed at a rate which is significantly higher than twice the input band. The ratio between the sampling rate and twice the band of interest is usually referred to as the oversampling factor, Ros. oversampling as large as many hundreds is often used.

As we will study in this chapter, the important effect of oversampling is that the power of the error associated with the quantization (quantization noise) is spread over a band that is much wider than the band of interest. Thus the digital output of an over sampled A/D converter contains a given fraction of the quantization noise that is pushed out of the band of interest. This fraction can be filtered out without affecting the informative content associated with the input signal. After filtering, the power of the quantization noise is reduced and, in turn, the resolution of the converter is increased.

Different categories of oversampling converters perform the spread (or the shaping) of the quantization noise in the band and outside of the band of interest with different efficiencies; consequently, they determine the need for post filtering at different levels of complexity. This chapter, after considering basic oversampling converters, concentrates on the sigma delta schemes.

7.2 Intuitive Introduction to Oversampling Data Converters

Before looking at oversampling converters analytically, it is important to gain an intuitive understanding of how they function. Most publications on oversampling converters start with very tedious analytical examinations of their functionality. This has given oversampling converters the reputation of being complicated, which is not true.

To help the reader gain an intuitive understanding of oversampling converters, this introduction starts by looking at digital to analogue conversion. Digital to analogue conversion has been chosen because it is understood and perceived more simply than analogue to digital conversion. However, the principles and problems of oversampling are similar for both D/A and A/D converters.

Let us consider to begin with the conversion system known as pulse width modulation, of which a possible generated waveform is shown in Figure 7.1 the output of the converter can have only one of two voltage levels which are either V_{ref} or ground. The conversion time T_c is long with respect to the clock period T_s. Consequently, there are a large number of clock cycles, say m, available for each conversion. The desired output voltage is approximated by the repetition of a pulse, whose duty–cycle is digitally controlled. The average level of the pulse corresponds to the output analogue signal. In the example shown in Figure 7.1, the output voltage is set to V_{ref} for four of the m possible time slots. Consequently, its average $V_{out,av}$ is equal to $(4/m)V_{ref}$. The average output voltage can be generated by either integrating the DAC output over the conversion time interval, T_C, or by a low–pass filtering of the pulse stream. The latter is more to be preferred.

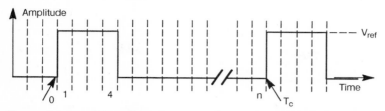

Figure 7.1 Possible pulse placement for an oversampling converter

As can be seen from the principle of operation the output voltage is always smaller than the reference; moreover, given m clock period for conversion, the precision of the output voltage turns out to be equal to one part per m, that corresponds to $\log_2(m)$ equivalent number of bits.

Since the clock frequency is much higher than the conversion frequency the described technique falls into the oversampling data conversion category. The oversampling ratio R_{os}, defined as the ratio of the sampling frequency f_s to the required Nyquist frequency, $f_{Nq} = f_C$, is equal to m. Consequently, the equivalent number of bits n_{eq} in such an oversampled converter is given by $\log_2(m)$.

This is an important result, since it represents a fundamental limit for the accuracy of pure oversampling modulators. Doubling the clock frequency while keeping the signal band constant results in a doubling of the number of time slots available per conversion and, in turn, allows a doubling of the achievable accuracy: the converter gains half a bit of accuracy for each doubling of the clock frequency.

It is now important to examine the spectral properties of the digital to analogue converter output. For this, consider a pulse width modulator with 32 time slots (5 bits) during 7 of which the output is high i.e. connected to V_{ref}. Since the output of the PWM is periodic, the spectrum will consist of discrete spectral lines spaced evenly at multiples of the conversion frequency f_c. The amplitude envelope of these spurs is given by the Fourier transform of the output during one period of the conversion. In the present case this results in $\sin(x)/x$ envelope. A plot of the Fourier transform for this example is shown in Figure 7.2. The $\sin(x)/x$ envelope and the DC level of 7/32 can be seen.

The value at DC is the desired signal, all other frequency components must be considered as disturbances. The disturbances with the largest magnitude are at lower frequencies. These noise components do not represent a problem when driving systems with long time constants or which have large momentum. This explains why pulse width modulators are popular when

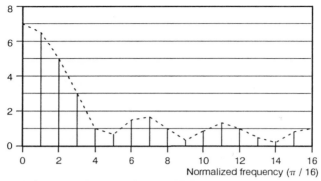

Figure 7.2 Output spectrum from a PWM for 7 from 32 high time pulses

driving motors and other mechanical devices. Furthermore, only one rising edge and one falling edge is required per conversion, thus making it possible to reduce the switching transients in power systems. However, for other applications the undesired discrete spectral components must be removed via a low pass filter. This requires a filter with a very steep transition from pass band to rejection band. Additionally, a large stop–band rejection is required because the first unwanted frequency components have almost the same amplitude as the desired signal.

The generation of a pulse width modulated signal can be implemented with simple circuitry. A possible solution is shown in Figure 7.3. The input data is

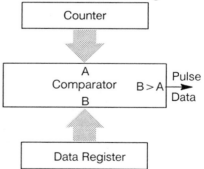

Figure 7.3 Pulse width modulator implementation

stored in a register, the output of which is compared with the output of a binary counter once per clock cycle. This is equivalent to comparing the data word to a sawtooth signal, as shown in Figure 7.4. The output bit is high so long as the data value is greater than that of the counter.

In order to avoid large disturbances at low frequency a different placement of the pulses required to construct the output signal within the m available

Over Sampling Converters

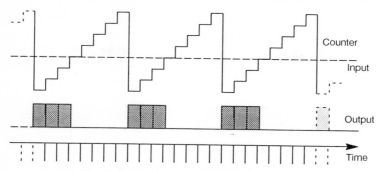

Figure 7.4 Pulse width modulation timing diagram

clock periods can be used. One possibility is shown in Figure 7.5, where the pulses are uniformly distributed in time. This pulse placement is often called "pulse density modulation" (PDM) and corresponds to the output from a "first order sigma–delta modulator" or a "charge balancing converter" as they are also known.

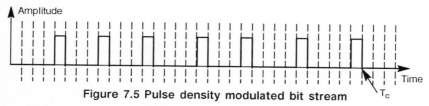

Figure 7.5 Pulse density modulated bit stream

The accuracy which can be achieved with this pulse placement is the same as for pulse width modulation, however, the spectral properties of the output signal are different. By evaluating the output spectrum of this pulse stream under the same conditions as for the PWM i.e. 7 from 32 pulses high, we come to the spectrum shown in Figure 7.6. The desired output DC level has

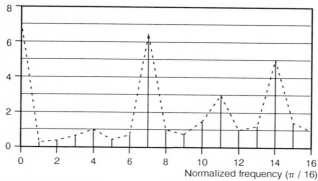

Figure 7.6 Output spectrum from a PDM for 7 from 32 high time pulses

been maintained, while the disturbances have been moved to higher frequencies. However, it is important to note that, since m time slots are used to determine the average output value, the achieved accuracy of the DC level remains unchanged. In other terms, the different modulation algorithms modify the signal to noise ratio but not accuracy.

The pulse density modulated bit stream can be generated using simple circuitry, see Figure 7.7. The input digital data is stored in a register (B) and

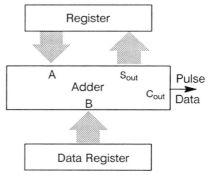

Figure 7.7 Pulse density modulator implementation
(first order sigma–delta DAC)

at every clock cycle it is summed up with the contents of register A, which, in turn, is the delayed version of the output of the accumulator. The system generates an output pulse whenever the accumulator reaches its full scale (carry out). It is easy to verify that for a full scale equal to 32 and 7 applied at the input, the full scale is obtained, assuming the accumulator at the beginning to be empty, after 5, 9, 13, 18, 22, 27, 32 clock periods.

The required pulses could also be placed in a random manner. Such modulators are known a "stochastic" converters. If the pulses are truly random then the quantization error is spread uniformly in the frequency domain. The output spectrum of a stochastic modulator (shown in Figure 7.8) is "whiter" than for the previous examples of PWM and PDM. These converters are advantageous in applications where the system is sensitive to disturbances with discrete frequency components. The implementation of the stochastic modulator is very similar to that shown in previous examples. Figure 7.9. A pseudo random (PN) generator is the key to stochastic converters, it must generate numbers which are truly random and having a uniform probability density function. Any deviation from these desired properties will result in an output spectrum where the noise is not white.

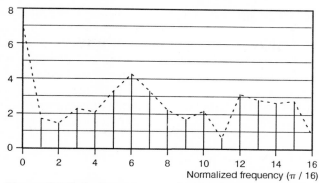

Figure 7.8 Output spectrum of a stochastic modulator for 7 from 32 high time pulses

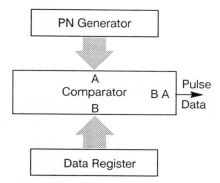

Figure 7.9 Implementation of a stochastic DAC

It is obvious that there is an almost infinite number of possibilities of placing the pulses. The three examples considered had the aim of giving the reader an insight into oversampling techniques. More generally the high order sigma–modulators which will be analyzed later are an extension of the above ideas.

7.3 Quantization

Quantization is the process of converting a signal with continuous amplitude to a signal with discrete amplitude levels [2][3]. During this process the signal is changed resulting in a modified spectrum. The change in signal power is referred to as quantization noise. However, the error due to the quantization process can be assimilated to noise only under a number of assumptions. These assumptions have certain weaknesses and it is important to examine them beforehand.

Assumption 1: All quantization levels are exercised with equal probability.

This assumption is almost never fulfilled. For a sine wave the quantization levels have a non uniform probability distribution function. For DC signals this condition is not even approximated. Only signals with uniform amplitude distribution are compatible with this assumption e.g. triangular waves with maximum amplitude.

Assumption 2: The quantization steps are uniform.

This requires an ideal analogue to digital converter. This assumption may be fulfilled for low resolution converters. If the steps are not uniform then the error is a function of the input signal and cannot be regarded as a simple additive noise component.

Assumption 3: The quantization error is not correlated with the input signal.

This is one of the weakest assumptions in the calculation of the signal to noise ratio. On the basis of this premise it is assumed that the quantization error can be handled as noise with a white spectrum. The correlated nature of the quantization can be demonstrated with a simple example of a sine wave and its 4–bit quantized equivalent, see Figure 7.10. It can be seen that the resulting quantization error, Figure 7.11, is correlated with the sine wave at the input.

Assumption 4: A large number of quantization levels are used.

This assumption is usually fulfilled. However in the case of one bit oversampling converters this assumption is violated. It is important to be aware of the above limitations when calculating the performance of a data conversion system. If all the above assumptions can be regarded as fulfilled then quantization can be dealt with as an additive process.

The transfer curve for an ADC and the resulting quantization error is shown in Figure 7.12. It corresponds to an ideal analogue to digital converter with the transfer characteristics shifted by half LSB; the error is limited to the interval $-Q/2 + Q/2$, where Q is the quantization step equivalent to one least significant bit.

The probability distribution function of the quantization error, $p(x)$, has the following limitation:

$$\int_{-\frac{Q}{2}}^{\frac{Q}{2}} p(x) \, dx = 1$$

$\boxed{7.1}$

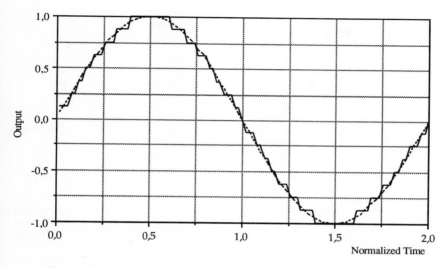

Figure 7.10 4–bit quantized sine wave and the original sine wave

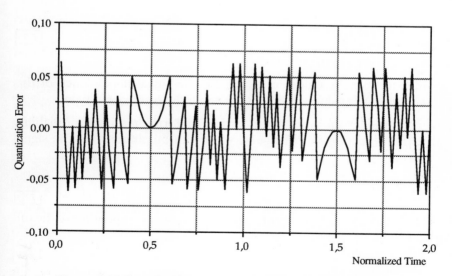

Figure 7.11 Quantization error for a 4–bit quantized sine wave

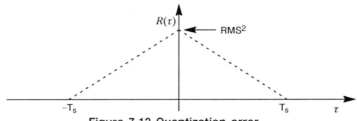

Figure 7.12 Quantization error

Now if we apply assumption 1 that all the allowed levels are exercised with equal probability we get:

$$p(x) = \frac{1}{Q} \text{ for } x \in [-\frac{Q}{2}, \frac{Q}{2}]$$

$$p(x) = 0 \qquad \qquad \textit{otherwise}$$

7.2

The time average power for the quantization error, N_Q, can now be calculated:

$$N_Q = \int_{\infty}^{\infty} x^2 \, p(x) \, dx = \int_{-\frac{Q}{2}}^{\frac{Q}{2}} \frac{x^2}{Q} dx = \frac{Q^2}{12}$$

7.3

The signal to noise ratio in dB is defined as:

$$SNR = 10 log_{10}(\frac{N_S}{N_Q}) \, dB$$

7.4

Where N_S is the signal power.

In order to give the reader a feeling for the signal dependence of this parameter let us evaluate the SNR for three different signal types.

Case 1: Consider an input signal with a uniform amplitude probability distribution function i.e. triangular wave with amplitude A_0. The signal power is given by:

$$N_S = \frac{A_0^2}{12}$$

7.5

If the triangular wave has maximum amplitude A0 corresponding to the full scale of the n bits ADC, we get:

$$N_S = \frac{[(2^n - 1)Q]^2}{12}$$

7.6

It follows that the signal to noise ratio is:

$$SNR = 20 \log_e (2^n - 1) \approx n \, 20 \log_e (2) = 6.02 \, n \, [dB]$$

7.7

Case 2: Consider a maximum amplitude sine wave. The signal power is:

$$N_S = \frac{[(2^n - 1)Q]^2}{8}$$

7.8

Proceeding as with the previous example:

$$SNR = 10 \log_e [\frac{12(2^n - 1)^2}{8}] \approx 6.02n + 1.76[dB]$$

7.9

Case 3: The input signal is a random variable with a Gaussian distribution where the four sigma points correspond to the ADC full scale range. This represents a very common requirement for an ADC, for example, a weighing scale. The signal power is:

$$N_S = \frac{[(2^n - 1)Q]^2}{64}$$

7.10

Once again expressing the SNR in dB:

$$SNR \approx 6.02_n - 7.27[dB]$$

7.11

Following the analysis of these three cases it is important to note that the signal to noise ratio at the output of an ADC is dependent on the input signal. Specifications which depend on signal to noise ratios should be handled very carefully as all boundary conditions must be specified.

The transfer curve for an ADC and the resulting quantisation error is shown in figure. 7.13.

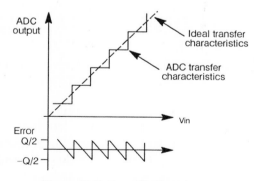

Figure 7.13 Quantisation error

7.3.1 Spectral properties of quantization noise

To analyse random signals it is now assumed that quantisation noise is a random, sampled data variable; there are two very important mathematical functions: the auto-correlation function and the power spectral density [4].

The auto-correlation function of a given variable v(t) is defined by:

$$R(\tau) = \lim_{T \to \infty} \int_{-\frac{T}{2}}^{\frac{T}{2}} \frac{v(t)\, v(t-\tau)}{T_s}\, d\tau \qquad \boxed{7.12}$$

it gives a measurement of the randomness of the variable v(t). From the auto-correlation function it is possible to derive the power spectral density for a signal.

$$G(f) = \int_{\infty}^{\infty} R(\tau) e - j^{\,2\pi f \tau}\, d\tau \qquad \boxed{7.13}$$

The power spectral density function is a statistical evaluation for the power of a random signal and its distribution in the frequency domain. For the assumed assumptions on the quantization noise ϵ_Q its auto correlation function is given by (Fig. 7.13):

$$R_Q(\tau) = RMS_Q^2 \left[1 - \frac{|\tau|}{T_s}\right] \quad \text{for } |\tau| \le T_s$$

$$R_Q(\tau) = 0 \qquad\qquad\qquad \text{for } |\tau| \ge T_s \qquad \boxed{7.14}$$

Where RMS_Q is the root mean square value of the noise and T_s is the sampling period. The assumptions which lead to this result are that the quantization error remains constant during one sampling period and that the quantization noise is uncorrelated from one sample to the next. For example, the auto correlation function is zero for τ greater than T_s.

The spectral density of the quantization noise $G_Q(f)$ can be calculated by making the Fourier transform of the auto-correlation function shown above.

$$G_Q(f) = RMS_Q^2\, T_s\, sinc^2(fT_s) \qquad \boxed{7.15}$$

Where $sinc(x) = sin(x) / x$.

Since this function will only be evaluated in a range where $f \le T_s/2$ a further simplification can be made by neglecting the action of the sinc function. Thus the noise level is independent of frequency i.e. the quantization noise has a white spectrum. Moreover, the noise floor is proportional to the sampling period, conversely the higher the sampling frequency the lower the noise floor.

The total noise power in a given band–width N_{QB} can now be calculated:

$$N_{QB} = \int_{-b}^{b} RMS_Q^2\, T_s\, df = 2B\, RMS_Q^2\, T_S \qquad \boxed{7.16}$$

and, using the oversampling ratio as previously defined:

$$N_{QB} = \frac{RMS_Q^2}{R_{os}}$$

7.17

This is a very useful form, it states that the quantization noise power is reduced by a factor of two for a doubling of the sampling frequency. That is, the resolution of the data converter could be increased by one half a bit by doubling the clock frequency. Note that this is only true for AC or so called busy signals, for DC signals this calculation is not valid because the quantization error is always the same and does not fulfil the requirement that the noise be random. Expressing the signal to noise ratio with this factor we obtain:

$$SNR = 10 \log \frac{N_s}{RMS_Q^2} + 10 \log [R_{os}]$$

7.18

Similar derivations to the ones given in this section are the basis for SNR analysis of noise shaping converters.

7.4 Noise shaping converters

In the previous sections we learnt that a way to improve the signal to noise ratio in data converters is to use oversampling. However, a mere increase in the sampling rate determines only a uniform spread of the power of the quantization noise over a wider bandwidth and, because the noise component is rejected out of band, only half a bit every doubling of the sampling frequency is gained. Thus, for example, to improve the resolution by 5 bits it is necessary to use an oversampling as large as $4^5 = 1024$.

Better efficiency is obtained if the quantization noise, spanning over the band $0 \div f_{ck}/2$, is shaped non–uniformly in such a way that the noise components falling in the band of interest $(0 \div f_{in,max} = B_{in})$ are greatly reduced. No matter if the noise out of the input bandwidth, B_{in}, is increased, a suitable post–filtering (digital for A/D converters, analog for A/D converters) will be capable of rejecting it.

A possible manner to shape the noise as desired is to put the quantizer into a feedback loop [5]. Figure 7.14a) shows the generic block diagram of a A/D converter implementing such a solution. The analogue input is subtracted from the feedback signal and it is passed through the z–transfer function H(z). Its digital conversion is re–transformed in the analog domain by a D/A converter and then passed through a second transfer function F(z) to give rise to the feedback signal. The cascade of the A/D and the D/A converter transforms an analogue signal with all the possible values into an analogue

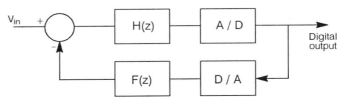

Figure 7.14 (a) General scheme of a niose shaper loop

signal with only discrete amplitudes. In order words, it superimposes an additive component onto the signal: quantization noise. Under the assumption that the addition of quantization noise ϵ_0 is an additive process the schematic in Figure 7.14a) can be transformed into its linearised version, shown in Figure 7.14b). The system has two inputs and

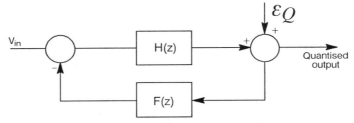

Figure 7.14 (b) linearized scheme

correspondingly two transfer functions: the signal transfer function, T(z), and the noise transfer function, N(z), they are given respectively by:

$$T(z) = \frac{Vout^{(z)}}{Vin^{(z)}} = \frac{H(z)}{1 + H(z)\ F(z)}$$

$$\boxed{7.19}$$

$$N(z) = \frac{Vout(z)}{\epsilon_Q(z)} = \frac{1}{1 + H(z)\ F(z)}$$

$$\boxed{7.20}$$

In order to get the desired effect, the signal transfer function should not insert any attenuation, at least in the band of interest, and by contrast, the noise transfer function should reject signals in the input band. A solution to this is obtained by using F(z) = 1 and H(z) of integrator–type. If H(z) is a simple integrator the modulator is referred to as a first order noise–shaper; for more complex functions, typically cascades or a more complex interconnection of integrators, second–order or high–order noise– shapers result.

When 1–bit A/D and D/A converts are used around the noise shaping loop the circuit is usually referred to as a "sigma delta modulator".

7.5 First order sigma delta modulators

A continuous time implementation of a first order sigma–delta modulator is shown in Figure 7.15. The 1–bit A/D converter is realized very simply by

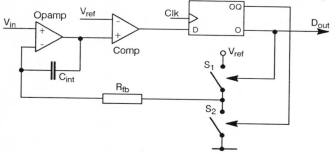

Figure 7.15 Continuous–time first order sigma delta modulator

means of a comparator and a latched flip– flop. The two complementary outputs of the flip–flop drive two switches which connect the non inverting input of the integrator to ground or to the reference voltage; in this way the 1–bit D/A converter is implemented. Depending on the value of the sign of the output of the analog integrator (measured with respect to V_{ref}) the current that is injected into the virtual ground is positive or negative; hence, determining a falling or a rising of the output of the op–amp. Now, since the op–amp is in the integrating configuration and the loop is assumed to be stable, the mean value of the injected current must be zero. Thus, assuming V_{in} constant (or varying very slowly in time), it must be, over a large number of clock cycles, n:

$$(V_{ref} - V_{in})\frac{k_1}{R_{fb}} \cong V_{in}\frac{K_0}{R_{fb}} = V_{in}\frac{n - k_1}{R_{fb}}$$

$$\boxed{7.21}$$

where k1 and k0 is the number of clock cycles for which the output of the flip–flop is 1 or 0 respectively. From 7.21 we get:

$$V_{in} \cong V_{ref}\frac{k_1}{n}$$

$$\boxed{7.22}$$

Thus, the digital pulses at the output of the modulator contain the information on the amplitude of the input signal. Moreover, after one or more 1 pulses the system will react with more or one 0 pulses to compensate the negative injected current with an equivalent positive one. Thus the pulses are uniformly distributed in time.

The switched capacitor implementation of a first order sigma delta modulator is shown in Figure 7.16. The integrator is realized by a conventional switched capacitor structure where the same value of

Over Sampling Converters

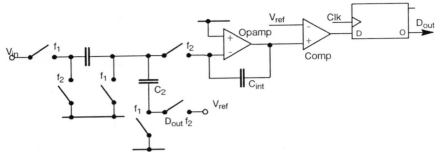

Figure 7.16 Switched capacitor first order sigma delta modulator

capacitance is used in the input structure and the feedback element. The linearised diagram of the circuit is shown in Figure 7.17. The quantization

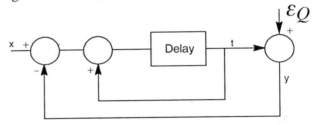

Figure 7.17 Linearized scheme of a first order sigma delta modulator

noise, as usual, is represented by the symbol ϵ_Q. In the time domain the system is described by [6]:

$$t(n + 1) = t(n) + [x(n) - y(n)]$$

$$y(n) = t(n) + \epsilon_Q(n) \hspace{2cm} \boxed{7.23a}$$

that, in the z–domain results in:

$$y(z) = x(z)z^{-1} + (1 - z^{-1})\epsilon_Q(z) \hspace{2cm} \boxed{7.23b}$$

The transfer function of the signal is only a simple delay (z–1); by contrast, the noise is modified by $N(z) = (1 - z–1)$ that corresponds to an high pass action:

$$N(\omega) = 1 - \epsilon^{-j\omega T} = 2je^{-j\omega T/2}\,\frac{e^{j\omega T/2} - e^{-j\omega T/2}}{2j} = 2je^{-j\omega T/2}\,\sin\frac{\omega T}{2} \hspace{1cm} \boxed{7.24}$$

where T is the sampling period.

The white spectrum of the quantization noise $S_Q(f)$, spread over the band $\pm f_S/2$, is:

$$S_q(f) = \frac{V_{ref}^2}{12} \frac{1}{f_S} = \frac{V_{ref}^2}{12} T$$

7.25

it is filtered by the square module of the noise transfer function, and results in the shaped noise spectrum:

$$S_N(f) = S_Q(f)(2\sin \pi f T)^2$$

7.26

If we assume that Bin $<<$ f$_S$, equation 7.26 can be approximated by using $\sin(x) \approx x$; in this case the total power in the bandwidth B$_{in}$ is given by:

$$N_{B,in} = \int_{-B_{in}}^{B_{in}} S_Q(f)\ (2\pi f T)^2\ df = \frac{V_{ref}^2}{2} \frac{\pi^2}{3}\ (2fT)^3 = \frac{V_{ref}^2}{12} \frac{\pi^2}{3}(R_{os})^3$$

7.27

This equation states that the quantization noise power is reduced by a factor of eight for a doubling of the sampling frequency. That is, the resolution of the data converter could be increased by one and a half bit by doubling the clock frequency. The advantage with respect to the simple oversampling operation is evident; however, the noise shaping transfer function, shown in Figure 7.18, amplifies the noise spectrum in the high frequency range. If

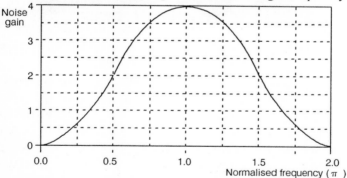

Figure 7.18 Noise shaping transer function of a first order modulator

only for this reason, the specifications of the digital filter used to reject the undesired noise will be more severe than for the one used in a simple oversampled converter: besides rejecting an amplified noise it must also avoid the folding in band base of noise power (due to aliasing) at a level which is negligible with respect to the one left by the more efficient shaping.

The SNR of a first order sigma delta converter, employing an ideal low pass digital filter with cut off frequency set at B$_{in}$, can be calculated by the use of equation 7.27. Assuming an input signal with uniform amplitude probability, we get:

$$SNR = 9.03 \log_2(R_{os}) - 5.2[dB] \qquad \boxed{7.28}$$

The above results have been obtained under the hypothesis that justifies the linearised model. In reality, since there are nonlinear elements around the loop, and since the assumptions taken for assimilating the quantization error to white noise have not been verified well, the derived equations give only an approximate description of the system's behaviour. The divergence between theory and reality can be shown by appropriate computer simulations [7][8]. Figure 7.19 shows the spectrum of the quantization noise

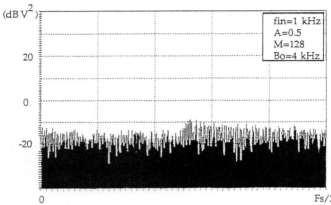

Fig.19 Spectrum of the quantization noise for a first order sigma delta modulator

for a first order sigma delta with a 1kHz sinewave applied at the input. It is obvious that its spectrum is not strictly white; however, in the low frequency range the spectrum is reasonably constant. Figure 7.20 shows the spectrum of the noise at the output of the first order modulator; it can be noted that the shaping follows the behaviour that was foreseen, however, discrete frequency components with relevant amplitude affect the spectrum.

7.6 Second order sigma delta modulator

The noise shaping of the modulator considered in the previous section can be further improved by using the cascade of two integrators around the noise shaper loop [9]. This configuration is referred to as "second order modulator"; its typical block diagram is shown in Figure 7.21. It consists of the cascade of two sampled data integrators, the first without delay, the second with delay. The second integrator, by the feedback connection from the 1–bit D/A converter, is dumped. This configuration avoids stability problems otherwise present when the cascade of two integrators is used in a feedback loop. The analysis of the schematic, in the z–domain, gives:

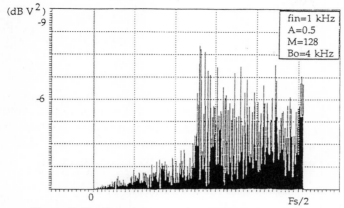

Figure 7.20 Spectrum of the noise at the output of a first order modulator

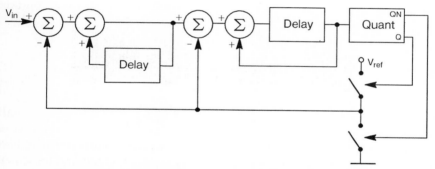

Figure 7.21 Block diagram of a second order modulator

$$s(z) = [x(z) - y(z)] \frac{1}{1 - z^{-1}}$$

$$t(z) = [s(z) - y(z)] \frac{z^{-1}}{1 - z^{-1}}$$

$$y(z) = t(z) + \epsilon_Q(z) \qquad \boxed{7.29}$$

by solving the system we get:

$$y(z) = x(z)z^{-1} + \epsilon_Q(z)(1 - z^{-1})^2 \qquad \boxed{7.30}$$

It should be noted that the signal is passed through a simple delay, while the quantization noise is passed through the square of the transfer function of the already analyzed first order modulator.

The same transfer functions expressed by 7.30 can also be obtained with different schematics; different architectures are often derived and used in order to optimize the dynamic range at the output of the two integrators. For the schematic shown in Figure 7.21 the voltage at the output of the first op–amp can extend up to $\pm 2V\text{ref}$ and the voltage of the second op–amp up to $\pm 4V\text{ref}$. Such large output swings can result in a limitation in practical circuits which can be removed by adopting a different modulator scheme.

As for the already considered first order modulator it is worthwhile calculating the in–band total power of the shaped quantization noise. Again using the approximation $\sin(x) \approx x$ (justified by the condition $B_{in} << f_S$) we get:

$$N_{B,in} = \int_{-B_{in}}^{B_{in}} S_Q(f) \ (2\pi f T)^4 \ df = \frac{V_{ref}^2}{12} \ \frac{\pi^4}{5} \ (2fT)^5 = \frac{V_{ref}^2}{12} \ \frac{\pi^4}{3} (R_{os})^5 \qquad \boxed{7.31}$$

The signal to noise ratio, SNR, for input signals with uniform amplitude probability

$(< v_{in} > = V_R / \sqrt{12})$

(is calculated by:

$$SRN = 15.05 \ \log_{2(R_{os})} - 12.9 \ [dB] \qquad \boxed{7.32}$$

Hence with a second order modulator the converter can gain two and a half bits of accuracy by doubling the clock frequency. The strong reduction in the quantization noise in the input band is achieved at the expense of increasing the requirements on the specifications of the digital filter which rejects the noise out of band. Since the output data must be decimated, the contribution that will be folded back in band base (aliasing) must be a negligible fraction of the term left in the band base by the shaping.

7.7 Multistage sigma–delta modulator

The modulator analyzed in the previous section was derived from a first order scheme by adding a second integrator to the feedback loop. The scheme of the modulator was designed in such a way that the signal transfer function is a simple delay while the noise transfer function becomes the squared of a sine shaping. The procedure can be generalized to improve the reduction of the noise in the low frequency band [10][11]. However, it is not advisable to put three integrators in a feedback loop because of the stability problems involved. Higher–order modulators are usually realized with a cascade of low– order modulators (1st or 2nd order). The outputs of the modulators are suitably combined to obtain a high–order noise transfer function. This approach is advantageous because it allows modularity in the architecture, improves the dynamic range and allows inconditional stability of the circuit. However, the high–order noise shaping is achieved by a perfect matching of elements. Errors due to mismatch rapidly degrade the performance of the modulator.

In order to explain the technique, let us consider the scheme shown in Figure 7.22. It is a third order modulator, popular known by its acronym

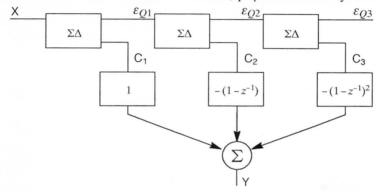

Figure 7.22 Third order modulator realized by the MASH architecture

MASH [12]. It is composed of three first order modulators which provide both the digital data and the quantization noise at their output (Fig. 7.23). The quantization noise of the first modulator is used as an input of the second modulator and the quantization noise of the second modulator as an input for the third one. The three outputs are then combined by a digital filter.

The architectures used in the first order modulators give rise to the following equations:

$$C_1 = X + (1-z^{-1})\epsilon_Q 1$$
$$C_2 = \epsilon_{Q1} + (1-z^{-1})\epsilon_{Q2}$$
$$C_3 = \epsilon_{Q2} + (1-z^{-1})\epsilon_{Q3} \qquad \boxed{7.33a}$$

the output of the digital filter is:

$$y = C_1 - (1-z^{-1)} C_2 - (1-z^{-1})^2 \ C_3 = X - (1-z^{-1})^3 \ \epsilon_{Q3} \qquad \boxed{7.33b}$$

The result is that the contributions of the quantization noise of the first and the second modulator are cancelled and the noise of the third integrator is shaped with the transfer function $(1-z^{-1})^3$.

The method described is also used with basic blocks made of second order modulators. As for the use of first order cells, it is necessary to generate the quantization noise that is then processed by a successive stage. From knowledge of the signal and noise transfer functions it is possible to find a given digital processing to cancel the effect of the quantization noise of the inner stages. It is evident that the weak point of the method lies in the generation of the quantization noise; as shown in the schematic of Figure 7.23 it is obtained by subtracting the output of the op–amp from the

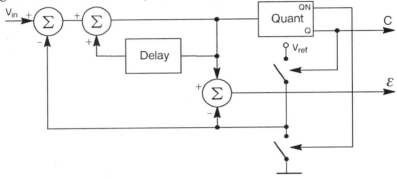

Figure 7.23 Schematic of the basic cell of the mash architecture

D/A converter data. Any error in this operation is reflected in an imperfect cancellation of the noise. For example, if in the first integrator in Figure 7.22 the generated quantization error is $\epsilon_{Q1}(1-\alpha)$ instead of the correct value ϵ_{Q1}, the residual part $\alpha\epsilon_{Q1}$ will be filtered out only by the noise transfer function $(1-z-1)$. Thus, in general, the residual quantization error of the first modulator is shaped only by a first–order transfer function, the residual error of the second integrator by a second order transfer function and so on. On the basis of the above observation it seems advisable to use second order modulators as elementary stage of the system.

7.8 Non ideal effects in sigma delta modulators

The basic blocks used in sigma delta modulators have been, up to now, considered as ideal elements. However, all the real elements are affected by limitations which reflect non idealistics in the system where they are utilised. We will concentrate on switched capacitor modulators, where operational amplifiers, comparators, switches and capacitors are employed.

A real operational amplifier is limited because its DC gain, bandwidth and slew–rate are finite; its output impedance is far from zero (very often transconductance operational amplifiers, OTAs, are utilised) and because the op–amp itself is noisy (the noise performances are described by an input referred noise generator). A comparator is affected by limitations similar to the ones of the op–amp, with, in addition, the presence of hysteresis in the input–output characteristic. Switches are limited by a finite on–resistance, at which a white noise source is associated. Moreover, for switches, a coupling between the driving logic and the analog section (clock–feedthrough effect) always affects the operation. The matching in integrated capacitors is quite good, however small inaccuracies, as well as non linearities and parasitic elements, can become significant for high resolution data converters [13].

The non idealistics recalled above determine some effect on the modulator performances. In this section the more important of them are considered.

7.8.1 Finite gain and leakage in integrators

Let us consider the conventional switched capacitor integrator shown in Figure 7.24. As known, in the ideal case, the injecting capacitor C_1 is

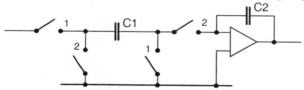

Figure 7.24 Switched capacitor integrator

pre–charged at the input signal (phase 1) and during the phase 2 it is completely discharged through the virtual ground. However, in a real case, because of the finite value of the DC gain the bandwidth and the slew rate of the op–amp, the charge is not completely transferred. Moreover, a mismatch between the capacitors C_1 and C_2 determines an additional error. If the above limits are taken into account, the integrator, in the time domain, is described by:

$$V_{out}(n + 1) = (1 - \delta)V_{out}(n) - \frac{C_1}{C_2}\beta V_{in}(n)$$

$$\boxed{7.34}$$

where δ and β are suitable parameters; in the z domain, assuming C_1 and C_2 nominally equal, it results:

$$H(z) = \frac{V_{out}(z)}{V_{in}(z)} = -\frac{(1 - a)z^{-1}}{1 - (1 - \delta)z^{-1}}$$

$$\boxed{7.35}$$

where $\dfrac{C_1}{C_2}\beta = (1 - a)$.

With respect to the ideal transfer function, equation 7.35 reveals a gain error α and a leakage error δ. A similar equation can be derived for the non inverting integrator.

Using the above result in the first order modulator shown in Figure 7.17 the output signal, as a function of the input $x(z)$ and the quantization noise ϵ_Q becomes:

$$y(z) = \frac{1}{1 + (\delta + a)z^{-1}} \left\{ x(z)z^{-1} + [(1 - z^{-1)} + \delta z^{-1}]\epsilon_Q \right\} \qquad \boxed{7.36}$$

at low frequency ($f < < f_S$) the effect of the denominator is negligible. Thus, beside the expected shaping, the quantization noise is also multiplied by the term δz^{-1}. It corresponds to an additional unshaped noise component proportional to the leakage error.

If gain and leakage errors are taken into account in the second order modulator we get:

$$y(z) = \frac{1}{D(Z)} \left\{ x(z) \, z^{-1} + [(1 - z^{-1}) + \delta_1 z^{-1}][(1 - z^{-1}) + \delta_2 z^{-1}]\epsilon_Q \right\} \qquad \boxed{7.37}$$

where $\delta 1$ and $\delta 2$ are the leakage factors of the two integrators and $D(z)$ is a second order function. Again, at low frequency, $D(z)$ has negligible effects. The resulting noise transfer function is made of three terms:

$$N(z) = (1 - z^{-1})^2 = (\delta_1 + \delta_2)z^{-1}(1 - z^{-1}) + \delta_1\delta_2 z^{-2} \qquad \boxed{7.38}$$

they corresponds to three additive components in the output noise. The first one is the same as in an ideal modulator. The second one corresponds to the quantization noise attenuated by $(\delta_1 + \delta_2)$ and shaped by a first order modulator; the third term is the quantization noise only attenuated by $\delta_1\delta_2$.

7.8.2 Electronic noise

The relevant noise sources in sigma delta modulators are the input referred noise generator of operational amplifiers and the noise source associated to the on–resistance of switches. Since the modulator is a discrete time network, all the noise sources are sampled every clock cycle. Then, their effect is transferred to the output by a proper sampled data (noise) transfer function. It is worthwhile to observe that, before the sampling, the noise generators are filtered by a continuous–time action. It takes place, very often, because of the finite on–resistance of switches and of finite bandwidth of the op–amps. However, the noise band is limited at frequencies that are well beyond the sampling frequency; consequently, the spectrum will be aliased into the band base. The simple case where the band limitation comes from the on–resistance of the switch and the sampling capacitor, is studied in this Book in the Chapter dealing on "Data Converters". For it, the thermal noise of the on–resistance results into a white sampled–data noise which power over the Nyquist band is kT/C (k is the Boltzman constant, T is the absolute temperature, C is the sampling capacitor) [14][15].

The sampled–data transfer functions acting on the noise critically depends on the point where the noise itself is injected. If the noise is applied immediately after the quantizer it will be shaped in the same way as the quantization noise. By contrast, if the noise is injected at the input of the modulator it will be treated like signal. It immediately turns out that the noise sources the the first stage of the modulator are much more critical than the ones in successive stages, since they are not favourably processed.

To be more specific, let us consider the first order modulator shown in Figure 7.25. The effect of the noise sources associated to the two switches

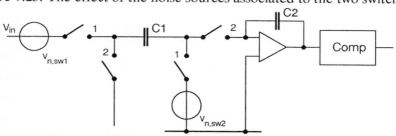

Figure 7.25 First order modulator with evidentiated the noise sources of two switches

exercised during the phase 1 is considered. As already pointed out, the continuous time filtering and the successive sampling results into a white sampled–data noise which power is kT/C_1. The related noise transfer function, being the switches in series with the input signal, is just the signal transfer function, z^{-1}. Therefore, the only rejection of the kT/C_1 power comes from the digital filter, used after the analog modulator. Assuming an ideal low pass response, as shown in Figure 7.25, the output noise that is contributed by the two considered switches is given by:

$$v_{out,sw}^2 = \frac{kT}{c_1}\frac{2B_{in}}{f_{CK}} = \frac{kT}{C_1 R_{os}}$$

$\boxed{7.39}$

for $C_1 = 1pF$ and $R_{os} = 128$, $v_{out,sw}$ is equal $5.7\mu V$, a value that becomes comparable to 1/2 LSB for a 16 bit ($V_R = 1V$) data converter. Thus, for high resolution it is necessary to use, at least in the input stage of the modulator, large capacitances.

Like the noise injected into inner points of the noise–shaping loop, outcomes of non idealistics, such as, for example hysteresis and threshold variations in the comparator are attenuated in the low frequency range.

7. 8.3 Sampling jitter

The input stage of a modulator samples the input analogue signal at the oversampled frequency. The exact sampling instant occurs when the sampling switch goes off. Since this switch is realized by an MOS transistor, it goes off when the gate voltage is larger (or smaller depending on the transistor type) than the input signal by just the threshold. If the falling edge (or the rising edge) of the driving signal is not negligible the sampling instant will depend on the input amplitude. Moreover, the driving signal can displace a jitter. These effects results in a non uniform sampling that is usually described with an additional noise source: the jitter noise, ϵ_{ji}. In order to estimate its relevance, let us consider the sampling of a sinusoidal input. If the sampling instant, T, is affected by an error, δ, its equivalent jitter noise, ϵ_{ji}, is given by:

$$\epsilon_{ji} = V_{in}(T + \delta) - V_{in}(T) = A\left\{sin[\omega_{in}(T + \omega)] - sin(\omega_{in}T)\right\} \cong$$
$$\cong \omega_{in}\,\delta\,A\,\cos\omega_{in}T \qquad \boxed{7.40}$$

which indicates that the random variable δ is modulated at the input frequency. However, if δ is a Gaussian process its spectrum is white and the modulation is irrelevant.

The jitter noise, introduced in the circuit, is transferred to the output in the same way as the electrical noise sources: it is passed through a suitable sampled–data transfer function. Remembering the results discussed for the electrical noise we can immediately state that the dominant spur arises from jitter in the input stage: the sampled data transfer function does not introduce any low frequency attenuation and the noise power is only reduced by the decimation process.

7. 9 Digital decimation

The purpose of a digital filter cascaded to a sigma delta modulator is twofold [16]:

• to remove the shaped quantization noise out of the base band

• to prevent the noise aliasing in the successive decimation After the decimation the rate of the digital data is twice the band of the input analogue signal and is represented by a given number of bits; they are completely meaningful only if the total residual noise is smaller than 1/2 LSB.

A very common architecture for digital decimators in sigma delta converters is shown in Figure 7.26. It is made of the cascade of many stages,

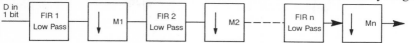

Figure 7.26   Architecture of a typical digital decimator

each of them introduces a given decimation factor. Usually, $M_1 > M_2 > ... > M_n$. The obtained decimation is, off course:

$$M = \prod_{i=1}^{n} M_i$$

$\boxed{7.41}$

Every decimator is preceded by a low pass filter for getting in each stage the two before mentioned actions.

In order to intuitively understand the low pass filtering needs, Figure 7.27 shows examples of the outcome of different filters followed by a by–4 decimator. The considered in put is the quantization noise shaped by a first–order sigma–delta modulator (Fig. 7.27a). Without any filtering, the decimation results in the band–base spectrum shown in Figure 7.27b. The noise is strongly dominated by the terms that are folded from the upper bands, b2, b3, b4; they are summed up in almost a white spectrum. It can be concluded that the advantage of using a sigma delta modulator is completely vanished. The filtering action shown in Figure 7.27c determines an improvement of the situation; thanks to a zero placed at fs/4 the noise that is folded at DC frequency is completely cancelled; however, in the low frequency spectrum the folded upper bands. The stronger filtering action shown in Figure 7.27e permits to push folded components well below the unfolded term, thus allowing full benefit by the modulator shaping.

As shown before it is necessary to have in the transfer function of the digital filter single or multiple zeros. This is get by the use of FIR architectures. Very often in the first stage of the decimator chain it is used a sinc–filter (comb–filter), since its realization does not need multipliers. The transfer function of a sinc–filter is expressed by:

$$D(z) = \frac{1}{N} \left[\frac{1-z^{-N}}{1-z} \right]^k$$

$\boxed{7.42}$

where k is the order of the sinc–filter. In order to have a negligible amount of aliased noise the order of the sinc–filter must be at least by 1 larger than the modulator order. The simple sinc–filter gives rise to some attenuation in the base band. The effect is usually corrected in the following stages.

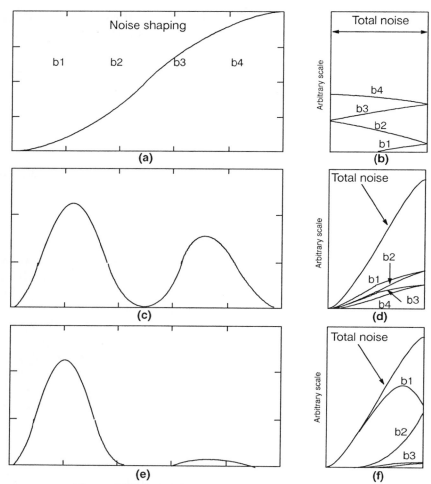

**Figure 7.27 Spectrum before and after decimation
for different filtering transfer functions**

7.10 References

[1]   M. Hauser, R. Brodersen, "Circuit and Technology Considerations for
      MOS Delta–Sigma A/D Converters", Proc. of the IEEE International
      Symposium on Circuits and Systems, pp. 250–253, 1986

[2]   A. Gersho, "Principles of Quantization", IEEE Trans. on Circuits and
      Systems, Vol. CAS–25, No. 7, pp. 427–436, July 1978

[3] D. Goodman, L. Greenstein, "Quantizing Noise of DM/PCM Coders", The Bell System Technical Journal, Vol. 52, No. 2, pp. 183–204, Feb. 1973

[4] A.V. Openheim, R.W. Schafer, "Digital Signal Processing", Prentice–Hall Inc., Englewood Cliffs, N.J.

[5] B. Agrawal, K. Shenoi, "Design Methodology for SDM", IEEE Trans. on Communications, Vol. COM–31, No. 3, pp. 360–369, March 1983

[6] J. Candy, O. Benjamin, "The Structure of Quantization Noise from Sigma–Delta modulation", IEEE Trans on Communications, Vol. COM–29, No. 9, pp. 1316–1323, Sept. 1981

[7] V.F. Dias, F. Maloberti, "Design and Performance Evaluation of High–Resolution Oversampling A/D Converters", The Fourth CSI/IEEE International Symposium on VLSI Design, New Deli, India, January 5–8 1991

[8] V.F. Dias, V. Liberali, F. Maloberti, "TOSCA: A Simulator for Oversampling Converters with Behavioural Modeling", CompEuro 1991, Bologna, Italy, May

[9] J. Candy, "A Use of Double Integration in Sigma–Delta Modulation", IEEE Trans. on Communications, Vol. COM–33, pp. 249–258, March 1985

[10] W. Chou, P. Wong, R.M. Gray, "Multistage Sigma–Delta Modulation", IEEE Trans. on Information Theory, Vol. IT–35, No. 4, pp. 784–796, July 1989

[11] W. Lee, C. Sodini, "A Topology for Higher Order Interpolative Coders", Proc. of the IEEE International Symposium on Circuits and Systems, pp. 459–462, 1987

[12] Y. Matsuya, K. Uchimura, A. Iwata, T. Kaneko, "A 17–bit Oversampling D–to–A Conversion Technology Using Multistage Noise–Shaping", IEEE J. Solid–State Circuits, Vol. SC–24, No. 4, pp. 969–975, Aug. 1989

[13] B. Boser, B. Wooley, "The Design of Sigma–Delta Modulation Analog–to–Digital Converters", IEEE J. Solid–State Circuits, Vol. SC–23, No. 6, pp. 1298–1308, Dec. 1988

[14] C.A. Gobet, A. Knob, "Noise Analysis of Switched–Capacitor Networks", IEEE Trans. on Circuits and Systems, Vol. CAS–30, pp. 96–102, Jan. 1983

[15] H. Fisher, "Noise Sources and Calculation Techniques for Switched–Capacitor Filters", IEEE J. of Solid–State Circuits, Vol. SC–17, pp. 85–95, Aug. 1982

[16] J. Candy, "Decimation for Sigma–Delta Modulation", IEEE Trans. on Communications, Vol. COM–34, pp. 72–76, Jan. 1986

Chapter 8
Self-calibrating and Algorithmic Converters
Joao Vital, Jose E. Franca, Carlos A. Leme
Instituto Superior Tecnico, Lisboa

8.1 Introduction

The relevance of data converters in modern ASIC designs combining both analogue and digital signal processing has motivated significant efforts worldwide devoted to the development of circuit techniques which can not only improve their basic specifications for conversion speed and resolution, but also increase their functionality and achieve more economic realisations with respect to power consumption and silicon area.

Data converter architectures are usually tailored to specific application areas ranging, for example, from low speed–high resolution for high quality audio band systems with signal bandwidths of 22kHz and 18–bit resolution, to high speed–moderate resolution for video band systems with signal bandwidths of the order of 4MHz and requiring not more than 10–bits of resolution. In between, there is a wide variety of system applications which may need conversion speeds up to the order of several hundreds of kHz, or even a few MHz, and conversion resolutions ranging from 8–bit to 16–bit. Some of the circuit techniques available for designing data converters in such a range of specifications are described in this Chapter, whereas elsewhere in the book we can see detailed descriptions of alternative data conversion techniques for lower speeds and higher resolutions as well as for much higher speeds and lower resolutions.

One of the most popular architectures for realising A/D converters with conversion resolution from 10–bits to 16–bits and conversion speeds between the audio and video frequency bands is described in Section 2. It is based on the well known combination of the successive approximation algorithm together with self–calibrating techniques. Section 3 addresses some types of algorithmic D/A converters which are designed to achieve very specific targets, for example with respect to the cost of manufacture and conversion characteristics. This is case of a low–cost quasi–passive D/A converters yielding up to 8–bit resolution and conversion speeds up to a few MHz, and of a capacitance–ratio–independent D/A converter which can achieve high conversion resolutions using operational amplifiers with relatively modest values of the DC gain. In Section 8.4, a further functional dimension is added to traditional algorithmic D/A converters by considering a built–in filtering function to shape the resulting output analogue signals, and thus realise a complete mixed–signal digital–to–analogue interface system. For some of the data converters described in Sections 8.2 to 8.4 we shall also present examples of a CMOS integrated circuit implementation. The final summary of this Chapter is given in Section 8.5.

8.2 Self–calibrated analogue–digital converters

8.2.1 Introductory remarks

Two alternative types of converters have become dominant for high–resolution A/D data conversion, namely the sigma–delta and the successive approximation with self–calibration. While sigma–delta converters, which are addressed in a separate Chapter, attract widespread interest for high performance digital audio, since they minimize the number of critical analogue components and make extensive use of digital signal processing techniques, self–calibrated successive approximation converters are still rather competitive for a large number of data acquisition applications requiring high resolution of conversion [1, 2]. At the expense of a relatively more complex digital micro-controller, such self–calibrated architectures are capable of determining and correcting their internal linearity errors. Moreover, since the calibration procedure can be periodically repeated, this type of A/D converter can maintain the specified accuracy over a wide range of operating conditions. The potential of self–calibrating A/D converters for high performance signal conversion applications has led to the development of a number of different types of architectures [3], some of which have already found considerable commercial interest [4]. Besides discussing the basic principles and circuit techniques of self–calibrated successive approximation A/D converters, we shall also describe in this Section the practical design of a CMOS converter for high resolution applications.

8.2.2 Architecture with segmented binary–weighted capacitor array

The architecture of a successive approximation A/D converter is illustrated in Figure 8.1, consisting of a Successive Approximation Register (SAR), a comparator and a binary–weighted capacitor array [4]. The capacitive array is segmented into a Main–Array (MA), for the M Most Significant Bits (MSB's) and a Sub–Array (SA), for the L Least Significant Bits (LSB's), such that $N = L + M$ is the resolution of the converter. Both arrays are interconnected through an active block giving an attenuation factor of 2^{-M}. The adoption of such a solution for the design of the overall capacitor array significantly reduces the overall capacitance spread and, consequently, the

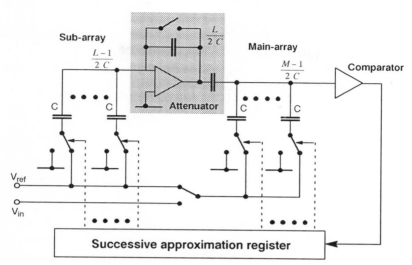

Figure 8.1 Architecture of a successive approximation A/D converter with segmented binary–weighted capacitor array.

input capacitive load and also the silicon area required for integration. The length L of the SA is determined such that the worst–case matching accuracy of the capacitance ratios guarantees the required resolution and linearity specifications of the converter. Since the matching accuracy which can be practically achieved for the capacitance ratios in the MA, with length M, may not guarantee the resolution nor the linearity specifications of the converter, the corresponding capacitors must be calibrated using the self–calibrating architecture and calibration cells described next.

8.2.3 Self–calibration technique and circuits

8.2.3.1 Principle of calibration

The calibration of the MA is accomplished by associating with each appropriate capacitor, Carray, a Calibrating Capacitor Array (CCA) together with the associated calibration register, as illustrated in Figure 8.2. The number of calibrating capacitors in each CCA depends on the maximum mismatch error of the corresponding capacitor in the array. In each CCA, the smallest calibrating capacitor introduces a minimum nominal bit weight of 1/4 LSB, whereas the largest calibrating capacitor is switched in opposition to C_{array} to introduce a negative bit weight which allows for both positive and negative mismatch errors. The number of capacitors of the MA which need calibration, together with the number of calibrating capacitors of the associated CCA's, can be determined by means

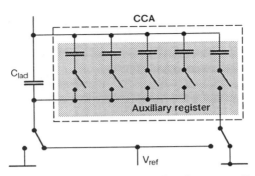

Figure 8.2 Architecture of the calibrating capacitor array.

of an automated design tool which takes into account the statistical variations of the matching accuracies determined by the technology [5]. An additional CCA may also be used for fine offset voltage compensation.

The basic operation for calibrating each capacitor C_{array} in the MA consists of testing whether its capacitance value is above or below the nominal value. This can be determined by considering the property of a binary weighted capacitor array where the nominal capacitance value of each capacitor is equal to the sum of all the remaining less significant capacitances plus a unit capacitance [6]. Figure 8.3 illustrates how to carry–out such test for an

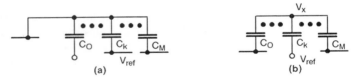

Figure 8.3 Generation of residual voltage associated with capacitance mismatch of Ck

arbitrary capacitor C_k. In a first step, shown in Figure 8.3a, the top plates of C_k and of the corresponding most significant capacitors are all connected to ground, whereas the reference voltage is sampled in all the less significant capacitors plus capacitor C_0 with unit capacitance value. Then, in Figure 8.3b, only capacitor C_k is connected to the reference voltage yielding an error voltage Vx at the bottom plate of the array. If the value of C_k is above the nominal value then V_x will be positive, in the opposite case V_x becomes negative. By carrying–out a successive approximation search, the calibrating capacitors associated with C_k are inserted in parallel to achieve the matching accuracy corresponding to the required resolution and linearity specifications of the converter. Afterwards, the final configurations (i.e. the connections of the switches of the CCA) are stored in the calibrating registers. The circuit is then ready to be operated by a classical successive approximation register to carry–out the A/D conversion.

8.2.3.2 Calibrating capacitors

Because of the segmented architecture of the binary–weighted capacitor array, the calibrating capacitors in the MA also have to be multiplied by the attenuation factor 2^{-M}. This can lead to such low capacitance values that they are impractical for IC implementation, since the corresponding geometries would violate design rule constraints determined by the technology. However, the problem can be overcome by realizing that the circuit variable that matters in both processes of calibration and conversion is a *packet of charge* defined by a product *voltage times capacitance*. Thus, if the minimum capacitance value is technology limited, the reference voltage can be scaled down as required to obtain the value for the minimum packet of charge which introduces the corresponding bit weight of 1/4 LSB. The CCA's in the Main–Array are, therefore, associations of minimum technology capacitors cp switched to an auxiliary reference voltage, Vrefaux, given by:

$$V_{ref\,aux} = V_{ref}.\frac{C/4}{cp}.2^{-4} \qquad \boxed{8.1}$$

where C is the unit capacitance value of the MA. The large capacitance ratio mismatches inherent to minimum technology capacitors limit the number of capacitors in a CCA if they are designed to follow a binary–weighting rule. However, since the value of each calibrating capacitor in any CCA must only satisfy the condition of its value never exceeding the sum of the remaining less significant ones, we can adopt instead the following weighting rule:

$$C_k = (1-\tau) \sum_{i=1}^{k-1} C_i + C_1 \qquad \boxed{8.2}$$

where τ is the tolerance of the capacitance ratio mismatch for the minimum geometry [7].

8.2.3.3 Calibrating registers

The calibrating registers associated with each CCA have one memory cell for each calibrating capacitor, and perform several operations. During the A/D conversion cycle, they only have to apply to the corresponding CCA's the calibrating configurations stored in their memory cells. During the calibration cycle, on the other hand, the calibrating registers must carry–out the operations of decoding the active memory cell, driving the calibrating capacitor and storing its state (either ON or OFF). Each memory cell employs the structure illustrated in Figure 8.4, where the SEL_BIT line

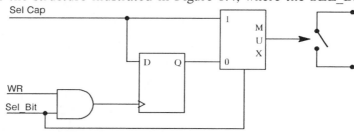

Figure 8.4 Block diagram of the memory cells.

selects the bit to be calibrated and the SEL_CAP line determines the configuration of the CCA. The signal WR, which is common to all the cells, stores in the register selected by SEL_BIT the configuration available in SEL_CAP . This structure can lead to a very compact calibration circuitry and a rather simple controller [10].

8.2.4 An example of a CMOS integrated circuit implementation

The photomicrograph of a $2\upsilon m$ CMOS self–calibrated successive approximation A/D converter employing the architecture and circuit techniques described above can be seen in Figure 8.5 [8]. In this example, the MA defines 8 bits whereas the SA defines another 7 bits. The placement of both arrays corresponds to *Zone B*. *Zone A* indicates the area taken up by the calibration circuitry which has been designed to cope with a maximum capacitance ratio error of 0.5% in the main array and of 20% in the calibrating arrays.The excellent layout compactation results on one hand from the very modular topology of the memory cells and on the other hand, from the extensive use of Metal–II lines running over the memory cells and which saved much interconnection area. *Zones C* through *F* identify, respectively, the switches, the comparator, the operational amplifier in the active interconnection block between both arrays, and the generator of the auxiliary voltage reference used in the calibrating arrays. A high performance comparator is employed to resolve an input voltage as low as $10\mu V$ in less than $1\mu s$ [9]. In this implementation example, a great care was devoted to the routing of the analogue signals using shielded channels, i.e., the analogue lines run in Metal–I sandwiched between two ground planes in Metal–II and Poly (Figure 8.6a). Where a crossover by digital lines is required, the channel structure is changed: the analogue lines run in Poly

Figure 8.5 A photomicrograph of a 2um CMOS self-calibrated successive approximation A/D converter

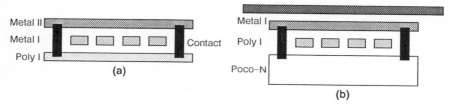

Figure 8.6 Schematic illustration of shielded channels for routing of the analogue signals.

and for the ground planes we employ Metal–I and an N–Tub, thus leaving the Metal–II level of interconnection for running the digital lines (Figure 8.6b). A typical result of the computer–simulated non–linearity performance of this A/D converter, for 15–bit resolution, is illustrated in Figure 8.7.

8.3 Algorithmic digital–analogue converters

8.3.1 Introductory remarks

Algorithmic D/A converters constitute an attractive alternative for applications covering a wide range of specifications because they occupy a relatively small silicon area and can also be programmed by digital means [10–14]. In one type of algorithmic D/A convertor the conversion algorithm is implemented by means of an active circuit employing, at least, one

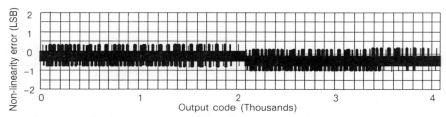

Figure 8.7 Typical computer-simulated non-linearity performance of a 15-bit A/D converter with self-calibration

operational amplifier [11, 12]. In quasi–passive algorithmic D/A convertors, on the other hand, signal conversion is performed entirely in the charge domain and therefore can potentially achieve much higher speeds of operation [13, 14]. In both cases, however, the conversion resolution is essentially limited by the capacitance matching accuracy which can be achieved in a given process technology. There is still another type of algorithmic convertors whose resolution is basically independent of capacitance ratio accuracies and, therefore, can achieve higher conversion resolution [11, 16].

This Section describes first a quasi–passive algorithmic D/A converter whose resolution is basically limited by the capacitance matching accuracy of 3 equal–valued capacitors, and then a capacitance–ratio–independent algorithmic D/A converter which employs only one operational amplifier. In both cases signal conversion is carried–out bit–by–bit, and the resolution as well as the speed of conversion can be easily programmed by digital means. Experimental results are presented to demonstrate the operation of an integrated 2.5μm CMOS D/A converter based on the quasi–passive algorithmic conversion technique.

8.3.2 Quasi–passive algorithmic D/A converter

8.3.2.1 Principle of conversion

The basic circuit diagram of a quasi–passive D/A converter is shown in Figure 8.8, where it is assumed that the bits of the input digital word are sequentially applied through a shift–register array to the logic circuitry associated with the gate terminals of transistors Q_1 and Q_2. Here, it is assumed that b_0 is the LSB and b_{n-1} the MSB. The circuit uses two capacitors of equal capacitance value and its operation requires three clock waveforms. The clock waveform with phase ϕ_0 defines the period T_c during which one full conversion cycle can be performed. The conversion algorithm is controlled by the non–overlapping clock waveforms with phases ϕ_1 and ϕ_2 and whose frequency must be N/T_c for an N–bit

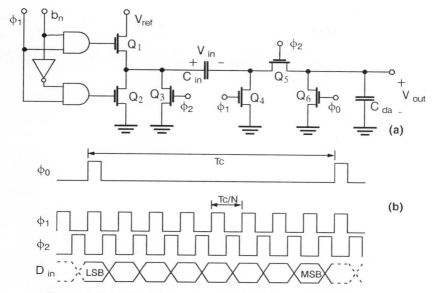

Figure 8.8 (a) Quasi–passive SC circuit for algorithmic D/A conversion and (b) switching waveforms.

conversion. Every conversion period starts by resetting the output capacitor C_{da} during the time when $\phi_0 = 1$ (switch Q_6 closes). When this clock phase goes low, switch Q_6 is cut–off and hence the output capacitor is enabled to partake, together with the input capacitor, in the conversion algorithm which is carried–out sequentially from the LSB to the MSB. One iteration of the conversion algorithm is accomplished during one period of the clock waveforms with phases ϕ_1 and ϕ_2. First, during the time when $\phi_1 = 1$, the input capacitor C_{in} acts as a sampling capacitor which samples either the voltage reference if $b_0 = 1$ (switch Q_1 closes and switch Q_2 opens), or the ground voltage if $b_0 = 0$ (switch Q_1 opens and switch Q_2 closes). When the clock phase ϕ_1 goes low the switches Q_1, Q_2 and Q_4 cut–off, and then the voltages across capacitors C_{in} and C_{da} are $V_{in} = V_{ref}b_0$ and $V_{da} = 0$, respectively. When the clock phase ϕ_2 goes high, the closing of switches Q_3 and Q_5 produces a toggle action upon capacitor C_{in} such that the negative of its charge will be redistributed through the parallel combination of C_{in} and C_{da}. When the clock phase ϕ_2 goes low, both switches Q_3 and Q_5 are cut–off and hence the final voltage across the output capacitor will be $V_{da} = -(1/2)(b_0)V_{ref}$.

The second conversion iteration starts when clock phase ϕ_1 goes high again. In the same fashion as before, the input capacitor samples either the reference voltage if the next least significant bit is at logic "one" ($b_1 = 1$), or the ground voltage if it is at logic "zero" ($b_1 = 0$). After the sampling phase the voltages across the two circuit capacitors will be $V_{da} = -(1/2)(b_0)V_{ref}$ and $V_{in} = V_{ref}b_1$, respectively for capacitors C_{da} and C_{in}. Then, clock phase ϕ_2 goes high again and the charges in capacitors C_{in} and C_{da} will be redistributed taking into account the toggle action upon capacitor C_{in}. At the end of clock phase ϕ_2 the voltage across the output capacitor will be $V_{da} = -(1/2)^2(b_0)V_{ref} - (1/2)(b_1)V_{ref}$. Clearly, the functions of the output capacitor are twofold. During one iteration it acts as a divide–by–two operator, whereas between two consecutive iterations it acts as the memory element which accumulates the results of the previous and present operations of dividing–by–two. So, after one full conversion cycle with N iterations the analogue voltage:

$$V_{da} = -V_{ref}.\sum_{n=0}^{n-1} b_n.2^{n-N}$$

$$\boxed{8.3}$$

available across capacitor C_{da} corresponds to the converted input digital word with N–bits.

8.3.2.2 D/A Converter with multiplexed SC branches and output voltage signal

The complete D/A converter employing the conversion technique described above is shown in Figure 8.9, where it is required that all three

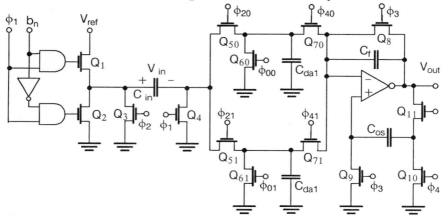

Figure 8.9 Quasi–passive algorithmic D/A converter with two multiplexed input SC branches and output voltage signal.

capacitors C_{in}, C_{da0} and C_{da1} have equal capacitance values. During one conversion cycle, digital–to–analogue conversion is carried–out by the combined operation of capacitors C_{in} and C_{da0}, while capacitor C_{da1} transfers the charge obtained at the end of the previous conversion cycle to the feedback capacitor C_f of the charge–to–voltage converter. During the next conversion cycle capacitors C_{da0} and C_{da1} interchange their roles. Now, C_{da1} combines with C_{in} to carry–out the digital–to–analogue conversion, while C_{da0} transfers the charge obtained at the end of the previous conversion period to the feedback capacitor C_f. With this arrangement, the total operating cycle of the complete D/A converter of Figure 8.9 is equal to the conversion cycle T_c of the basic D/A converter of Figure 8.8. The amplifier and associated feedback branches are used to transfer the charge accumulated in capacitors C_{da} into the feedback capacitor C_f and produce a buffered output voltage capable of driving a wide range of output loads, both capacitive and resistive. The feedback SC network with capacitor C_{os} is used here to render the overall charge–to–voltage transmission factor of the circuit ideally independent of both the finite DC gain and offset voltage of the amplifier [14].

The non–ideal effects associated with the basic circuit elements can affect the performance behaviour of the D/A converter of Figure 8.9. As described in [14] and [17], such non–ideal effects include the amplifier finite DC gain and offset voltage.as well as the non–ideal effects associated with both the capacitors and the switches. The switches are mainly responsible for clock feed-through effects although they may also give rise to errors resulting from incomplete charge transfer due to non–zero time constants. The capacitors produce capacitance mismatch errors. The parasitic capacitances which are associated with the floating nodes of capacitors C_{da0} and C_{da1} result from grounded parasitic capacitances of both the capacitors and switches, and can also become an important cause of disturbance of the operation of the circuit. Despite the influence of such non–ideal effects on the operation of the quasi–passive D/A converter shown in Figure 8.9, it is possible to achieve in practice 8–bit conversion resolution and conversion speeds in the MHz range, and thus cover a broad span of specifications as required in many practical applications.

8.3.2.3 An example of a CMOS integrated circuit implementation

The photomicrograph of a prototype $2.5\mu m$ CMOS D/A converter employing the architecture of Figure 8.9 is shown in Figure 8.10 (analogue section alone), where the active silicon area without pads is a mere $0.25mm^2$. The digital circuitry which is necessary to control the operation of this converter comprises a master clock generator, a counter, a shift register and some auxiliary glue logic, as schematically illustrated in the block

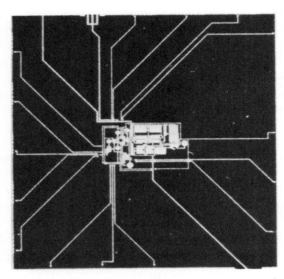

Figure 8.10: Photomicrograph of an integrated quasi–passive D/A converter.

diagram of Figure 8.11a. For an 8–bit convertor, Figure 8.11b illustrates a complete set of switching waveforms generated during two consecutive conversion periods.

A number of prototype chips have been tested to verify the conversion linearity versus the speed of operation of the integrated circuit of Figure 8.10, and some of the results obtained are summarised in Figure 8.12. The typical conversion errors of the integrated circuit operating as an 8–bit D/A converter at a 100kHz conversion rate are shown in Figure 8.12a. Figure 8.12b, on the other hand, shows the conversion errors of the integrated circuit operating as a 6–bit DAC at conversion frequency of 1MHz.

8.3.4 Capacitance–ratio–independent D/A converter

8.3.4.1 Principle of conversion

In Figure 8.13a we present the basic schematic of a D/A converter where we have replaced the traditional binary–weighted capacitor array [18] by a single SC branch operated by a set of switching waveforms realising a binary–weighted time slot array, as represented in Figure 8.13b for the case of $N = 4$. The switches S_A, S_B and S_C are operated by switching waveforms ϕ_0 and ϕ_1 whose switching period $1/[(2^N-1)F_s]$ is determined not only by the conversion rate F_s of the converter but also by the bit resolution N. On the other hand, the switches S_0, S_1, ..., S_i, ..., S_{N-1}, which are either connected to

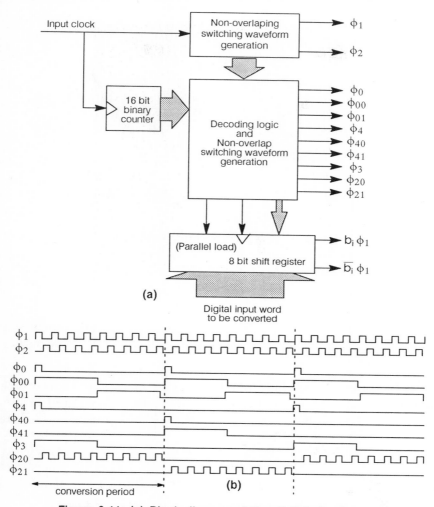

Figure 8.11: (a) Block diagram of the digital circuitry and (b) switching waveforms for 8-bit conversion.

the reference voltage V_R or to ground, depending on the bit level, are controlled, respectively, by switching waveforms A_0, A_1, ..., A_i, ..., A_{N-1} operating at the conversion rate F_s, and each of which has a total of 2^i *time slots* that are synchronous with ϕ_1. At the beginning of each conversion period $1/F_s$, the feedback capacitor C_F in the integrator of Figure 8.13a) is pre–charged to the offset voltage of the operational amplifier by the switch S_R, controlled by A_0. The D/A conversion is performed sequentially from bit b_0 to bit b_{N-1}, and the equivalent bit voltage V_i corresponding to each bit

Self-calibrated and algorithmic converters

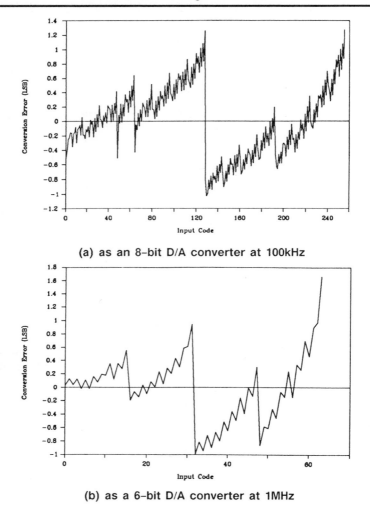

(a) as an 8–bit D/A converter at 100kHz

(b) as a 6–bit D/A converter at 1MHz

Figure 8.12: Conversion performance of the integrated circuit

of the input digital word $[b_0 \, b_1 \, ... \, b_i \, ... \, b_{N-1}]$ is determined by the *number* of time slots of the corresponding switching waveform A_i. Hence, it can easily be seen that the resulting total output voltage of an N–bit digital word is given by:

$$V_{OUT} = \frac{C}{C_F} V_R \sum_{i=0}^{n-1} b_i \, 2^i$$

8.4

where $[(C/C_F)V_R]$ is the incremental output voltage variation corresponding to the least significant bit. Since the above D/A converter

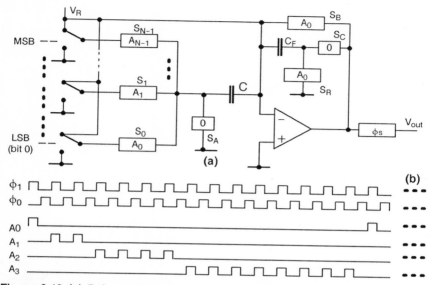

Figure 8.13 (a) D/A converter based on a binary–weighted time slot array and (b) switching waveforms for N=4

requires a long period for conversion, but its resolution does not depend on the accuracy of the capacitance ratio C/C_F, it can be particularly useful for low speed and very high resolution applications.

8.3.4.2 Effect of the Amplifier Finite DC Gain

Although the conversion linearity of the D/A converter of Figure 8.13 can be affected by the finite DC gain of the amplifier, it can be shown that high linearity of conversion can be achieved for high resolution D/A converters even using amplifiers with relatively low DC gain [16]. This is illustrated in Figure 8.14a showing the conversion error obtained by the behavioral simulation of a 13–bit D/A using an amplifier with a mere 40dB of DC gain, and from Figure 8.14b we can see that 19–bit resolution could be reached using an amplifier with 60dB of DC gain which can be easily achieved using current CMOS technology.

8.4 Algorithmic digital–analogue converter with FIR filtering

8.4.1 Introductory remarks

Besides the D/A conversion function described in the previous Section, many interfacing systems additionally require a filtering function to shape the output analogue signals. Often the filtering function must be of the FIR type (Finite Impulse Response) because of the stringent requirements with respect to the phase response characteristic. For such applications, we

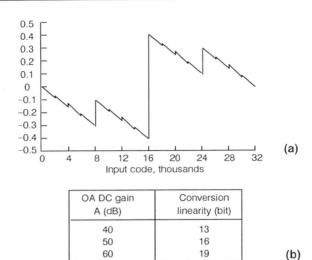

(a)

OA DC gain A (dB)	Conversion linearity (bit)
40	13
50	16
60	19

(b)

Figure 8.14 (a) Conversion error obtained for a 13–bit D/A converter with a 40dB amplifier; (b) Conversion resolution achieved with different values of the amplifier DC gain.

describe next a mixed–signal analogue–digital building block which is capable of realizing simultaneously a D/A conversion function together with a FIR filtering function whose impulse response may have an arbitrary length [19–20]. In the DAFIC (Digital–Analogue–FIlter–Converter) building block, the filter multiplications and the addition are both realized in the analogue domain simultaneously with the D/A conversion, whereas the delay line alone is realized in the digital domain. An algorithmic D/A converter employing a single amplifier offers the best solution to implement the DAFIC building block when considering such important aspects as the complexity of both the digital and the analogue parts as well as the capacitance spread, speed of operation and accuracy of the conversion and filtering functions. This is illustrated in this section by considering the practical design and experimental evaluation of a 8–bit/4–tap DAFIC using a 3μm single–metal/double–poly CMOS process.

8.4.2 Algorithmic DAFIC architecture

As mentioned in the previous Section, algorithmic DAC's are usually utilized to minimize silicon area in applications where a lower speed of operation is not a limiting factor. When an algorithmic conversion technique is applied to the DAFIC this also leads to significant reductions in silicon area and circuit complexity in comparison with alternative architectures [20]. Figure 8.15 represents the general architecture of an

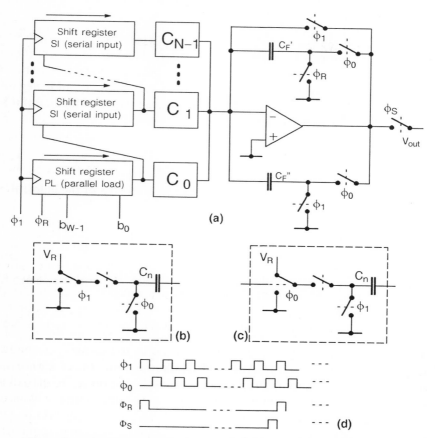

Figure 8.15 (a) General architecture of an algorithmic DAFIC and corresponding switching arrangements for (b) positive and (c) negative impulse response coefficients. (d) Switching waveforms.

algorithmic DAFIC comprising a digital delay line, realized by serially associated shift registers, together with an analogue output block where the delayed words are converted according to the coefficients h_n ($n = 0$ to $N-1$) of the FIR transfer function and added together. In this case, the input digital word associated with the SC branch with capacitance value C_n ($n = 0$ to $N-1$) produces an analogue voltage, V_{OUT_n}, that results from the super-position of the output voltages produced bit–by–bit according to the conversion algorithm [12]. This process of conversion is carried–out sequentially from the least significant bit $b0$ to the most significant bit b_{W-1}, and it can be shown that at the end of the conversion period the contribution of each bit is given by:

$$V_{OUT_n}(b_i) = \pm \frac{C_n}{C_F} V_R \, 2^{-W+i} b_i \qquad \boxed{8.5}$$

where W represents the conversion resolution. The sign of the coefficient above depends on the switching of the input capacitor as indicated in Figure 8.15b and in Figure 8.15c, respectively, for positive and negative coefficients. Therefore, the total contribution associated with one input SC branch alone can be expressed as:

$$V_{OUT_n} = \pm \frac{C_n}{C_F} \sum_{i=0}^{W-1} V_R \, 2^{-W+i} b_i \qquad \boxed{8.6}$$

yielding the following expression for the output voltage of the algorithmic DAFIC

$$V_{OUT} = \sum_{n=0}^{n-1} \pm \frac{C_n}{C_F} \sum_{i=o}^{W-1} V_R \, 2^{-W+i} b_i \qquad \boxed{8.7}$$

In this architecture, we must have $C_n = |h_n| C_F$ in order to preserve the DC gain factor of the FIR transfer function. As illustrated in Figure 8.15d, the operation of this circuit requires 4 switching waveforms, ϕ_1, ϕ_0, ϕ_R and ϕ_S, and each conversion is completed in W cycles of switching waveform ϕ_1.

The accuracy of the DAFIC can be measured by the linearity of the D/A conversion and by the accuracy of the FIR filtering function, which are both affected by the capacitance–ratio accuracy and by the finite DC gain of the amplifier. The contribution of each capacitor to the final output voltage can be evaluated according to the equation [20]

$$V_O(n) = \frac{C_F(1+\frac{1}{A})}{\underset{(1+\frac{1}{A})(C_F+C_F'')+\frac{j}{A}}{\sum C_j}} V_O(n-1) + \frac{C_i}{\underset{(1+\frac{1}{a})(C_F+C_F'')+\frac{j}{A}}{\sum C_J}} V_R b_i \qquad \boxed{8.8}$$

Numerical analysis of the above equation shows that the amplifier must have a DC gain equal or greater $1.25*2^{W+1}$ for $\pm 1/2$LSB linearity error. The conversion resolution is limited to 9–10 bit due to the accuracy that can be practically obtained for the capacitance ratios and which, in turn, also determine the precision of the FIR filtering function.

8.4.3 An Example of a CMOS Integrated Circuit Implementation

An experimental prototype chip of the algorithmic DAFIC of Figure 8.15, implemented in a 3μm CMOS technology, was designed in order to realize an 8–bit D/A conversion together with an FIR transfer function with 4 equally weighted taps and unity DC gain factor. This corresponds to a z–transfer function given by $H(z) = 1/4(1 + z^{-1} + z^{-2} + z^{-3})$, whose amplitude response notches are located at $F_s/4$, $F_s/2$ and $3F_s/4$.

The analogue section of the integrated DAFIC is basically a SC circuit comprising an amplifier and several switches and capacitors. For a modular layout, the analogue switch cell is implemented using a pair of CMOS transmission gates. High quality double–poly capacitors are implemented as an association of unit capacitors to improve matching. The digital circuitry is formed by 4 serially associated shift registers, each of which having $W = 8$ cells. The basic register cell is illustrated in Figure 8.16a. At the end of the shift register there is an interface cell, shown in Figure 8.16b, to control the switching of the corresponding SC branch to the reference voltage according to the corresponding logic bit value.

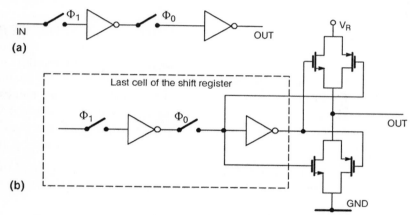

Figure 8.16 (a) Dynamic shift–register cell and (b) its interface with the SC circuit.

Being a mixed–signal analogue–digital IC, the chip floor-planning should be carefully studied to minimize coupling of digital signals onto critical analogue lines. The chip employs independent power supplies to the digital and analogue sections, and the N–type substrate under the analogue section is biased with a surrounding ring connected to the analogue V_{DD}. This technique keeps the inner substrate at a nearly constant potential, thus preventing the propagation of digital spikes via the substrate. In Figure 8.17 we present a photomicrograph of the chip whose active silicon area occupies a mere 0.6mm^2. In the layout of the amplifier, the MOS transistors are implemented in an interleaved form (stacked implementation) to improve matching. The capacitors are laid–out as an association of unit capacitors, and for capacitors C_F these are also interleaved to maximize their matching.

Self-calibrated and algorithmic converters

Figure 8.17 Photomicrograph of the integrated algorithmic DAFIC.

In order to measure the linearity of the D/A conversion alone, an 8–bit digital word is applied serially to the input of a single tap of the DAFIC. In Figure 8.18 we present a plot of the conversion error obtained from the difference between the D/A conversion characteristic and the best adapted straight line. While the analysis of the plot shows that the error is always between ±0.6 LSB, the best straight line indicates a gain deviation of 2% from the theoretical value of 1/4, and an output offset voltage of only –2mV (–0.3LSB). A computer based analysis of the performance behaviour of this circuit indicated that such gain and linearity errors may be caused by an amplifier DC gain lower than its expected worst case value. The overall

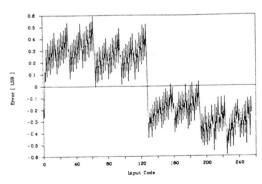

Figure 8.18 D/A conversion error from a single tap of the integrated DAFIC.

frequency response of the DAFIC, on the other hand, can be measured by applying a continuous–time input signal to a 12–bit A/D converter, and then taking the resulting 8 most significant bits and applying them serially to the input of the DAFIC. This is represented in Figure 8.19, where the 20dB attenuation is due to the difference between the reference voltage of the

DAFIC and the reference voltage of the auxiliary ADC. In the same Figure 8.19, the computer simulated amplitude response is also superimposed (broken line) to show the correctness of the filtering function of the integrated DAFIC.

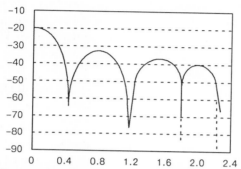

Figure 8.19 Measured and computer simulated (broken line) amplitude responses of the filtering function of the integrated DAFIC.

8.5 Conclusions

This Chapter described a number of data conversion circuit techniques, both for A/D and D/A, which can be employed in a wide variety of system applications which may need conversion speeds up to the order of several hundreds of kHz, or even few MHz, and conversion resolutions ranging from 8–bit to 16–bit. Firstly, we presented one of the most popular architectures for realising A/D converters with conversion resolution from 10–bits to 16–bits and conversion speeds between the audio and the video frequency bands and which is based on the well known combination of the successive approximation algorithm together with self–calibrating techniques. Then, we described a low–cost quasi–passive D/A converter yielding up to 8–bit resolution and conversion speeds up to a few MHz, and a capacitance–ratio–independent D/A converter which can achieve high conversion resolutions using operational amplifiers with relatively modest values of the DC gain. Both types of converters can be easily programmed by digital means to achieve different conversion characteristics, both for speed and resolution. Finally, we considered an algorithmic D/A converter with a built–in filtering function to shape the resulting output analogue signals, as required in a complete mixed–signal digital–to–analogue interface system. Various examples of a CMOS integrated circuit implementation were also presented to illustrate the circuit design techniques described for the self–calibrated and algorithmic converters.

8.6 Acknowledgements:

The realisation of the prototype chips presented herein was supported by Austria Mikro System, the Technical Research Centre of Finland, the University of Pavia, SGS, and the Instituto Nacional de Investigation Cientifica. We would like to thank Franco Maloberti, Paul O'Leary, Helena Pohjonen, and Gebert Melcher for their support and critical suggestions during the realisation of the various chips. Thanks are also due to Vitor Dias and Jorge Guilherme who colaborated in the practical realisations and testing of some of the chips.

8.7 References

[1] Hae–Seung Lee and David A. Hodges, "Self–Calibration Technique for A/D Converters", IEEE trans. Circuits Syst., pp.188–190, March 1983

[2] H.Lee, D. Hodges, P. Gray, "A Self–Calibrating 15 bit CMOS A/D Converter", IEEE J.Solid–S.Circuits, vol.SC–19, pp.813–819, Dec.1984

[3] C. A. Leme, J. E. Franca, "An Overview and Novel Solutions for High Resolution Self–Calibrated Analogue–Digital Converters", in Proceedings International Symposium on Signals, Systems, and Electronics, Erlangen, Germany, Sept. 18–20, 1989, pp.815–819.

[4] J.Croteau, D. Kerth, D.Welland, "Autocalibration Cements 16–Bit Performance", Electronic Design, Sept. 4, 1986, pp.101–106.

[5] N. C. Horta, J. E. Franca, C. A. Leme, "Framework for Architecture Synthesis of Data Conversion Systems Employing Binary–Weighted Capacitor Arrays", in Proceedings International Symposium on Circuits and Systems, Singapore, June 11–14, 1991, pp. 1789– 1792.

[6] J. McCreary and P. Gray, "All–MOS Charge Redistribution Analog–to–Digital Conversion Techniques – Part I", IEEE J. Solid–State Circuits, vol. SC–10, pp. 371–379, Dec. 1975.

[7] J–B. Shyu, G. C. Temes, K. Yao, "Random Errors in MOS Capacitors", IEEE Journal Solid State Circuits, Vol.SC–17, No.6, pp.1070–1076, Dec. 1982.

[8] C. A. Leme, J. E. Franca, "A 20's 16–Bit Successive Approximation CMOS Analogue–Digital converter", in Proceedings Mideuropean Conference on Custom Integrated Circuits,

[9] C. A. Leme, J.E.Franca, "High Performance CMOS Comparator for Analogue–Digital Convertors", Electronics Letters, Vol.26, No.20, 27th. Sept. 1990, pp.1725–1726.

[10] C. A. Leme, J. E. Franca, "Efficient Calibration of Binary–Weighted Networks Using a Mixed Analogue–Digital RAM", in Proceedings International Symposium on Circuits and Systems, Singapore, June 11–14, 1991, pp.1545–1548.

[11] K. Nagaraj, "High–Resolution Switched–Capacitor Algorithmic Digital–to–Analogue Convertor", IEE Proceedings, Vol.132, Pt.G, Oct. 1985, pp.200–204.

[12] H. Matsumoto, K. Watanabe, "Switched–Capacitor Algorithmic Digital–to–Analog Converters", IEEE Trans. Circuits and Syst., Vol.CAS–33, No.7, pp.721–724, July 1986.

[13] F–J. Wang, G. C. Temes, S. Law, "A Quasi–Passive CMOS Pipeline D/A Converter", in Proceedings Custom Integrated Circuits Conference 1989, U.S.A., May 1988.

[14] V. F. Dias, J.E. Franca, J.C.Vital, "High–Speed Digital–to–Analogue Convertor Using Passive SC Algorithmic Conversion", Electronics Letters, 1988, May 10th..

[15] P. W. Li, M. J. Chin, P. R. Gray, R. Castello, "A Ratio–Independent Algorithmic Analog– to–Digital Conversion Technique", IEEE Journal Solid State Circuits, Vol.SC–19, pp.828–836, Dec. 1984.

[16] J. C. Vital, J. E. Franca, "Novel Capacitance–Ratio–Independent Switched–Capacitor Digital– Analogue Convertor", Electronics Letters, Vol.25, No.20, 28th. Sept. 1989, pp.1362– 1363.

[17] J. Vital, J. E. Franca, "Programmable Parasitic–Compensated Algorithmic Digital–Analogue Convertor", submitted for publication.

[18] Roubik Gregorian, "High–Resolution Switched–Capacitor D/A Converter", Microelectronics Journal Vol.12 No. 2 1981 Mackintosh Publications.

[19] J. E. Franca, J. Vital, C. A. Leme, "Filtered DAC", United States Patent No.5,008,674, April 16, 1991.

[20] J. Vital, J. E. Franca, F. Maloberti, "Integrated Mixed–Mode Digital–Analogue Filter Converters", IEEE Journal Solid–State Circuits, Vol. SC–25, No.3, June 1990, pp.660– 668.

Chapter 9
A High Flexibility BiCMOS Standard Cell Library for mixed Analogue / Digital Application

Christian Caillon SGS – Thomson, Grenoble, France

9.1 Introduction

Originating as an extension of digital standard cells, mixed analogue/digital standard cell technology is now beginning to mature. Increasingly, analogue and digital cells can be mixed on the same die resulting in a continuing increase in integration density and reliability and concomitant decrease in external component count and overall cost. Some years ago the only means of integrating analogue functions was the full custom approach which was only relevant for high volume production. Initially in an attempt to enlarge the use of analogue functions in ICs, a small number of standard cells appeared in digital CMOS libraries. However the use of these cells was limited due to a lack of flexibility in two particular areas. These were firstly the problem of integrating resistors, capacitors and current sources of different values with standard cells and secondly the problem of integrating complex analogue functionality with only a limited range of standard cells. The first obstacle was overcome with the introduction of the programmable or parametrizable standard cell concept, managed by specific CAD tools such as analogue cell compilers or leaf cell compilers. This cleared the way for the design of entire ASICs from a standard library, included in a secure design environment, implemented on a work station with design flow checked from schematic capture to layout using automated CAD tools. For example SGS–Thomson's TSGSM mixed analogue digital CMOS library contains more than 250 cells, many of which are programmable. This allows implementation of many types of design for most segments of the analogue–digital ASIC market

With the introduction of programmable analogue standard cells, the next step was to improve the flexibility and ergonomics of mixed A/D cell libraries. To achieve this, two strategic routes were selected. Firstly in the area of technology; it was recognized that the best way of mixing high performance linear functions in terms of preserving accuracy, the ability to operate at high speed and consume small amounts of power, with a high density of digital functions was to choose a BiCMOS technology. Therefore an innovative silicon process especially dedicated to mixed A/D applications was developed. The process HF2CMOS offers 6GHz bipolar NPN and 2.5GHz vertical PNP transistors with a 2.5μm channel length for

CMOS and two poly–silicon and two metal layers. The second strategic route involved introducing greater flexibility. Several improvements had to be introduced to take account of constraints imposed by the analogue functions. For example multiple power supplies, power down capabilities, centralized biasing blocks automatically managed by CAD tools, simplified analogue symbols, alternative management of cells, circuit floorplan management and more parameterized cells. Based on these two strategies a new mixed A/D standards cell library was developed, the STKM2000; a merge between two processing technologies, CMOS and Bipolar, and two worlds analogue and digital. For example a system from sensor to actuator in a data processing chain requires either CMOS or Bipolor processing for its various component parts. (Figure 9.1).

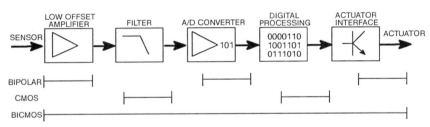

**Figure 9.1 The most appropriate silicon technologies
for mixed analogue / digital applications**

9.2 A BiCMOS process dedicated to mixed A/D applications

At SGS-Thomson a new BiCMOS process, the HF2CMOS, has been developed for mixed analogue digital applications and optimised for telecommunication, video processing and analogue semi-custom applications. It is a double–poly, double–metal Bipolar/CMOS technology. It offers five different kinds of transistors viz. 15V, 6GHz NPN, 15V, 20MHz lateral PNP, 15V, 2.5GHz, vertical PNP, NMOS and PMOS. The CMOS part is similar to an N–Well technology with the possibility of fabricating isolated NMOS transistors. The two processes were optimised for 5V and 10V applications with a different thickness of gate oxide. The minimum channel length in the 5V process is 1.8μm. A range of seven resistor types is available with values from 20 ohms/square to 8 Kohms/square. It is also possible to implement two kinds of capacitor viz. poly/diffusion capacitors with a specific capacitance of 500pf/mm^2 and poly/poly capacitors also with a specific capacitance of 500pf/mm^2. Low resistance buried layers have been used to reduce risk of latch up effects. Figure 9.2. below gives the cross section of the HF2CMOS process.

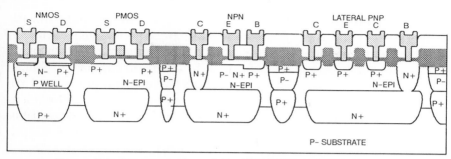

Figure 9.2   Cross section of the HF2CMOS BiCMOS Process

9.3 Cell libraries

9.3.1 Analogue libraries

The basic principle behind the provision of the analogue library is to offer the maximum flexibility for a large range of applications. The primary mechanism for achieving this goal has been cell programmability. Programmability has been implemented at different levels, from the selectability of a resistor value to a filter compiler passing through a leaf cell compilation process. Step by step adjustability is provided, for example, for resistor values, supply voltages, comparator speeds, operational amplifier performance features etc. Even the phase margins of the amplifiers are adjustable to meet the needs of particular applications. This comprehensive use of programmability allows the designer many of the advantages of the full–custom approach with the integrity and short turnaround times associated with semi–custom methodologies. Programmable values captured during the schematic phase with attributes are managed by an integrated CAD tool which checks all property values and automatically generates the cell required by the user. The circuit design of the analogue library cells takes full advantage of the BiCMOS processing technology. For example for high–speed or low–offset, bipolar transistors are used. For a high input impedance amplifier, CMOS input transistors are chosen. For many other applications bipolar and MOS transistors are used in the same cell as and when appropriate. Whereas CMOS is required for analogue switches, multiplexers and switched capacitor filters, bipolar transistors are essential for bandgap reference voltages, low noise amplifiers and output current drivers. Most of the analogue library operates from 2.7 Volts to 11 Volts. All analogue functions have a power down mode and the facility for adjustment of the power consumption. Biasing currents and voltages required by individual analogue functions are automatically managed by the CAD tools. Another important feature of this cell library is the possibility of managing multiple supplies automatically and it is straightforward to separate supply rails between different blocks in the same chip with a view to reducing crosstalk. Hence it is a global property which fixes the supply voltage for a part of a sub-circuit in the schematic capture. Figure 9.3 shows an example of the schematic capture of a simple triangular/square wave generator with all component properties. An important part of the analogue library is a set of macro–functions such as converters and filters. In the STKM2000 library a switched capacitor filter compiler allows the designer to generate any kind of filter up to twelfth order; including low–pass, high–pass, band–pass, band–reject and all pass networks, implementing any type of mathematical approximation. All tasks from the specification up to layout and creation of simulation models are taken care of by the compiler which is one of the most powerful currently available. Typically, cut of frequencies of up to 150KHz are achieved. Figure 9.4 shows a 6th order Butterworth SC filter issued from the compiler. A range of A/D and D/A converters are available. At present 8 bit A/D converters operating between 200KHz and 20MHz and a 13 bit operating at 80KHz are available. Table 9.1 lists a number of elements of the STKM2000 analogue cell library.

Table 9.1 STKM2000 Analogue library, abstract

Operational Amplifier	: OPB81 (BICMOS Video Amplifier) : G.BP = 80MHz (20pF ; 75 Ohms) : Slew rate = 200V/μs (20pF ; 75 Ohms)
Comparator	: CMP31 (BICMOS High Speed) : Tp = 80nSec Typ. (5mV overdrive) : Offset = $\pm$1mV Typ.
Bandgap Reference Voltage	: VREF11 (REF to Vss) : VREF = 1.125 $\pm$2% : Temperature drift = 30ppm/^oC
Oscillator	: OSC31 (One pin IC Oscillator) : Frequency range - 2 to 200KHz (Programmable) : Stability versus supply voltage - 0.5%/Volt
Power ON Reset (POR11)	: Adjustable threshold - 2.8 to 7.3 Volts : Adjustable hysteresis - 0 to 50%
Power NPN Bipolar : BN3E	: Iout = 100mA (VCE SAT = 0.2V) : BETA = 80
Convertor	: AD12 (12 bit + sign A/D converter) : Resolution - 12 bit. : Conversion time 12.5μs : Integral linearity $\pm$ 1 LSB : Differential linearity $\pm$ 0.5 LSB

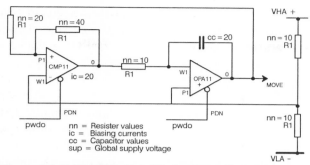

**Figure 9.3 STKM2000 Simplified schematic capture of an
analogue function**

9.3.2 The digital cell library

Two digital libraries are made available; one dedicated to 5 Volt applications and operating at frequencies up to 30MHz and a second one dedicated to 10 Volts, where a single supply and high voltages are required. The selection between the two libraries is automatically managed by the CAD tools depending on the supply voltage and application. The digital library is composed of 60 internal hard macro cells such as standard gates, flip flops, translators, multiplexers etc. and several types of input and output buffers, some of which are programmable, depending on loads or current requirements. In order to simplify the schematic capture of a digital system, a set of macro cell generators is available, including N–bit synchronous and N–bit serial or parallel registers. Recent additions include SRAMs, ROMs and PLA macro–block generators. (table 9.2).

Table 9.2 Main features of the digital macrocell generators

PLA Generator	: Limited only by the silicon area and timing
	: Example - 10 Inputs, 10 Minterms, 10 Outputs
	    Area = 0.4mm^2
	    Operating frequency - 30MHz.

SRAM Generator	: Maximum complexity - 8KBit : Maximum operating frequency = 30MHz
ROM Generator	: Maximum complexity - 64KBit : Maximum operating frequency = 30MHz

9.4 CAD tools

To manage the design of chips with the STKM2000 library efficiently novel CAD tools have been developed which offer flexibility and integrity of analogue designs and reduce the overall time to design. SGS–Thomson's Analogue Design System (ADS) has been developed as an integrated system which offers coherent solutions from schematic capture through simulation to mask generation. ADS covers all the standard features of such CAD systems such as delay evaluation and design rule checking and in addition offers proper library management of mixed analogue and digital cells.

9.4.1 The CAD capability.

The STKM2000 product is one of the best recent examples of the synergy between design and CAD. It has been achieved by developing CAD tools to fit the requirements of the library while at the same time the library has been developed to be easily handled by software tools. The final results are greater flexibility and reliability. Schematic capture is effected by use of a reduced set of analogue icons in order ease selection from the library. The required parameters are then added to customize the function to meet the needs of the design. The software will generate for each particular instance the following: propagation times for digital behavioral models; transistor level netlists for accurate electrical simulation; cell footprints to be used by the automatic placement and routing program; netlists for layout–versus–schematic checks and GDS2 layout representation. For example an operational amplifier may be generated according to two parameters. Firstly the bias current which controls the major performance parameters such as gain, bandwidth, slew rate, power consumption and secondly frequency compensation which allows adjustment and optimisation of the dynamic parameters versus the capacitive and resistive load. Op–amp layout is compiled by automatic selection and abutment of leaf cells, using a "TELESCOPIC CELL" feature as shown in figures 9.4.

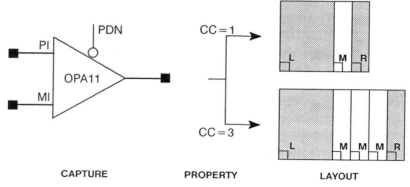

| CAPTURE | PROPERTY | LAYOUT |

Figure 9.4 Telescopic cells

Another type of programmability is offered by the PARAMETRIZABLE feature indicated in figures 9.5 The user selects a property such as design resistance or capacitance. A cell is generated to fit this value with a

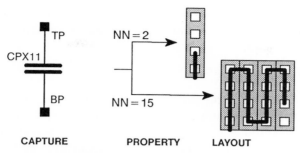

CAPTURE   PROPERTY   LAYOUT

Figure 9.5 Parametrizable cells

resolution of 0.1pF for capacitance and 300 ohms for resistance. To facilitate generation of switched capacitor systems or resistive dividers the device is generated using a repetitive structure. A field of unit devices is created as a "telescopic" cell and then automatically personalized by metal lines to arrive at the required value. Yet another kind of programmable feature is the "ADJUSTABLE CELL" shown in figures 9.6 This is used for functions containing a device that must be adjusted to tune a parameter R_{on} of a switch, percentage hysteresis of a power–on–reset, or current in a bipolar transistor. The property ADJ corresponds to this feature.

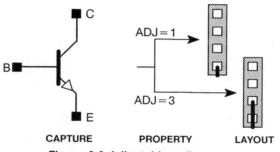

CAPTURE   PROPERTY   LAYOUT

Figure 9.6 Adjustable cells

9.4.1.1 Telescopic Cells

The TELESCOPIC Cell concept is used to compile digital output buffers, operational amplifiers, monolithic capacitors, MOS transistors etc. The cell is generated by abutment of 3 leaf cells: Left side subcell, one of several middle subcells to reach the value selected by the property, and finally one right side subcell.

9.4.1.2 Parametrizable cells

Parameterable cells are used to compile resistors or capacitors from basic elementary devices. The cell is assembled in two steps. Firstly an array of devices is generated by abutment of leaf cells, each made up of basic components. During the second step, a metal wire is added to customize this field of devices. ("gate-array-like" methodology).

9.4.1.3 Adjustable cells

This approach is used within the STKM 2000 library to generate associative bipolar transistors, variable Ron switches, pull-up/pull-down resistors, adjustable power-on-reset. The cell is generated from a slice of basic elementary devices, and an aluminium interconnection layer.

9.4.2 Automatic cell biasing and power down

Most analogue cells need to be biased by a current. As discussed previously the biasing value is given for each cell by the property IC. The design is scanned to evaluate global design needs in biasing currents and a synthesis is performed to calculate the reference current to be generated. This current is then reproduced and multiplied using current mirroring techniques. A centralized block is generated by abutment of a bias generator sub-cell with the current mirror cells required to bias every analogue cell. Connections from the biasing block to each analogue cell are automatically added during placement and routing. Figure 9.7 shows an example of centralized biasing

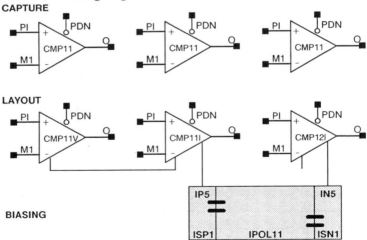

Figure 9.7 Sink/Source biasing selection

management, and table 9.3 an example of the main features of an operational amplifier versus IC property. Analogue chips often have more than one pair of power pads. To facilitate the design of such chips and to avoid connecting each cell to power nets during the schematic capture a novel "property" based approach has been developed. Capture is effected with icons having no supply pins and then a property is given by the user on parts of the design to be supplied by an extra pads property SUP.

Simulation netlists, both electrical and digital are generated according to these properties. Finally the floor–planning is prepared by cell partitioning in different classes to allow an easy separation between parts of the chip supplied with separate power nets. A typical application of this feature is the splitting of analogue and digital supplies in order to reduce crosstalk effects. (figure 9.8 illustrates the multiple–supply management of mixed

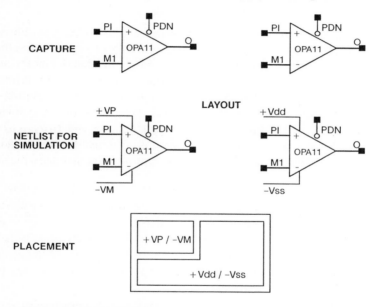

Figure 9.8 Multiple Supply management using "SUP" property

configuration circuits. Most analogue cells have an optional power down mode, driven by a command available on the icon. When these pins are put in the active state, the power consumption of the respective cell is suppressed. A typical application of this feature is the design of chips having a "watch" mode and supplied from a battery. If the power down mode is not used in a chip the command pin may be left unconnected at the design capture stage. The program scans the design to check connectivity of these pins. When left unconnected they are tied to inactive state lines wired inside the outline of the cell. All such tasks are performed by CORAIL, the heart of the SGS–Thomson Microelectronics analogue ASIC CAD system.

**Table 9.3 Example of bias programing of the OPB81
BiCMOS Operation Amplifier**

Biasing current programing	IC = 10μA	IC = 60μA	IC = 120μA
Open loop gain	78dB	69dB	56dB
Gain bandwidth	15MHz	55MHz	80MHz
Slew rate	10V/μSec	150V/μSec	600V/μSec
Output current	15mA	40mA	56mA
Current consumption	0.5mA	6mA	12mA
All intermediate values between 10μA and 120μA are possible in steps of 1μA.			

In order to organize the chip placement, a set of hierarchical properties have been created. Three different attributes allow one to create sub-chips ("CHIP" property) block cells ("AREA" property) and hard–macro cells ("MACRO" property). Used during the schematic capture step, it will allow one to drive the place and route software. "CHIP" can be used to split analogue and digital parts, "AREA" can be used for critical functions using several cells (AGC, PLL etc) and "MACRO" allows one to create a hard-macro cell from basic cells (accurate gain, Sallen and Key continuous filters). Figure 9.9 illustrates the ADS floor planning management.

9.4.5 ADS (Analog Design System) An environment for mixed signal design

ADS consists of a set of generic tools such as schematic entry, simulators, placement and rooters and specific tools such as CORAIL in charge of the library management, or macrocell compilers. The design flow is fully integrated, thus ensuring maximum compatibility at each step of the design (figure 9.10 shows the general ADS design flow). CORAIL undertakes the following tasks:

1. A provisional evaluation of the routing capacitance. This depends on the connectivity.

2. A check on the load driven by the output of a cell against the maximum allowed value.

3. A check of the electrical design rules such as parallel outputs or unconnected outputs, for the digital parts.

4. Similarly for the analogue parts, a check on rules of current or supply compatibility etc.

5. The transfer of design information to circuit level, digital and mixed–mode simulators; to the placement and routing programs and to the layout–versus–schematic checking program.

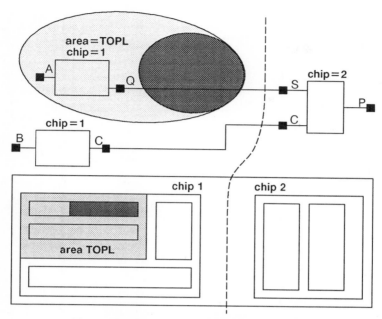

Figure 9.9 ADS Floor planing management

6. Extraction of the exact routing capacitance and the process of back annotation after completion of the placement and routing phases. CORAIL is integrated in ADS. ADS checks that each step is performed at the correct time only if the previous task has been completed without error. ADS is available on various platforms e.g. SUN/UNIX™ with CADENCE™ software, MENTOR™ and VIEWLOGIC.™

Depending on the simulation requirements, a set of simulators are available allowing different alternatives when designing: MOZART multiple level digital simulator for the digital part. ST-SPICE and ELDO™ for electrical and accurate simulation, MOZART MIXED MODE (MOZART and ELDO coupling allowing analog and digital multi-level simulation (pin-to-pin) top level simulators

9.4.6 Analogue/digital multi-level mixed mode simulations

To meet the simulation requirements, ADS and MOZART MIXED MODE offer one of the most flexible approaches to analogue and digital multi-level mixed mode simulation. This tool which is the link between two multi-level simulators, allows simulation of any application. A specific

A High Flexibility BiCMOS Standard Cell Library
for mixed Analogue/Digital Applications

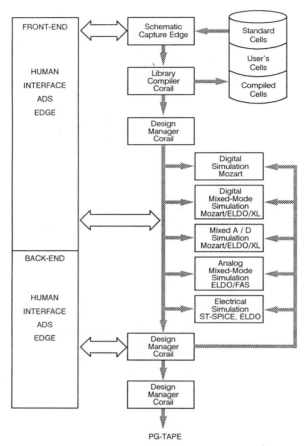

Figure 9.10 ADS environment
(XL and FAS are options of the ELDO simulator)

simulation property (SIM Property) allows one to handle the splitting of analogue and digital netlists for the modeling simulations, and the level: transistor level, accurate, macro-model or speedy macro-model. Figure 9.11 shows the different possibilities of model selection of each cell.

9.5 Application examples

The association of a flexible library, a powerful CAD tool and a true BiCMOS process allows one to address various applications for most segments of the market, for example:

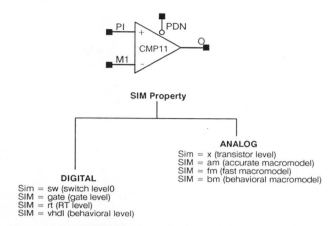

Figure 9.11 "SIM" Property for A / D netlist partitioning
and multi-level simulation

Market Sement	Application
Computers	Peripherals, Disc drives, Data communications
Telecommunications	Modems, Mobile phones, Radio communications
Industrial	Energy motors, Remote control, Battery charging
Automotive	Dash board control, Engine Management, Battery charge control
Consumer	Household electricals, Magnetic card readers

9.5.1 Example 1: Infra red receiver with decoder and actuator

Main electrical functions:

▶ Low level input signal amplifier: $30\mu V$ sensitivity

▶ Continuous time bandpass filter: fc = 50KHz Q = 4

▶ High gain loop: 7000

▶ 50Hz Synchronization mode

▶ Internal shunt regulation: 220V direct supply

▶ Processing logic: 1.2K gates

▶ LED display: 10mA outputs

▶ Triac driver

Applications:

▶   Any remote control unit.

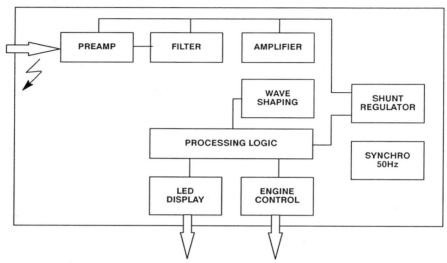

Figure 9.13 Remote control of energy distribution lines

9.5.2 Example 2: Remote control

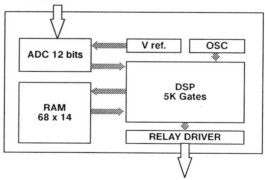

Figure 9.13 Remote control of energy distribution lines

Function: Information decoder and actuator

Main electrical functions:

▶   12bits ADC: ±0.5 LBS Integral non-linearity

▶   RAM

▶   Digital signal precessing: 5K gates

▶ Reference voltage for 12bits converter

▶ Hysteresis comparator

Applications:

▶ Telecommunication relay

9.6 Conclusions and future trends.

The STKM2000 library is a comprehensive library arrived at from a merger between bipolar and CMOS technologies, between analogue and digital functions and between various CAD and design concepts. This merged library is able to respond to the needs of systems between sensors and actuators; from low frequencies to video frequencies; from 2.7V to 11V applications. The availability of accurate functions which combine the ability to fabricate 10K gates together with many analogue functions opens the door towards the one chip solution to many requirements in equipment design. Its availability on a majority of standard workstations with a "friendly" human interface and design manager allows any system designer to undertake the design of a mixed analogue/digital custom circuit.

Chapter 10

Practical Aspects of Mixed Analogue and Digital Design

Paul O'Leary
Joannuem Research, steyrergasse 17,
A – 8010 Graz, Austria

10.1 Introduction

In previous chapters some of the background required for the design of mixed analogue and digital circuits has been discussed. Typical, integrated circuit consist of complex interacting building blocks containing a mixture of digital, sampled analogue (e.g switched capacitor circuits, analogue to digital and digital to analogue converters) and continuous time analogue circuits (e.g anti–aliasing and smoothing filters). To achieve the performance which has been theoretically predicted it is necessary to be aware of certain parasitic effects and practical limits.

The topics discussed in this chapter make it possible to minimize the standard deviation of parameters which are effected by random process variations. This maximizes the yield of critical circuits. In certain cases the required performance can only be achieved with the use of such structures. All the theoretical calculations can not help if these practical aspects are neglected. The topics discussed here should not be considered as trivial or unimportant side effects, but as essential knowledge required for the design of high performance mixed analogue and digital circuits.

This chapter concentrates on parasitic effects and practical limitations. Methods are presented to minimize the consequences of such undesired effects, With optimal circuit layout and or block inter-connection it is possible to reduce or eliminate many disturbances.

10.2 Element matching

A thorough understanding of matching and the structures with which good matching can be achieved is essential for the design of high performance integrated circuits. Element matching is important in almost all analogue circuits. To demonstrate this point a simple amplifier circuit is analysed (see Figure 10.1).

The gain of this circuit is given by:

$$Gain = -\frac{R_1}{R_2}$$

10.1

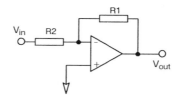

Figure 10.1 Simple amplifier circuit

If unit resistors are used to ensure good matching then the gain is given by:

$$Gain = -\frac{mR}{nR}$$

10.2

Where: R is the unit resistor value, m and n are the number of unit resistors required for R_1 and R_2 respectivly.

It must be remembered that the value of each resistor has a statistical variation. If for simplicity, we assume that the variation has a Gaussian distribution then it is relatively simple to calculate the maximum acceptable standard deviation of the resistor value using the following approximation:

$$\frac{\Delta G}{G} = \frac{\Delta mR}{mR} + \frac{\Delta nR}{nR}$$

10.3

This approximation is only valid for small Δ's.

To guarantee good yield, the gain of the amplifier should remain within the required specification for a three sigma (3σ) variation of the resistances. A 3σ boundary by a Gaussian distribution corresponds to a yield loss of 0.13% for this element alone. If the circuit consists of very many elements then an extremely high yield is required for each individual circuit component. Now using this boundary condition with equation 10.3 gives:

$$\delta_{r\,max} = \frac{m.n}{3\sqrt{m^2 + n^2}}\,\frac{\Delta G}{G}$$

10.4

Now consider the following specification :

$$Gain = 0dB \quad \pm 0.1dB$$

10.5

This is a typical specification for an amplifier circuit and not an extreme example. A gain of 0dB requires a resistor ratio of 1:1.

$$0.1dB \equiv 1.16\%$$

10.6

Using these values in equation 4 gives:

$$\delta_{r\,max} = 0.27\%$$

10.7

The required standard deviation of the resistor value to fulfill the specification with a good yield in production is 0.27%. Such a matching can be achieved in a standard CMOS process but only with the use of optimal layout structures. With this simple example the importance of understanding matching is demonstrated.

The nominal element ratio is defined by the geometries drawn in layout. The actual ratio produced is a random variable with a mean in the vicinity of the desired value and a standard deviation between 0.1% and 10%. The standard deviation is strongly dependent on the layout used. The mismatch is influenced by a number of factors including: local process variations, global lithographic variations, local lithographic variations and process gradients [1–4]. This is true for all elements (Transistors, Capacitors, Resistors) and therefore, similar techniques may be used to optimize the layout of all elements.

10.2.1 Local process variations

All process parameter display local variations (i.e. the region affected by the variation is small, typically $< 1 \upsilon m^2$). These variations are randomly distributed over the whole die and their placement can not be predicted. For example, the capacitor oxide thickness varies and etching has a random variation at the edge, both these factors result in a capacitor value with a random distribution. Similar effects are also present for resistors and transistors. Additionally, resistors and transistors are also dependent on doping implants and diffusions. In general, better matching can be achieved for capacitors than for either resistors or transistors (with the same area), because there are less influences governing the capacitor matching.

The only effective method of combatting local process errors, is to ensure that the precision elements are so large that the local process variations only play a minor role in the achievable matching. McCreary [2] in one of the most comprehensive reports on capacitor matching concluded, for the process which he evaluated, the local oxide thickness variations are negligible if the capacitors are larger than $25\mu m \times 25\mu m$.

For resistors the following rules of thumb can be used. Precision poly–silicon and diffusion resistors should not be laid out with a width smaller than $8\upsilon m$ and for well resistors a minimum width of $10\mu m$ should be used. The well junction is deeper than normal diffusion junctions, making the device more susceptible to three dimensional errors. This is the reason for requiring wider well resistors. Current sources which require precision matching should never use minimum geometries. This data has been observed for a 3μm CMOS process. It is clear that such data is process dependent, but the above guidelines can be used in the absence of better data.

10.2.2 Global process variations

One of the main causes of global process variations are errors during the photo-lithographic processing and etching of the wafers. Similar to all photographic processes, the photo-lithographic processing of silicon is prone to over or under exposure, over or under development etc.. The etching steps which follow the photo-lithographic process are also prone to over or under etching. Characteristics of such errors is that they are constant over the complete wafer. Error in the etching and photo lithographic processing cause degradation of the circuit element matching.

To examine the consequences of such errors and how the resulting mismatch can be minimized, a pair of matched capacitors with the desired capacitor ratio K_d are examined (see Figure 10.2).

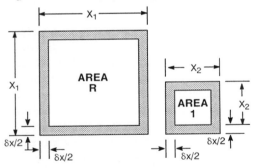

Figure 10.2 Capacitor ratio example

The area ratio for the capacitors drawn during layout is:

$$K_d = \frac{X_1^2}{X_2^2}$$

$\boxed{10.8}$

Where: K_d is the desired capacitor ratio.

If the poly-silicon is over–etched by δx (such errors may occur due to variations in the etching time or changing saturation of the etching fluid from one wafer batch to the next) then the capacitor ratio is given by:

$$K_R = \frac{(X_1 - \delta x)^2}{(X_2 - \delta x)^2}$$

$\boxed{10.9}$

Where: K_R is the real capacitor ratio.

Using equations 10.8 and 10.9 it is possible to calculate a relative error which can be used to estimate the achievable matching.

$$E = \frac{K_d - K_R}{K_d}$$

10.10

Approximating this equation up to terms in δx^2 gives:

$$E \simeq \frac{2\delta x}{x_1}\left\{\sqrt{K_d} - 1\right\}$$

10.11

To gain a feeling for the consequences of such errors it is interesting to look at a typical example. If a capacitor with dimension $30\mu m \times 30\mu m$ is used and typical etching variation $\delta x \simeq 0.1\mu m$ must be accepted. The wavelength of blue light is approximatly 460nm, An error of 100nm is less than one quater of the wavelength. Despite the small magnitude of this error it can result in serious problems in high precision circuits. Then for a capacitor ratio of 4:1 equation 10.11 gives the following error:

$$E = 0.66\%$$

10.12

If these two capacitors were being used in a data converter, then this error would correspond to 6.2bits ±0.5bit of linearity. Clearly the use of such simple structures is not appropriate for the design of precision circuits.

This error can be eliminated or at least dramatically reduced by the use of unit capacitors (see Figure 10.3). In the case of unit capacitors, the area of

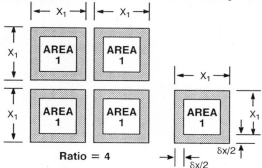

Figure 10.3 Use of unit capacitors

each capacitor is reduced by the same factor and so, the ratio remains unchanged. Using unit elements requires that the lowest common denominator of the desired ratio is feasible to implement.

If the required ratio can not be achieved with a reasonable number of unit elements then a non unit capacitor (or element) must be used. For example, a ratio of 11:13.3 would require 243 unit capacitors, which is not realistic. If non–unit capacitors are used then the matching has a sensitivity to global lithographic variations. Consequently, the capacitor ratio is dependent on such factors as over or under etching. The magnitude of these errors can be minimized with the use of optimum dimensioning (Figure 10.4).

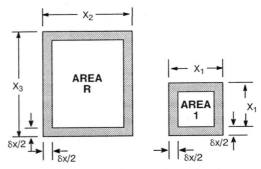

Figure 10.4 Non-unit capacitor structure

For a non-unit capacitor:

$$K_d = \frac{X_2 \cdot X_3}{X_1^2}$$

$\boxed{10.13}$

Where K_d is the desired capacitor ratio.

Over or under etching this structure introduces an error resulting in the following ratio:

$$K_R = \frac{(X_2 - \delta x)\,(X_3 - \delta x)}{(X_1 - \delta x)^2}$$

$\boxed{10.14}$

Where: K_R is the real capacitor ratio.

Once again calculating the relative error and approximating to terms in δx^2 gives:

$$E \approx\, = \frac{2\,\delta x}{X_2^2}K_d \left\{ 2X_1 - \frac{X_1^2\,(X_2 + X_3)}{X_2\,X_3} \right\}$$

$\boxed{10.15}$

Equation 10.15 gives a good approximation for the mismatch occurring due to etching errors. It is now desirable to find the dimensions for X_2 and X_3 which give the minimum sensitivity of E to δx. This can be found by taking the first derivative of E with respect to δx. and setting the equation equal to zero.

$$\frac{dE}{d\ \delta x} = \underbrace{\frac{2X_1}{(X_2 + X_3)}}_{\text{Term 1}} - \underbrace{\frac{X_1^2}{X_2\ X_3}}_{\text{Term 2}} = 0$$

10.16

Term 1 of equation 10.16 is the ratio of the perimeters and term 2 the ratio of the areas. If equation 10.16 is equal to zero, then the ratio of perimeters should be equal to the ratio of the areas. Now solving equation 10.16 to find the optimum dimensioning for the capacitors gives:

$$X_2^2 - 2\ K_d\ X_2 + K_d = 0$$

10.17

Now solving this quadratic equation:

$$X_2 = K_d \pm \sqrt{K_d\ (K_d - 1)}$$

10.18

One restriction on the non-unit capacitors can be seen from equation 10.18: if the solutions to this equation are to be real, then the ratio K_d must be greater than 1. This means the smaller capacitor should correspond to the unit capacitor and the lager capacitor should be the non-unit element. If equation 10.18 is used to dimension the non-unit capacitors, then a layout is achieved with minimum sensitivity to photo–lithographic errors and etching variations.

Taking the example which was previously discussed of a capacitor ratio 10.13.3. This ratio would now be implemented using 11 unit capacitors for one capacitor and 12 units plus one non-unit for the second capacitor.

Unfortunately this solution assumes an isotropic etching. Equation 10.18 results in very long and narrow capacitors for ratios $K_d = 1 + \Delta$ or $K_d = 2 - \Delta$. Where Δ is small such capacitors are very sensitive to anisotropic etching and photo-lithographic processing (anisotropic processing is the norm rather than the exception for small geometry processes). The capacitors with value $K_d = 1 + \Delta$ should be realized with a square capacitor, similarly the capacitors in the range $K_d = 2 - \Delta$ should be implemented as one unit capacitor and one capacitor of value $1 - \Delta$ The increased sensitivity of these structures to isotropic errors is more than compensated by the improved performance with respect to anisotropic processing.

10.2.3 Process gradients

Most process parameters are effected by process gradients. Process gradients are systematic variations of a parameter over the wafer or die. For example the oxide thickness may vary systematically from one side of the die to the other, similarly for a transistor the threshold voltage may have a systematic variation. Figure 10.5 shows the unit capacitors discussed

Practical Aspects of Mixed Analogue and Digital Design

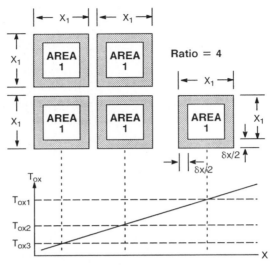

Figure 10.5 Process gradient for capacitors

previously, a possible oxide thickness gradient is shown below the capacitors. Due to the gradient the average oxide thickness for the four capacitors on the left is different form that of the single capacitor on the right. This would results in a capacitor ratio error. A solution to process gradients can be achieved with the use of common centroid layout techniques.

For capacitors the use of a common centroid layout, see Figure 10.6, offers

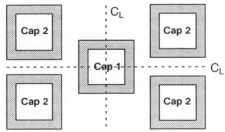

Figure 10.6 Unit capacitor with common centroid layout

a relatively simple solution (if the process gradient is approximately linear). The capacitors are laid out so that they have the same centroid (1st. moment of area. As a result of the common centroid layout the capacitors have the same effective gate thickness. To ensure that process gradients can be approximated linearly, the matched elements should be placed as close to one an other as possible.

If the input offset voltage of an operational amplifier is to be minimized, then a common centroid layout of the input differential transistors is essential. With optimum layout and design structures, operational amplifiers with a maximum input offset of 3–4mV (3σ) have been demonstrated [5]. One possible common centroid layout solution for an input differential pair is shown in Figure 10.7 [6]. The two transistors have been split into four equal parts and are connected in parallel in a suitable manner. This structure has the advantage over other solutions, that through the interleaved layout of the transistors, the common source and both drains can be connected in metal and require no crossovers. This structure is particularity well suited for input stages with wide input transistors.

The transistor layout in Figure 10.7 leads to a very compact layout with a

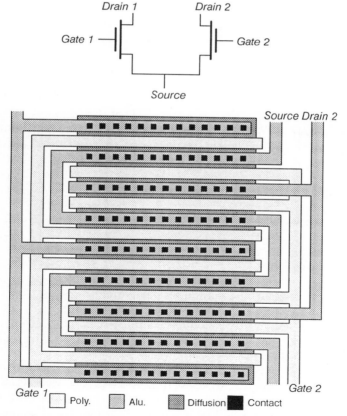

Figure 10.7 Common centroid layout for a differential pair

common centroid for the two input transistors. A good rule to keep in mind

Practical Aspects of Mixed Analogue and Digital Design

when dimensioning input transistors is; that the transistor matching is proportional to the square root of the gate area, this was determined by Jyn Bang Shyu in a report on achievable matching in CMOS [7]. The equivalent input noise of the transistor is inversely proportional to the gate area.

In the case where more than the two elements have to be matched, a common centroid layout can become cumbersome and impractical. An example in case is a current switching DAC using a thermometer decoding (see Figure 10.8).

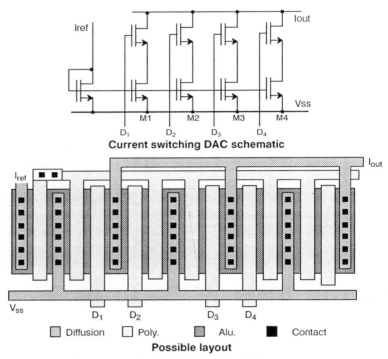

Current switching DAC schematic

Possible layout

Figure 10.8 Current switching DAC structure

The transistors M1–M4 are the unit current sources of a current switching DAC. To ensure a good differential and integral nonlinearity an optimum matching is required between the transistors.

A common centroid layout is not possible for the transistors M1–M4 without extreme penalties in area. The undesirable effects of the gradient can however be minimized by design. In the case of a DAC, if we know which parameter is the most important, differential or integral nonlinearity, then it is possible to optimize the decoding.

If the differential nonlinearity is the most critical parameter, then the decoding should be done in a sequential manner, that is, the transistors are switched on one after the other, starting at one end of the structure and ending at the other. The minimum difference from one current source to the next occurs between neighboring transistors. As a result sequential decoding gives the minimum differential nonlinearity. If however, the integral nonlinearity is more important then a scatter decoding should be used [8]. To produce the minimum integral nonlinearity, such decoding schemes can be optimized to average the process gradient. Some degradation of the differential nonlinearity must however be accepted in this case. Using such decoding techniques integral nonlinearities equivalent to 1 LSB at 10 bits have been achieved with conversion rates of 20MHz [9].

The use of common centroid layout of resistors is a debatable subject. In general the resistor structures require more contacts when laid–out in a common centroid manner. These contacts must also carry current and as a result the contact resistance (which is not negligible) effects the resistor matching. Figure 10.9 shows a precision resistor structure used in an ADC.

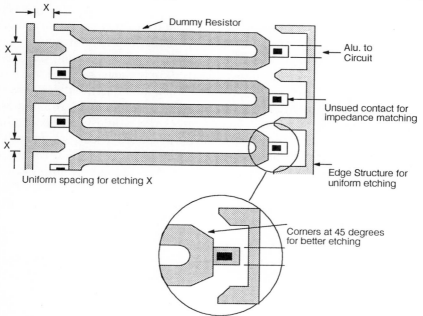

Figure 10.9 Precision resistor structure

This resistor is not laid out using a common centroid technique but still demonstrated in a worst case matching better than 0.2% . We believe the errors due to contact impedance in a common centroid layout structure would have been comparable.

Most analysis and proposed solutions [4, 8] assume that the process gradient is linear. Practical experience confirms that this assumption is sufficient.

10.2.4 Boundary effects.

With shrinking process geometries boundary effects are becoming more and more important. Most boundary effects are due to fringing or anisotropic photo-lithographic processing. Structures are regarded as two dimensional during design and layout. However, in reality the structures are three dimensional. Fringing effects are particularly acute for transistors and well resistors, whereas, anisotropic photo-lithographic processing is a general problem.

With larger geometry processes ($>5\mu$m) such boundary effects could be ignored, but with modern process dimensions (1.0 to 2μm) these effects cause major errors.

Figure 10.10 shows a cross section of four transistors from a precision current mirror. The field oxide between the transistors can be marginally thinner than the field oxide at the edge of the transistor structure. This results from differential growth rates for narrow strips of field oxide and large field oxide areas (The drawing is purposely exaggerated to make the effect visible). The thinner field oxide between the transistors, would result in an effective width difference between the transistors M2, M3 and M1, M4. For precision matched current sources this would be unacceptable. Transistors M1 and M4 would deliver a different current from the transistors M2 and M3. This effect can be eliminated with the use of dummy structures. The matching of the transistors M1–M4 can be improved by physically placing one transistor to the left and one to the right of this structures and not using these transistors in the circuit. Each transistor then has the same boundary conditions.

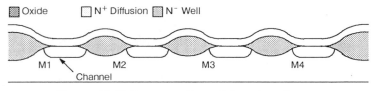

Figure 10.10 Cross section of four transistors

The use of dummy structures for resistors and capacitors should also be encouraged. Dummy elements also serve to prohibit or at least reduce the effects of anisotropic processing. All active elements which are used in a circuit should have the same boundary conditions. The dummy structures require very little area and give a dramatic improvement in the element matching achieved.

Figure 10.11 shows two possible layouts for a well resistor. The first implementation would function correctly, if the well could be considered to be a two dimensional structure for layout.

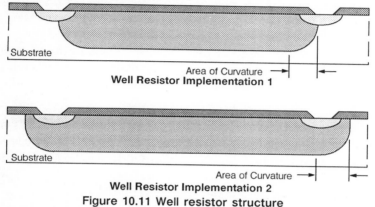

Substrate

Area of Curvature ⟶
Well Resistor Implementation 1

Substrate

Area of Curvature ⟶
Well Resistor Implementation 2
Figure 10.11 Well resistor structure

The impedance of the well in Figure 10.11 is given by:

$$R = \frac{L}{W} * R_\square$$

10.19

Where: $R_\square$... is the resistance per square.

However the well is a three dimensional structure. In a typical 3μm CMOS process the well is 3–3.5μm's deep. If the side diffusion is approximately 70%, then the area of curvature is roughly 2.5μm long. The doping concentration in the area of curvature is lower than in the normal well area. This results in a higher impedance at the end of each well resistor. The impedance can be as high as 1.5 times the normal impedance per square. If such a layout is used, then equation 10.19 cannot be used to calculate the resistance value. The structure has a second disadvantage. A misalignment of the N^+ diffusion well contact with respect to the well causes the contact to move further into the area of curvature, resulting in a change in the resistor value. This problem is easily solved by extending the well sufficiently beyond the N^+ to ensure, that with misalignment the N^+ is not over the area of curvature.

The orientation of precision elements is also important. Matched elements should always be drawn with the same orientation. They should not be rotated with respect to one another. Pressure on silicon results in an anisotropic stress distribution due to the crystal structure. Figure 10.12

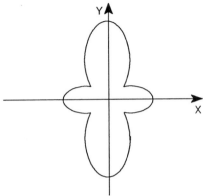

Figure 10.12 Stress distribution in 100 silicon due to pressure

shows the stress distribution of pressure for 100 silicon. It can be seen that the pressure distortion in X and Y is different. If matched elements were laid out with different orientations, they would experience different degrees of stress from pressure on the die. This differential stress gives rise to an effective mismatch of the elements. The pressure on the silicon die in a plastic package can be as high as 200 bar [10]. Effective mismatching of elements up to 5% have been observed under such conditions.

The orientation of matched elements with respect to power dissipation elements on an integrated circuit is also of great importance. Power dissipation on the chip causes temperature gradients. For elements with a temperature dependence (Transistors and resistors etc.) this gradient causes a change in the element value. To avoid a mismatch of the elements due to the temperature gradient the element should be laid out symmetric to the main power dissipation elements. Figure 10.13 shows two groups of resistors laid out with respect to a power dissipation element. The resistor group A will have a mismatch because the two resistors are at different temperatures. Whereas the resistors in group B which are symmetrically laid out with respect to the power dissipation have the same average temperature and as a result, no error due to temperature occurs.

The multiplicity of boundary effects is so large that it is impossible to describe all such effects. The aim here is to make the reader aware of such problems and to encourage him to think about possible solutions during the design and layout phase of a project. The understanding and prediction of such effects requires a good knowledge of the process being used.

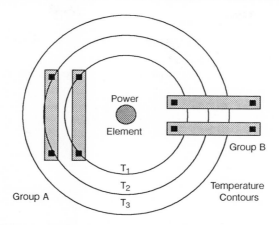

Figure 10.13 Matched elements with temperature contours

10.3.0 Noise coupling

One of the major problems of mixed analogue and digital designs is noise coupling. The noise generated in the digital sections of the circuit is coupled into the analogue circuits through the power supplies and over the substrate. Routing an analogue signal parallel to a digital line results in a capacitive coupling between the two lines.

Noise generated in a digital circuit is particularity unpleasant in precision analogue applications because it is not purely random. If the digital circuit contains counters or synchronous logic, then discrete spectral components are generated corresponding to the clock frequency and its harmonics, the sub-harmonics are also present. Such discrete spectral lines are often more disturbing than random (white) noise sources. Switched capacitor circuits are particularity sensitive to such disturbances. The original spectral line may be outside the used frequency band but due to the sampling in the switched capacitor stages the spectral lines may be folded into the pass-band. The anti–aliasing filter in front of the switched capacitor only prevents aliasing of the signal at the input of the filter, however the on chip noise may be coupled into any stage of the structure.

The techniques used to minimize noise coupling can also be used for circuits where very low levels of cross-talk are required such as modem front ends.

10.3.1 Power supply coupling

The first source of noise coupling or cross-talk is over the power supply connections. Figure 10.14 shows a section of a mixed analogue and digital circuit with a common power supply. The voltage change at point a is given by:

$$\Delta V = R_c \cdot I_{dd} + L_b \frac{dI_{dd}}{dt}$$

$\boxed{10.20}$

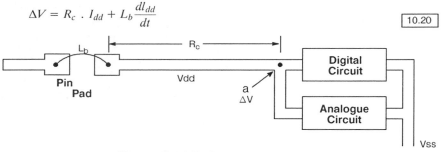

Figure 10.14 Bad power connection

The bus driver shown in Figure 10.15 can be used to demonstrate just how severe this problem can be. Considering an 8–bit data bus with a rising edge requirement of 50nSec and a load of 100pF. The power supply connection to the bus driver is estimated at 1mm long and 20μm wide. At the initial switching of the buffer a current slew as high as $5*10^6$ A/s can occur, this is followed by a capacitor charging current of approximately 10–11mA. Using these values in equation 20 gives a voltage variation of 880mV. Such a noisy power supply would be totally unacceptable for use in a precision analogue block. Such a power supply is also undesirable for digital circuits because it would reduce the noise margin to an unacceptable level.

The first and most obvious solution to the problem of noise coupling, is to use separate power supply lines for the digital and analogue sections of the circuit (Figure 10.16). This solution unfortunately requires extra pins which are not always available. Care must be taken when using two separate power supplies because of the danger of latch-up. For an N–well process the V_{ss} must be common external to the circuit and V_{dd} must be common for a P–Well process, if latch-up is to be avoided. For twin well processes the doping of the substrate must be considered.

If separate pins are not available, then the next best solution is to use separate power supplies on the die coming to two bonding pads. These two pads are then bonded to the same pin (Figure 10.17). This type of connection separates some of the common power supply connection but some degree of coupling still remains. If a pad has already got a double bond

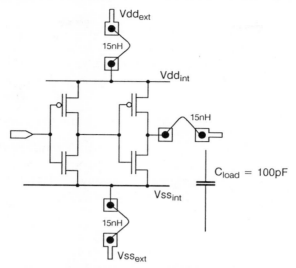

Figure 10.15 High speed data-bus driver

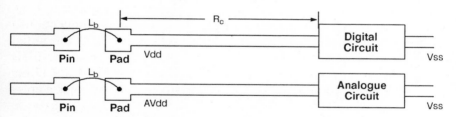

Figure 10.16 Separate pins for analogue and digital supplies

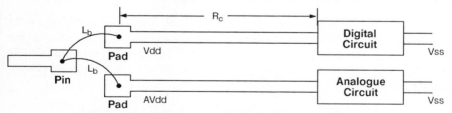

**Figure 10.17 Separate bonding pads for analogue and
digital power supplies**

then a third bond is usually not permitted. In such cases, the only possibility
which remains is to use separate power supplies on the chip which are then
connected together at the bonding pad (Figure 10.18). Cross-talk figures of
better than –90/–110dB at 10kHz have been achieved [11] using separate
power supplies on the die, connected to one pad.

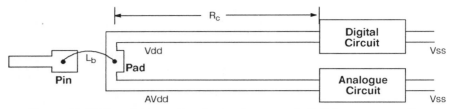

Figure 10.18 Separate routing for analogue and digital power supplies

In addition to the above measures, the pins with the minimum inductance should be used for the power supplies. The pin inductance consists of two portions. The bonding wire inductance (usually ≤3nH) and the lead frame inductance (see Figure 10.19). The lead frame inductance is due to the

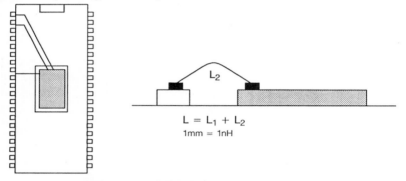

$$L = L_1 + L_2$$
$$1mm = 1nH$$

Figure 10.19 Bonding wire inductance

connection from the bonding point to the pin and is particularity large for dual in line packages with more than forty pins. A good rule to estimate the inductance of a connection is:

$$1mm = 1nH \qquad \boxed{10.21}$$

This rule can be used with relatively little error for almost all printed circuit board connections.

Table 10.1 Bonding wire inductance and capacitance
for a 40 pin DIL package

Package	Pin	Bond L	Bond C
40 Pin Plastic	1,2 10,11	15nH 4.4nH	2.4pF 0.7pF
40 Pin PLastic with Socket	1,2 10,11	18.6nH 7.6nH	2.6nH 0.8nH
40 Pin Ceramic	1,2 10,11	20.9nH 9.0nH	2.7pF 0.8pF

Table 10.1 gives the measured inductance and capacitance for the pins of a forty pin dual in line package. The pins at the corner of the package have longer connections than the pins in the middle of the package and as a result have a higher inductance. The difference can be as large as a factor of three. This corresponds to 9.5dB lower noise spikes.

The pads near the middle of the package should be used for power supplies and critical signals.

If a high speed data bus is required, then the bus drivers should have a separate power supply pin. This pin should be placed between the data bus pins to minimize the power supply interconnection. In general digital connections should be kept as far from the analogue circuit sections as possible.

Rather than trying to minimize the problem of high speed data busses, it is desirable to try and reduce or prevent the problem from occurring in the first place. This can be achieved with the use of controlled edge buffers [12, 13]. These buffers are designed to control the speed of the rising and falling edges of the output lines. The maximum slew rate at the output pad is limited so as not to exceed a tolerable limit. This limits the dynamic loading of the power supplies, reducing the disturbances on the chip.

10.3.2 Substrate noise coupling

The second source of noise coupling is over the substrate. Because the substrate offers a relatively high impedance to ground, noise generated in one section of the circuit can be propagated through the substrate to other parts of the chip. This noise is then capacitively coupled into the active part of an analogue circuit. High impedance nodes, such as the output nodes of an operational transconductance amplifiers (OTA's) or the compensation node of an internally compensated operational amplifier, as well as the input nodes are particularity sensitive to capacitive coupling.

The capacitive coupling into circuit elements (resistors and capacitors) and sensitive signal lines, can be reduced with the use of shielding layers. To provide the maximum possible de-coupling, the shielding layer should have the lowest possible impedance to a quiet power supply.

The most commonly used shielding is to place a well under capacitors and resistors. Figure 10.20 shows the implementation of a resistor and a capacitor using a well as shielding layer. The well under the resistor and capacitor served to attenuate the noise coupling from the substrate to poly–silicon or metal layer.

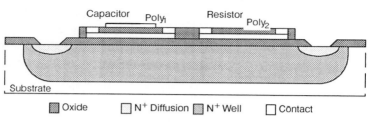

Figure 10.20 Well as shield

Sensitive analogue lines should be connected, where ever possible, in aluminium. This permits the use of a poly–silicon layer under the metal as a shielding layer. The poly–silicon having a lower impedance than the well, provides a better attenuation of capacitivly coupled noise.

The prudent use of well and substrate contacts helps to keep the silicon free from noise in the analogue sections of the circuit. Some time should be spent considering where and when to use such contacts. For very sensitive applications the use of three bus digital cells is to be recommended [14] (see Figure 10.21). The cells are designed with two Vss lines (for an N–well

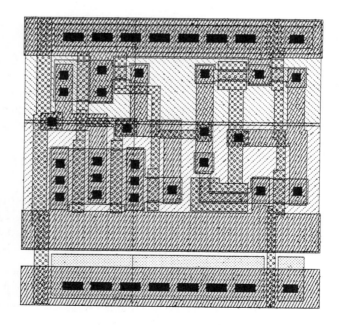

Figure 10.21 Example of three bus digital cell

process). One Vss line is used to supply the cell with current and the second line is used to connect the substrate. The two supplies are then led to extra pads, this helps to prevent noise on the digital supply lines from entering the substrate.

10.3.3 Signal noise coupling

Noise can also be coupled into a signal line. If an analogue signal is routed parallel to a digital signal, then noise is coupled from the digital to the analogue signal through the mutual capacitance. Figure 10.22a shows such a case. To prevent this coupling horizontal shielding should be used. The placement of an analogue ground between the digital and analogue signal functions to decouple the two lines (see Figure 10.22b). To minimize the dangers of parasitic coupling the analogue interconnections should be kept as short as possible.

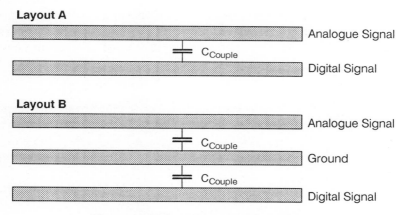

Figure 10.22 interconnection coupling

For switched capacitor circuits, the charge injection through the switches is a major problem. Figure 10.23 shows a double switch as used in switched capacitor applications.

Figure 10.24 shows an improved layout for this cell. The dimensioning of the transistors has been reduced to the minimum acceptable value. The smaller the transistor dimensioning, the smaller is the injected charge. Unfortunately the switch transistors cannot always be used with minimum geometries because of the resulting RC delay in the switched capacitor circuit.

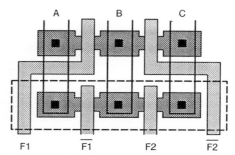

Figure 10.23 Switched capacitor switch version 1.

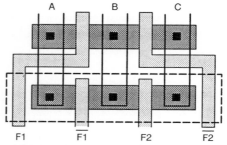

Figure 10.24 Switched capacitor switch version 3.

Finally Figure 10.25 shows a further improvement of the layout with respect to clock feed through. The clock lines F1 and F2 are now routed over the first switches and not over the analogue lines A and C, as is the case in Figure 10.24. This reduces the parasitic capacitance between the clock and the analogue signals.

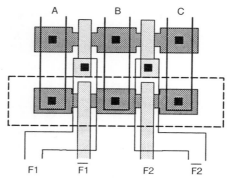

Figure 10.25 Switched capacitor switch version 2.

10.4 Examples of optimized structures.

After discussing all the above effects, it is now interesting to look at some precision structures implemented for mixed analogue and digital integrated circuits.

Photo 1 shows a capacitor array used in an 8–bit analogue to digital converter. The capacitors have been implemented using a chess board pattern layout. With a chess board layout, the elements are laid out corresponding to the black and white squares. This interleaved structure leads to optimum conditions for a common centroid layout. Dummy structures have been used to ensure uniform boundary conditions for all capacitors.At the edge of the capacitor array and surrounding the two contacts in the middle of the array. Examining this structure, it can be seen that all corners are at 45 degrees. This improves the uniformity of etching. Measurements show a worst case matching of 0.2% for the capacitors. Finally, to give maximum noise blocking, a well has been placed under the capacitors. The well is contacted on all four sides and is connected to a quiet analogue power supply.

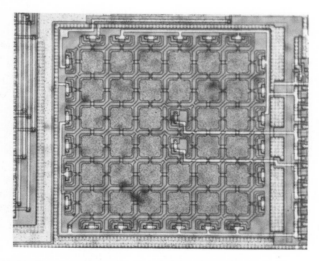

Photo 1 Capacitor array used in an 8–bit analogue to digital converter

A second example from the same ADC is shown in photo 2. The precision resistor is used to divide the reference voltage to intermediate levels. The resistor string consists of unit elements, at the top and bottom of the structure dummy elements are used. Similarly to the capacitors all corners are at 45 degrees. For the implementation of the ADC only eight contacts are required to the resistor string. Contacts have been placed at the end of every resistor, to eliminate any possible impedance change and resulting mismatch due to aluminium contacts to the poly–silicon. Thus ensures uniform impedance for all resistors. Because the poly–silicon is continuous, the contacts of the resistor chain can only carry a transient current while charging a capacitor. In the steady state the resistor contacts do not carry current, as a results the contact impedance plays no role in the resistor matching.

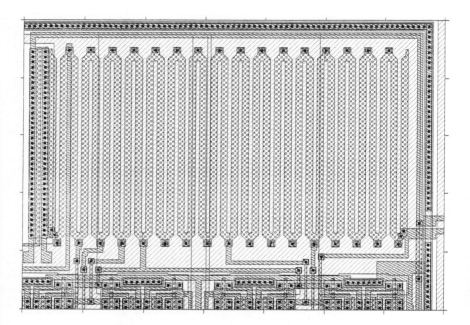

Photo 2. A second example from the same ADC showing precision resistor used to divide the reference voltage to intermediate levels

10.5 Conclusions

This chapter has presented methods to improve element matching for MOS integrated circuits. A high degree of matching is required if a good yield is to be guaranteed in production. Methods to reduce the coupling of noise from the substrate and power supplies have also been presented. Hopefully the reader has been made aware of some of the practical aspects of mixed analogue and digital design. He should now be motivated to consider such problems for himself.

10.6 References

[1]   Gregorian R and Temes G, "Analogue MOS Integrated Circuits for Signal Processing", J. Wiley 1986, Chapter 7 "Nonideal effects in switched capacitor circuits."

[2]   McCreary J, et al. "Matching properties, Voltage and Temperature Dependence of MOS Capacitors", IEEE Journal of Solid State Circuits, Vol. SC–16, December 1981.

[3]   Gray P, Castello R, "Design of MOS VLSI Circuits for Telecommunications.", Editors Tsividis Y and Antognetti P, Prentice Hall, 1985. Chapter 10, "Performance limitations in switched capacitor filters."

[4]   Lakshmikumar K, Copeland M, and Hadaway R, "Characteriztion and Modelling of mismatch in MOS Transistors for precision Analog Design.", IEEE Journal of solid–state circuits, Vol. SC–21, No. 6, pp.1057–1066, Dec. 1986.

[5]   Monticelli D, "A quad CMOS single–supply Op–Amp with rail–to–rail output swing.", IEEE Journal of Solid State Circuits, Vol. SC–21, No. 6, Dec. 1986.

[6]   Gatti U, Maloberti F and Liberali V, "Full stacked layout of analogue cells.", Proc. IEEE, Int. Symp, Ccts. and Sys. ISCAS, Portland 1989.

[7]   Shyu J, B, "The obtainable accuracy of analogue metal oxide semiconductor integrated circuit elements.", Ph.D thesis, Univ. of California, Los Angeles, 1984.

[8]   Conroy C.S.G and Lane W.A, "Statistical Modelling and simulation for DAC design.", Proc. 12th. ESSIRC, pp.214–216, 1986.

[9]   Internal work done at ITT Intermetall, Freiburg FDR.

[10] Maloberti F, "Design of CMOS analogue integrated circuits.", internal course at Austrian Mikro Systems, Graz, September 1988.

[11] Melcher G, et al., "A 3um CMOS analogue audio circuit for mobile telephone applications.", Proceedings of the 15th. ESSIRC, Vienna, Austria, September 1989.

[12] Raver N, "Open–loop gain limitations for push pull off chip drivers.", IEEE Journal of solid–state circuits, Vol, SC–22, No. 2, pp. 145–150, April 1987.

[13] Wong S and Salama C, "An efficient CMOS buffer for driving large capacitive loads.", IEEE Journal of solid state circuits, Vol. SC–21 June 1986, pp.464–469.

[14] "High performance CMOS analogue and digital cell library", AMS Data book, 1988.

[15] Allstot D.J and Black W.C, "Technological design considerations for monolithic MOS switched capacitor filtering systems.", Proceedings of IEEE, Vol.71, August 1983.

[16] Tsividis Y, et al., "Continuous–time MOSFET C–filters in VLSI.", IEEE, Jou. of Solid State Circuits, pp. 15–30, Vol. SC–21, No.1, Feb. 1986.

Chapter 11

Some Applications of Mixed Signal ASICs

Randeep Singh Soin. GenRad UK Ltd.,
Steve Morris, Dialog Semiconductor,
Franco Maloberti, University of Pavia.

11.1 Introduction

Although the main thrust in the development of VLSI incorporating ever larger numbers of components on single integrated circuits has been directed at digital functions, mixed signal processing is by no means a new development. Many of the phenomena which characterise the physical world are represented by continuous, i.e. analogue functions of time. As a consequence many electronic systems involve some analogue signal processing, even if this only manifests itself in the interfacing functions, e.g at inputs and outputs. For reasons of reliability, cost and technical merit it is desirable to take the integration further to include interfacing to the physical world. Therefore processing technologies, circuit design techniques, circuit building blocks, and design tools are to be developed developed which facilitate the mixing of analogue and digital signal processing on a common chip.

This particular chapter describes several specific applications of mixed analogue digital ASICs. Although some of these examples essentially perform digital signal processing with analogue circuits to provide the interfacing functions, others are more general. The development of many recent methods in the area of analogue circuit design have been exploited to produce systems where a significant amount of the signal processing is carried out in analogue form before conversion to the digital. This leads to much more flexible and cost effective solutions. The flexibility of the analogue functions is greatly enhanced by the fact that digital feedback circuitry can be used to obtain adaptive capabilities, and to provide greater precision in performance.

The present chapter and the one following it will deal with several applications of mixed signal chips. The material in this chapter is organised as follows. The first section gives very brief descriptions of a number of chips from various segments of the market. The basic intention is to emphasise to the reader, the widespread relevance of such ASICs. The section following this is organised by way of a number of case studies which discuss some examples in greater detail. The number of possible examples in development or already in the market is very great. We have chosen several from the telecommunications and automotive sector. These examples are not intended to represent the most complex, efficient or necessarily best solutions in any one area; but are simply chips with which one or other of the authors of this chapter has been directly involved. It is hoped that the examples serve to arouse the reader's interest and demonstrate the applicability of some the ideas discussed elsewhere in this book.

11.2 Applications areas

Since most segments of the electronic market involve analogue and digital processing, the areas of application of mixed signal ASICs are very broad. This fact is illustrated by Table 11.1 which lists several examples of mixed analogue digital chips in various market segments. These chips were all designed and fabricated for external customers by Dialog Semiconductor, a company specialising in mixed signal ASICs.

Table 11.1 Examples of mixed signal ASICs form various market segments.

Telecomms	Industrial
Mobile telephone IC's for cellular systems Radio pager Modems Echo canceller ISDN devices Telephone handset IC DTMF circuits Speech scrambler/descrambler Extended band CODEC RF circuit controller Voice over modem	Switched mode power supply Utilities receiver/decoder Bar code reader Weigh scale IC LCD display Light controller
	Automotive ASICs
	Panel LCD display Gear selector Airbag controller
Consumer	**Military**
Electronic organ sound generator Music synthesiser Car alarm ICs Television picture-in-picture colour palette Video DAC Satellite sound decoder Freezer temperature controller.	Programmable fuse timer Multichannel sensor Flight attitude analyser Control surface servo controller Detonator Flight path analyser/controller
Medical	**Computer**
Hearing aid circuits ECG monitor X-ray CAT scanner analogue multichannel processor X-ray CAT scanner charge amplifier IC Remote control for hearing aid	Video graphics controller Speech Recognition Disc controller Ethernet SIA

Some of the above examples still represent only part of the system which they serve while others incorporate levels of integration which would qualify as "systems on a chip". An excellent example of a system on a chip implemented using mixed signal techniques was developed by General Electric and Yokogawa Electric Corporation [1]. This is a CMOS ASIC which measures several parameters which characterise ac power. The inputs to the chip are six analogue signals from three pairs of voltage and current sensors. The signals are processed in both analogue and digital modes . The outputs which represent rms (root mean square) voltages and currents, active, reactive and apparent power, and power factor, phase angle and frequency are available in two forms. A bit serial digital form for further processing by a digital computer or as pulse width modulated (PWM) analogue form to drive analogue display meters. See Figure 11.1.

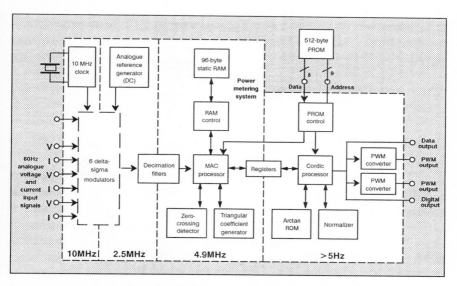

Figure 11.1 An example of a "system on a chip"

Amongst its various functional blocks, the chip incorporates six delta sigma A/D converters, decimation filters, DC reference generators, D/A converters whose output is in PWM form; RAMs, and various digital registers. At the heart of the system is a multiply–accumulate (MAC) processor which performs calibration, cross–products and low pass functions and a processor which implements the CORDIC [2] trigonmetric computing technique to calculate the output primitive functions. As techniques, tools and technologies in this field develop, more and more analogue digital systems on a chip will begin to appear.

11.3 Brief examples

This section gives brief descriptions of some mixed signal ASICs from various market segments. Chip floor plans or functional block diagrams are shown as a method of listing the analogue and digital functions of the chip and to give an idea of their relative silicon area occupation.

11.3.1 Some medical applications

The monitoring and measurement of biomedical signals usually involves the detection of low amplitude emf's which are often buried in noise. The situation is further exacerbated by the presence of important low frequency signal content. This makes CMOS (MOSFET) devices difficult to use because of their dominant $1/f$ noise content. However CMOS does offer low power operation and this is important for ambulatory applications. Innovative design has, to a large extent, overcome the noise problem in CMOS firstly by use of parasitic bipolar elements on any bulk CMOS process and secondly by use of oversampling or commutative methods (these methods also reduce offset voltages very significantly). Also switched capacitor filters may be used which enable filtering to be carried out at frequencies of 1Hz or even lower. High resolution data converters can now be included on a CMOS IC using technologies such as delta–sigma oversampling [15] or, reference refreshed cyclic converters. Thus very complex DSP can also be included on chips to perform Fourier analysis, high order filtering and/or noise reduction using correlation techniques. While mixed signal ASICs have been designed for enhancement of performance in Xray, ECG, CAT and other diagnostic medical equipment, there is a growing requirement for their use in ambulatory situations such as hearing aids, rf monitoring links and heart pacemakers. In this latter category of patient worn devices, the driving force is for low power and low voltage operation to yield increased performance. A good example is hearing aids where the user now expects to get "intelligent" information, conveyed at a comfortable level by a small non–intrusive device which can operate on a small battery for several weeks (not days!). The control of this device needs to be virtually "hands off", so that requirements on dexterity are minimised. An ASIC meeting this goal is described later in this section.

11.3.1.1. A heart rate meter

Figure 11.2 shows the floor plan of a chip whose main function is the monitoring of the condition of the heart. It directly reads and stores measured heart beat rate from chest electrodes and stores this information in an on chip RAM. The data can then be downloaded optically into an external Microprocessor unit. In addition to the digital functions, the chip features low noise and very low frequency (typically 20Hz) filtering and detection of analogue signals. The circuit consumes very low current ($<100\mu$A), which by use of a power down mode feature has an average supply current of less than 70μA. In summary the chip incorporates a 4k RAM; 1100 logic gates; an LED optical communication driver; low frequency bandpass switched capacitor filtering; very low noise and low power amplifiers together with a peak–clamped window detector. The chip was fabricated in a single metal, double poly, 3μm , low threshold, CMOS process.

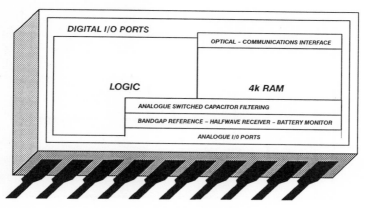

DIGITAL I/O PORTS

OPTICAL - COMMUNICATIONS INTERFACE

LOGIC

4k RAM

ANALOGUE SWITCHED CAPACITOR FILTERING

BANDGAP REFERENCE – HALFWAVE RECEIVER - BATTERY MONITOR

ANALOGUE I/O PORTS

Figure 11.2 Chip floor plan for ECG/EKG processor

11.3.1.2. Hearing aid ASIC

Figure 11.3 shows the floor plan for another medical application, an analogue/digital signal processing chip employed to aid hearing in subjects with certain forms of hearing impairment. The primary function of this chip is the recognition and attenuation of noise signals while passing speech signals. This function is performed by the use of a digital signal processing circuit whose input is a sample of the incoming analogue signal after A/D conversion. The outputs of the digital processing circuit are used to attenuate the noise signals from the composite input signal. The chip was fabricated in a single metal, double poly 3um low threshold CMOS process. A voltage doubler was included on chip to allow the device to be used with power sources down to 1 Volt.

11.3.2 Some consumer applications

Consumer applications are characterised by high production volumes. Therefore the primary impetus to integrating analogue and digital functions

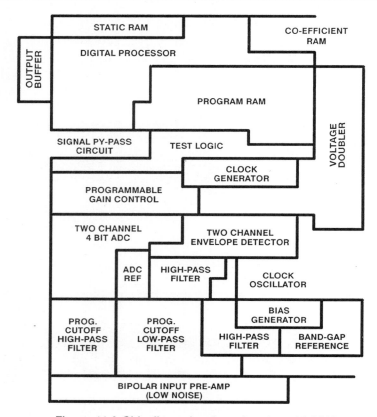

Figure 11.3 Chip floor plan for a hearing aid ASIC

onto single chips is the need to reduce the component count and thus reduce unit costs.

11.3.2.1. Sound and rhythm generator

Figure 11.4 shows the floor plan for a sound and rhythm generator ASIC for a musical organ. This was produced in 2um CMOS process. The main advantage of integration was that of cost as compared to the alternative solutions. Integration enabled an electronic organ/synthesiser to be made with just the ASIC, a microprocessor, memory and a power amplifier. Two versions of the chip were produced. One was for use on its own and a second more expensive versions was for use with an external microprocessor. The latter implemented more features and addressed the middle range of this market.

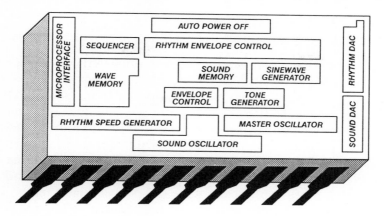

Figure 11.4 Floor plan for a sound and rhythm generator

11.3.2.2 TV picture in picture processor

A recent example of a mixed signal ASIC for the consumer market is a chip for a Television picture–in–picture facility. The chip floor plan is shown in Figure 11.5. The ASIC was fabricated in a 1.2u CMOS process and incorporates high speed Video DACs and ADCs. It provided a low cost alternative to a standard solution which involved six chips and 2 delay lines. This application enabled the display of up to four small pictures in the

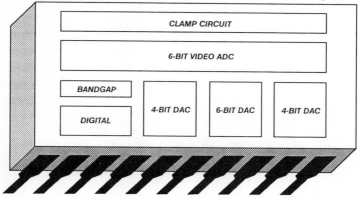

Figure 11.5 Floor plan for TV "picture–in–picture" ASIC

corners of a consumer TV screen in addition to the main picture. Of the four additional pictures only two were required to display motion while the other two could be used for a frozen image. The 4bit DACs offer sufficient resolution for the two colour difference signals while the 6–bit DAC is used for the luminance signal.

11.3.3 Some telecommunications examples

11.3.3.1. A multi-standard modem

Figure 11.6 shows the floor plan for a mixed signal ASIC to provide an integrated solution to the analogue front end of a duplex multi standard modem . The analogue portions include various programmable and fully differential switched capacitor filters. The programmability feature which is achieved by virtue of the digital control circuits, allows the same integrated circuit to meet the requirements of a number of different transmission rates and standards. An on chip crystal controlled oscillator provides a master clock which is processed by the digital section to provide the various frequencies and phases which control the switched capacitor filters. The digital section also provides an asynchronous "micro" interface to an external microprocessor which is used to control the programmability aspects of the chip. The "DSP" interface controls D/A and ADC converters and performs data preparation for an external digital signal processing IC. In the next phase of development the ASIC will contain sample/hold, ADC and DAC circuitry as well as a programmable gain amplifier (PGA), and summers, so that full duplex operation can be catered for.

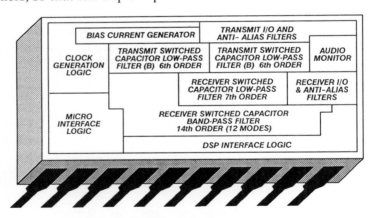

Figure 11.6 Floor plan for a modem frontend

11.3.3.2. A speech scrambler de-scrambler

Figure 11.7 shows the block diagram of an IC which handles the security aspects of speech transmission in commercial mobile radio systems [6]. It is a dual speech scrambler which can be used both in a base–station and in mobile transceivers. The transceivers can be hand held or vehicle mounted. The chip is included in the speech path of both base–station and mobile transceivers. The speech signal is spectrally split into four bands before passing through a digitally controlled scrambling (de-scrambling) process. The same chip scrambles transmitted audio signal and de-scrambles received signals.

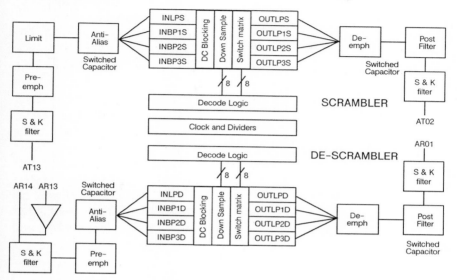

Figure 11.7 Block diagram of a speech scrambler/de-scrambler

11.3.4 An automotive example

Figure 11.8 shows the block diagram of a knock detector for spark ignition engines [3]. This ASIC forms part of a knock control system which can be used in a spark ignition automobile engine. Such systems can be used dynamically to adjust appropriate parameters to improve the performance and fuel efficiency of the engine and control pollutants emitted in the atmosphere.

Engine knock constitutes rapid combustion initiated by self ignition in the wall of the combustion chamber. This causes high frequency, pressure oscillations in the chamber. The frequency of oscillation corresponds to the acoustic resonances of the cavity. The oscillations are mechanically transmitted to the outside of the chamber and can be picked up by a suitable acoustic sensor. The sensed signal is heavily corrupted with background noise injected by the normal mechanical vibrations of the engine. The basic problem solved by the signal processing chip is the discrimination of signal components arising out of any possible engine knock, from the background noise. The frequency of the knock signal can vary with temperature and with other effects such as ageing of the engine. Hence the filtering functions

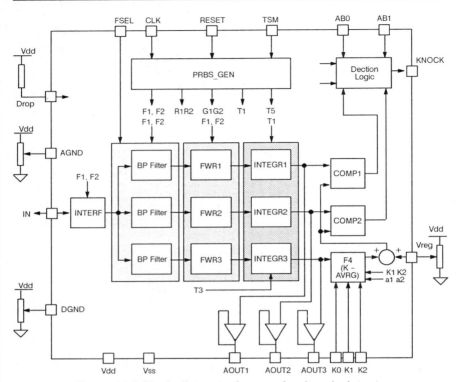

Figure 11.8 Block diagram of an engine knock detector

need to be tunable and adaptive. The use of switched capacitor filters whose frequency responses can be varied by changes in switching frequency allow programmability to be built into the detections system. The chip operates in a hostile environment where induced voltages could easily breakdown the MOS gate capacitances. Hence very special input protection circuits are required.

11.4 Case studies

This section constitutes some case studies of mixed signal ASICs from the telecommunications and automotive sectors of the market.

11.4.1 Mobile radio applications [6]

11.4.1.1. Mobile radio systems

One fertile area of the application of mixed signal ASICs is for the mobile radio market which, to stretch a point, has its origins as far back as 1932, when the New York Police department first used it for the control of dispatch cars. Such facilities have experienced a very rapid growth in recent times. It is estimated that by 1992 there will be 100 million users in Europe alone. Currently a number of different and mutually exclusive analogue cellular mobile radio systems are in operation in different countries in the world. Hence many different designs have been required to meet the non–compatible standards in use. Future developments will be geared towards the use of a common digital cellular system which can handle speech telephony and data transmission, and be compatible with the ISDN or PSTN.

Transceiver units built for any of the systems rely heavily on IC technologies for the complex VLSI digital systems and on low power analogue circuits. Increasing component integration in these portable system enables equipment manufacturers to reduce size, weight, current consumption and cost, while improving overall reliability. For convenience and customer acceptance, portable telephones require the battery, antenna and equipment body to be integrated into a single unit. Hence restrictions on weight and current consumption are severe.

11.4.1.2. System partioning

This case study briefly describes two mixed signal ASICs for use in the NMT (Nordic Mobile Telephone) and TACS (Total Access Communications System, U.K) analogue cellular systems, Firstly, to put their function into context, a brief discussion is provided of the partition of cellular radio systems with respect to IC realisation. Figure 11.9 shows in a much simplified form, the main sections of a mobile telephone transceiver. The radio section contains all the radio circuits. The receiver incorporates double frequency conversion where the frequency of the local oscillator (LO) is controlled by a frequency synthesiser. The synthesiser is itself controlled by a microprocessor. The second LO frequency is determined by a fixed crystal oscillator. A radio signal strength voltage (RSSI), whose strength is proportional to that of the of the received signal is obtained from the receiver. This is effectively measured by an analogue to digital converter (ADC) and used for the base station signal acquisition function.

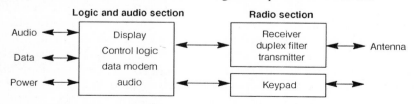

Figure 11.9 Simplified block diagram of a mobile telephone transceiver

The radio section of a transceiver is shown in Figure 11.10. A temperature compensated crystal oscillator TCXO provides the frequency reference for both transmitter and receiver circuits. The transmitted frequency generator consists of a frequency synthesiser and a VCO in a similar arrangement to the receive section. The VCO is modulated by an audio signal from the audio processing section. The correct transmission frequency is set by a micro–processor in the Logic section and is controlled via a digital–to–analogue converter (DAC). The output power from the power amplifier (PA) can be set to different levels depending on commands from the base station. The control circuits incorporate feedback to sense and report the output power level. This level is also controlled via a DAC, by the microprocessor in the logic section.

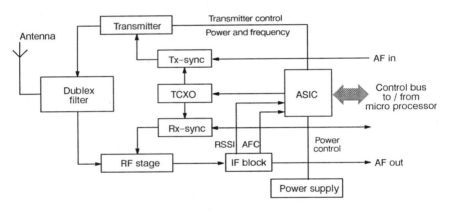

Figure 11.10 Radio section of transceiver

11.4.1.3. Mixed Mode ASIC for radio section control

This section briefly discusses a mixed mode ASIC (referred to as ASIC1), which performs the controlling functions of the radio section of a transceiver (Figure 11.10). This asic performs a number of tasks. It handles the detection and control of transmitted output power ; A/D conversion of receiver signal strength (derived from the 455kHz IF section), frequency pre-scaling for data modem use; automatic frequency control (AFC) based on the signal data and A/D conversion of the battery voltage. It also handles control of voltage regulators (power supply) so that the power consumption of the telephone can be reduced during standby mode, i.e when there is no transmit power or when the telephone is in sleep mode. It also incorporates a DTMF(Dual Tone Multi–frequency) generator and miscellaneous analogue blocks.

This mixed mode ASIC uses serial data to communicate with the microprocessor in the logic section. Figure 11.11 shows a sectional chip layout (chip floor plan). The ADC is multiplexed and implemented as an 8 bit charge redistribution circuit [14]. Both DACS are 8–bit and use the resistor–string/transmission gate principle with outputs buffered for high drive capability. The outputs are capable of driving capacitive loads with levels from 50mV to (VDD –50mV) while maintaining a 0.5 LSB settling time of 10uSec. The Bandgap reference used in this CMOS ASIC for purposes of circuit bias generation and the setting up of the DAC output levels employs lateral npn transistors (parasitic devices) for a low temperature coefficient (less than 30ppm/degC) and an absolute value of 1.26 volts (+/– 10mV). The digital section of ASIC1 consists of interface logic coding/decoding, control for the ADC and DAC, and the DTMF digital control circuitry, and clock generators.

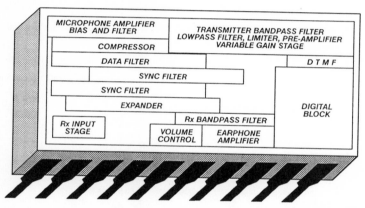

Figure 11.11 Floor plan of an ASIC for the radio section control

11.4.1.4. A mixed mode ASIC for the logic and audio section.

The logic and audio section (Figure 11.9) of the telephone also contains both digital and analogue functions. The digital part controls and supervises transmission on the radio channel by using control circuits in the radio section. This part also handles the keypad, the display and the transmission between the telephone and the switching centre. The analogue part handles audio signals for interfacing to the earphone, microphone and modem. The main logic resides in a microprocessor system with its program stored in EPROM, EEPROM and RAM, and a VLSI chip (ASIC2) of around 6000 gates. ASIC2 has a control interface to ASIC1, through which the radio parameters are controlled. It also has an audio control interface to another mixed mode ASIC, ASIC3. This chip contains the transmit and receive audio filters and amplifiers for speech circuits as well as supervisory audio tone filtering, detection and digital data signal processing. It satisfies the requirements for NMT and TACS standards and is intended for use in several countries. The chip incorporates all the filtering required for the two system standards, and is controlled by a serial data word generated in the local micro-controller. All filters which require controlled cutoff frequencies are implemented using switched capacitor techniques. Where appropriate continuous time anti–aliasing and smoothing filters have been used to band limit the filter input or remove unwanted clock signals.

Figure 11.12 shows a chip floor plan for ASIC3. It is based on an earlier chip designed for analogue cellular telephone handsets. The earlier device contained only RX, TX and SAT filters (anti–alias and Switched Capacitor) and was part of the first generation of mixed signal integration in analogue cellular systems and is likely to represent the last generation of analogue cellular systems as digital standards such as GSM, ADC and JDC will soon replace their analogue counterparts. Nevertheless there will remain the

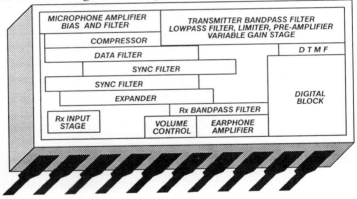

Figure 11.12 Floor plan of an ASIC for the logic and radio section

need for high performance analogue and mixed analogue–digital circuits in the overall system. Indeed mixed signal ASICS are already beginning to appear for the new standards [16]. Some important features of ASIC3:

- Switchable between NMT and TACS under external control.

- DTMF tone generator contained on chip with filters.

- Variable gain stages.

- Audio frequency signal compander

- Earphone driver.

- Supervisory Audio Tone (SAT) filters.

- Data filter for use with modem.

- Power down facility.

- Microphone amplifier and bias circuit.

The choice of TACS or NMT mode is made under external control. This determines the filter responses and signal levels throughout the IC. ASIC3 also includes a DTMF tone generator which covers the standard range of tones recommended by the CCITT. The tones are digitally generated, and are composed of 8 voltage levels with 14 steps/cycle. The different signal levels required are provided using digitally controlled variable gain stages. These cells are used to select the appropriate levels for NMT/TACS operation and for user defined functions such as volume control. These variable gain stages are realised using either switched resistors (continuous time), or switched capacitors (sampled data).

A signal compander has been included to improve noise performance in the transmit and receive channels. Signals in the transmit path are processed by the compressor before being filtered prior to transmission. The received signals are first filtered, and then expanded before being amplified in the audio output stage. The 2:1 compressor gives a 1dB change in signal level for every 2dB change in input level. The unaffected level is 40mV for NMT and 24mV for TACS.

The compander included on this ASIC is of a novel design using delta–sigma oversampling and switched capacitor techniques for attack/decay time constants, full wave rectification and filters used within the expander and compressor blocks. Using this approach it is possible to use MOSFET devices throughout despite their inferiority to bipolar components in terms of poor transistor threshold voltage matching and limited driving capability. The compressor consists of a multiplier to perform the automatic gain control function. The gain of the multiplier is controlled by the rms value of the input signal to provide signal compression. The analogue input signal is first coded into a string of digital pulses whose density reflects the input signal amplitude. The multiplication of an analogue signal with such a string is then achieved by using the digital string to control a switch gating the analogue signal. The resulting pulse amplitude modulated signal is then converted back to continuous time form by a low pass filter and this is the product of the two inputs. In this implementation, a second order integrator delta–sigma coder is used [15] in conjunction with a switched capacitor third order low pass filter and a very efficient SC rectifier.

The audio output circuit of ASIC3 includes a volume control and a circuit to drive an external earphone. The earphone type can be dynamic or piezo–electric and is controlled from a serial data word. The piezo–electric load can be driven to 2(Vdd–Vss–1) Volts into a 2k2 Ohms in series with 50nF. This is achieved by using two amplifiers in a bridge configuration. Alternatively, the circuit can drive a dynamic load to within 0.2 Volys of the rails, into a load of 180 Ohms and 100 pF. In this configuration, one of the bridge amolifiers is powered down, and a single amplifier drives the load.

ASIC3 includes filters to process both receive and transmit SAT (Supervisory Audio Tone) signals. The transmit SAT signal is generated externally and filtered before being summed into the transmit path. The received SAT signal is also filtered before being converted to a digital signal for synchronising received data. These circuits are 4th. order bandpass filters. The centre frequency of the transmit filter is 6kHz. The receive filter has a switchable centre frequency of 4kHz or 6kHz. The receive filter is followed by a switched capacitor Schmitt trigger circuit to produce a digital output.

A low pass filter has been included in ASIC3 for filtering signals generated by an external modem. The response of this filter can be adjusted, under external control to give a cutoff frequency of 4kHz or 20kHz.

This ASIC was realised in a 3um double poly single–metal CMOS technology and used a mixture of full custom and standard cell compilation techniques. The chip size is approximately 6.5mm x 6.5mm and the device is packaged in a low profile QFP. As power consumption is of prime importance there are complex power down facilities on this chip. Under external control, most parts of the circuit may be powered down individually, when not required. Total power consumption with everything working from the nominal 5V supply at 25°C is less than 10mA and in power down this drops to 0.5mA.

11.4.2 An actuator example [9]

This section describes a mixed analogue digital chip, whose function is to perform signal processing and direct driving of a resistive actuator. The application was initially developed for the automotive market, but could be applied more generally.

11.4.2.1. Introduction

In many applications, in addition to the analogue "sensed" input signals, the output drive or actuator signals are also required to be in analogue form. Nevertheless it is desirable to perform much of the processing of the signal in digital form thus exploiting the flexibility and noise immunity advantages of digital signal processing. Figure 11.13 shows a generic block diagram for such an arrangement. In this processing chain the analogue interface is the most critical, as typically it is quite sensitive to noise and electromagnetic interference. The general trend is to use an oversampling approach [7] which can be further enhanced by use of noise shaping techniques. Signal oversampling spreads the power of the noise over a wider spectrum, so that a lower amount of noise power falls inside the signal bandwidth. At the other end of this generic signal processing chain, the D/A converter should deliver an output signal with enough power to drive the actuator directly.

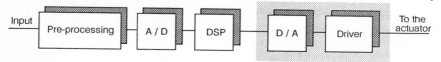

Figure 11.13 Generic block diagram for a signal processing chain

This would enhance system reliability and reduce cost. In many applications the output signal is not to be used to drive a machine member but is simply to be used for driving a display instrument. Nevertheless analogue displays using mechanical pointers are often preferred to digital displays.

11.4.2.2. System Description

The example described here is a CMOS IC dedicated to the processing and direct driving of resistive actuators. Specifically, a system for driving the crossed coils of a measurement and indication instrument, a tachometer is considered. The display section of a mechanical tachometer is made up of two fixed, crossed, coils in a mutually orthogonal arrangement. The display pointer is fixed to a permanent rotating magnet placed in the area common to the coils. The currents flowing through the coils give rise to two orthogonal strength components acting on the magnet, thus determining the angular deviation of the pointer. Normally the maximum required deviation of the pointer is 270 degrees with a resolution of 1 degree. This implies the requirement of 8 bit accuracy.

The schematic diagram of the proposed system is shown in Figure 11.14. The input signal consists of a train of pulses generated by a magnetic sensor which detects the rotation of a transmission shaft. The time interval between two successive pulses carries the instantaneous speed information. The interval is measured, and after processing, the result is displayed by means of the crossed–coil instrument. The use of digital processing allows

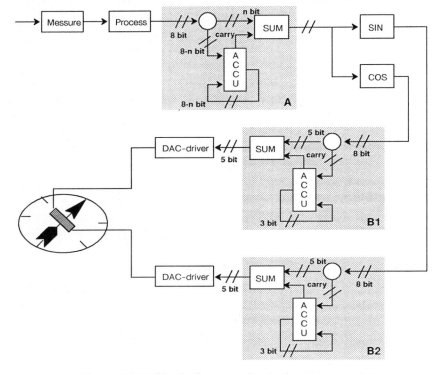

Figure 11.14 Block diagram of a tachometer system

noise immunity and accuracy to be enhanced. It also facilitates the convenient implementation of additional features such as jitter prevention of the pointer in the case of a constant or slowly changing speed. This is achieved by introducing hysteresis. The process of updating the display can also be provided with damping to avoid sudden variations in the deviation of the pointer in response to fast changes in the measured speed. The amount of damping can be made variable according to whether speed is increasing or decreasing.

To obtain an angular deviation directly proportional to the value of the displayed variable, the two orthogonal components of the strength acting on the rotating magnet of the instrument have to be proportional to the sine and the cosine of the desired angular deviation . The digital word containing the speed value is transcoded according to sine and cosine laws to generate the analogue currents which are fed to the two coils. The transcoding is performed in the DSP domain by means of two ROMs. The use of ROMs gives a higher degree of flexibility in the transcoding law to be employed, and allows easy changes to the design when the application is to be directed at a different system requirement. For example for the implementation of a fuel level indicating instrument, the transcoding law can be determined in accordance with the geometrical shape of the fuel tank.

11.4.2.3. Implementation

A straight implementation of the system would have required, 8 bit digital signal processing, i.e an 8 bit input and output for the transcoding ROMs and an 8 bit output from the D/A. In fact, to reduce the amount of silicon required, the digital transcoding is performed with n bits, where $n < 8$. This is done while still maintaining an 8–bit output from the ROMs. The required resolution is obtained in the transcoder by use of a linear interpolation technique performed at an oversampling rate (Figure 11.14). Only the n most significant bits of the digital word are used to perform transcoding. The 8–n least significant bits are iteratively summed at the oversampling rate by means of an accumulator. When a carry is generated, it is summed with the n–bit word which is in the process of being transcoded. This is then fed to the ROM input. The error introduced by this linear interpolation technique is small. For example for $n = 4$ and $n = 5$, the respective errors are $= \pm 0.5\%$ and $\pm 0.15\%$ of the full–scale value. This level of error is acceptable since it is smaller or comparable to the quantisation error of straight 8 bit conversion, while using significantly less silicon area.

To reduce the complexity of the output D/A converters, an oversampling approach is followed, as indicated in blocks B1 and B2 in Figure 11.14. Oversampling is implemented with a circuit identical to the one used to perform the linear interpolation of the word to be transcoded. This instance of the oversampling approach can be regarded as a first order sigma–delta modulation [7] and results in a lower resolution requirement for the D/A converter. Current switched converters [8] are the most suitable DACs for driving current–controlled actuators such as coils. In these converters the output variable is a current proportional to a fixed unity–current source Iu. The proportionality factor is given by the digital value to be converted and the output current is physically obtained by means of a matched replica of the elementary source. Figure 11.15 shows the circuit diagram of the current sink section of the CMOS D/A converter developed for the proposed ASIC [9]. The output current sources are binary weighted replicas of a unity current source Iu. Further implementation details and discussion can be found in reference [9].

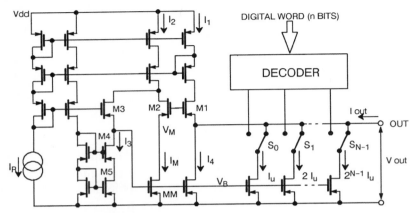

Figure 11.15 Output current sink section of the D/A converter

11.4.3 A receiver IC for a digital transmission system [4]

This application illustrates some architectural and circuit ideas which can profitably be employed in mixed signal ASICs. In particular the chip illustrates the power and versatility of sampled data analogue signal processing when combined with digital circuitry used for control and adaptability. Apart from the conventional filtering functions, switched capacitor techniques were used in a variety of analogue signal processing roles.

11.4.3.1. System architecture

The basic function of the chip is to perform a pcm receive function for a 96kbit/Sec digital channel. The digital signal is superimposed on the conventional analogue voice channel between a telephone subscriber and a local exchange. Thus, this so called "1 + 1 system", provides a second channel for use either as an additional audio circuit or for data transmission. Such a system uses the existing subscriber line, while only adding equipment at the exchange and subscriber ends,(Figure 11.16).

The receiver chip is part of the Line Termination Equipment (LTE). A WAL2 line code is used to code the digital 1 and 0 signals into an analogue waveform for transmission over the telephone line between exchange and subscriber. The definition of WAL2 can be understood by considering Figure 11.17a. This particular code is selected because it has low spectral energy in the audio band which is important to avoid interference with the

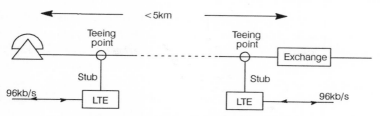

Figure 11.16 System configuration

co–channel audio signal. Figure 11.17b shows a typical eye diagram for a WAL2 line code. An eye diagram is a standard method of displaying signals in a digital transmission system. It essentially involves limiting the horizontal axis to the time interval of one digital symbol (one transmitted bit), and repeatedly superimposing the received signal in this interval. It is very useful for indicating the degree of inter-symbol interference and noise in a digital communication channel. In the case of a WAL2 code the system is confronted with a true eye corresponding to the the central part of each symbol (whether representing 1 or 0) and a false eye representing the two side lobes of the symbol.

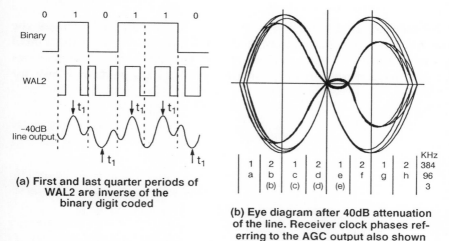

(a) First and last quarter periods of WAL2 are inverse of the binary digit coded

(b) Eye diagram after 40dB attenuation of the line. Receiver clock phases referring to the AGC output also shown

Figure 11.17 WAL2 coding used in the system

The LTE simultaneously transmits and receives. So the input signal to the receiver is a combination of the desired received signal with an unwanted coupling (echo) from the transmitted signal at the near end. Although a conventional "hybrid" transformer is used to help separate these two signal components, the action of this device is less than perfect. Consequently at the input to the receiver, the unwanted transmitted signal can be as much as

30db larger than the wanted received signal. To overcome this problem echo cancellation is employed. The LTE unit contains a digital echo cancellation chip in addition to the receiver chip. The echo canceller takes 4 samples per bit period (384 Ksamples/s) from both the scrambled transmitted data stream and the amplified composite signal (signal + echo) input data stream. The echo canceller processes these signals and the output is an accurate reconstruction of the unwanted echo which is subsequently subtracted from the composite signal to leave only the wanted component. The echo canceller employs purely digital processing whereas the receiver uses sampled data analogue, implemented in the form of switched capacitor (SC) circuits, together with digital circuits extensively used to control and regulate the sampled data analogue functions. Some examples of these will be described presently. The interface from the receiver to the echo canceler is via A/D and D/A converters. The receiver chip was developed for an experimental system and hence separate chips were developed for the echo cancellation functions. A further development would have endeavoured to combine these functions on the same chip.

11.4.3.2. Receiver architecture

The receiver in conjunction with the echo canceller regenerates the incoming data transmitted from the far end of the line. A block diagram of the receiver chip is shown in Figure 11.18. The signal processing functions of the receiver are performed in the sampled data domain. Hence for

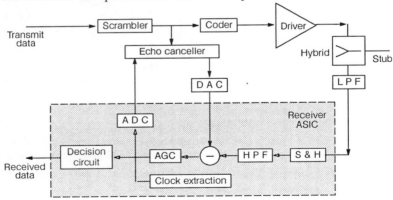

Figure 11.18 Block diagram of line interface unit

purposes of anti–aliasing, the incoming signal has to be band-limited by the use of a continuous time low pass filter, which also helps limit wide band noise. The sample and hold takes four samples per bit period All four samples are used by the echo canceler. When the samples are taken at instants t1 – t4 (Figure 11.18b) then the sample at t1 is from the most open part of the "true eye" and can subsequently be used by the decision circuit to determine the polarity of the data bit.

The sampled data stream is first high pass filtered to remove any residual audio band signal and then the echo replica generated by the echo canceler is subtracted from the composite incoming signal. The signal is then restored to a predetermined level by the automatic gain control (AGC) prior to decision making, A–D conversion and clock extraction. The ADC provides the necessary input to the echo canceler.

11.4.3.3. Automatic gain control

Figure 11.19 shows the general arrangement of the AGC system comprising analogue and digital parts. The AGC is required to provide a stable output level despite a variation of up to 40dB, introduced by the subscriber line. The configuration incorporates a three stage forward gain path, a switched capacitor (SC) level detector and digital control circuit. The level detector examines the signal samples at the end of the forward path over 32 consecutive bit periods and if the peak value of this set is less (more) than the reference voltage than the counters in the digital control are made to increment (decrement) and so increase (decrease) the forward gain. The gain, step size and the control operating frequency of 3Khz determine the AGC's convergence time and immunity to impulsive noise.

The forward gain path employs three cascaded SC gain setting stages to cover the required 40–dB range. The stages operate at 384kHz and the gain is determined by the setting of the programmable binary weighted capacitor arrays. The first two stages give a combined gain setting of 1, 2, 4,...., 64 while the third stage gives a finer gain setting in the range of 1–2 in 32 increments. Hence an overall gain ranging from 1 to 128 (43dB) with a maximum increment of 3 percent is achieved.

The first stage input incorporates a sub-tractor which suppresses the echo. The final stage is configured as a switchable gain first order high–pass filter filter with a cut off frequency of 6kHz. The function of the filtering is to remove accumulated offset voltages from the first two sections. The AGC control logic comprises a 5–bit up/down counter which controls switch states S7–S11, counter control logic and a 3–bit counter and decoder which performs like a reversible shift register and which after retiming controls switch states S1–S6. The counter control logic latches the UP control from the level detector, checks for an end–stop state (i.e all 0's or 1's), and steps the counters accordingly. If an end–stop state is inadvertently reached (as indicated by S1 or S6) then the counter control logic prevents anomalous counting. The multi-phase 3kHz clock (b),(c), (d), and (e) (Figure 11.19) is needed to ensue that the output states S1–S11 only change on the reserved phases of the gain setting stages. Simulated waveforms of the data regeneration path , from the receiver input, through the HPF, subtractor, three AGC gain stages, to the decision circuit, are shown in Figure 11.20

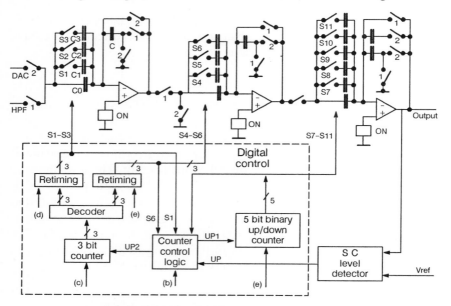

Figure 11.19 Automatic gain control (AGC) system

11.4.3.4. Clock extraction

Clock extraction is a pervasive problem of digital transmission systems. The problem is essentially that of generating local clocks to lock in phase and frequency to the clock used to time the line signals transmitted at the far end of the communication link. In this system the function of the clock extraction circuit is to ensure that the sample used for making a decision as to whether a 1 or 0 was transmitted be taken at an instant close to the centre of the true eye, i.e at t1 (Figure 11.17b). This is achieved by aligning t2 to a zero crossing using a phase–locked loop.

The situation where the phase–locked loop aligns the clock a 1/2 bit period later to the zero crossing of the false eye, is detected by an ambiguity detector . This enables the timing to jump by a 1/2 bit period to the correct condition. The SC clock extraction system is shown in block diagram form in Figure 11.21. The lower half of the circuit consists of a phase sensitive

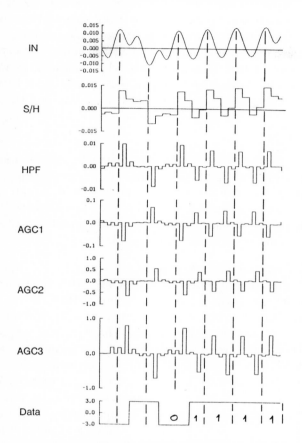

**Figure 11.20 Simulated waveform from receiver input
to the output of the decision circuit**

detector, a loop filter, and a sample and hold circuit which drives an off chip
voltage–controlled–oscillator (VCO).

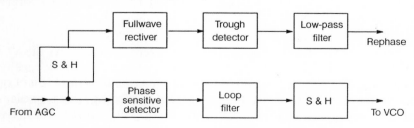

Figure 11.21 Digitally controlled SC clock extraction system

The top half the circuit consists of a cascade of a full wave rectifier, a trough detector and rephase circuit. The rectifier converts the (amplified) signal into a unipolar form. The trough detector, detects whether the system has locked to the false eye, and if it has the rephase circuit issues a command to make the decision circuit jump by half a bit period and hence lock to the centre of the true eye.

11.4.3.5. Simulation

To ensure a "right first time" outcome, extensive simulations were performed. The choice of digital simulation, circuit level simulation and switched capacitor frequency domain simulation was useful for the verification of the design of the relevant building blocks. However because of the size of the overall circuit, these conventional simulators offered no solution to the general system simulation problem.

Sampled data analogue circuits are inherently "event driven". That is to say although signal values from a continuous range need to be considered, the essential signal changes occur at discrete events in time. In the case of switched capacitor and clocked digital circuits, these correspond to the appropriate edges of various clock signals. This fact was employed to develop an ad-hoc program which described the individual switched capacitor functions in terms of transfer functions expressed as difference equations. Similarly conventional high–level behavioural descriptions were written for the digital blocks. The entire chip was described in this manner using a hierarchical, modular approach. This inherently made use of the event driven and latency principles. That is, the various parts of the chip were only evaluated when the relevant switching signal was active. Latent parts of the circuit were not (unnecessarily) evaluated. A further description can be found in reference [5]

The power of this approach and the results of a system level simulation of the entire chip, over 200,000 cycles of the basic clock frequency (384kHz) are shown in Figure 11.22. This required only 4 minutes of CPU time on an IBM 370/4381 computer. The diagram shows the value of the AGC gain as it counted up by the digital control circuit from zero. It also shows the VCO control voltage as it successfully locks to the correct phase of the received signal. That is it simulates and confirms the operation of the Phase Lock Loop. Figure 11.23 shows the behaviour of the system in the presence of impulsive noise. The AGC control decreases the value of the overall gain and the phase locked loop looses lock. However the system recovers after the noise pulse has disappeared.

Some Applications of Mixed Signal ASICs

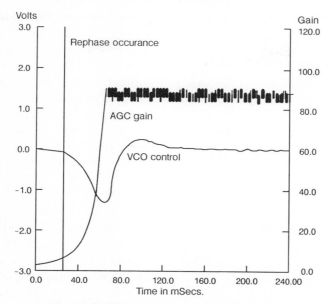

Figure 11.22 Results of system simulation

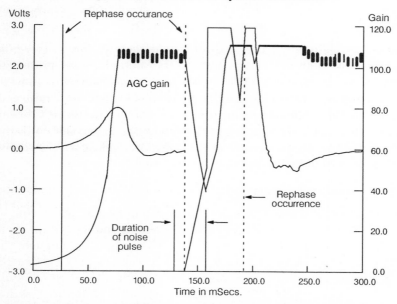

Figure 11.23 Results of system simulation (in presence of noise pulse)

The simulation program allows verification of the basic functionality of the system and provides various measures of system performance such as the recovery time after loosing phase and immunity or otherwise to impulsive noise etc.

11.4.4 A programmable tone receiver and generator. [10]

This section describes an integrated full–duplex tone receiver and generator used for purposes of signaling in mobile radio communication systems. The chip constitutes a very flexible mixed analogue digital single chip solution to this requirement. The resonant frequency, the Q factor of the receiver filter and the tone frequency are all digitally programmable. The analogue parts of the system are realised using sample–data switched capacitor (SC) circuits.

11.4.4.1. The application

Modern communications systems make extensive use of tones in the audio and sub-audio bands for signalling functions. In typically applications e.g in a DTMF a small number of different frequencies are required. However there is an important class of specification where the required tone frequencies are different from one application to another. One example of this is the situation where the same ASIC has been designed for multi–standard use. These applications demand tone receivers and generators with widely programmable parameters.

An important feature of a tone receiver is its selectivity. This is determined by a trade–off between the minimum transmitted signal power, the internally generated noise, and the maximum allowable detection time. For instance a high selectivity increases the signal–to–channel–noise ratio at the output to the filter, but at the same time implies longer detection times and higher receiver noise. The basic specification on the tone generator is that, the tones must be programmable at the centre frequencies of the receiver and be as free from higher harmonic components as possible.

11.4.4.2. System Description

Figure 11.24 shows a block diagram of the overall system, which contains two sections; one for the analogue signal processing and one for the digital signal processing. In the analogue section the signal is applied to a sampled–data band-pass filter realised with a pseudo N–path SC structure [11,12]. The pseudo N–path operation introduces pass-bands at DC and at even multiple integers of the fundamental frequency fs/N. To suppress these pass bands a notch filter is cascaded with the bandpass filter.

Tone detection is accomplished by the tone detector block (Figure 11.24) formed by a full wave rectifier, low–pass filter and a Schmitt Trigger. On the

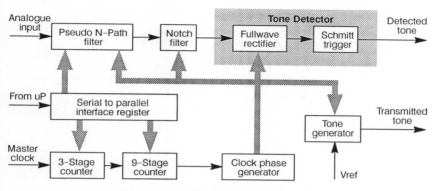

Figure 11.24 System block diagram

other hand the transmitted tone is generated within a block which employs the same clock phases as those used in the receiver pseudo N–path filter. The tone generator synthesises a sampled pseudo sine wave signal with N samples per period.

The digital section contains two programmable counters, a clock phase generator and a 17 bit shift register. The first 12 bits stored in the shift register control the two counters and establish the centre frequency of the receiver filter. The other 5 bits are used to program the quality factor "Q" of the N path filter.

11.4.4.3. Implementation

The pseudo N path SC filter is shown on Figure 11.25. The transfer function of this filter, in the Z–domain is given in expression 11.1.

$$H(z) = \frac{C_1/C_3}{z^8(1 + C_2/C_3) - 1}$$

$$\boxed{11.1}$$

It can be shown that the frequency response of the pseudo N path SC filter is a periodic function with a period equal to one–eighth of the sampling frequency fs. As a consequence of this characteristic the tone signalling frequency would be inconveniently limited because the second pass band would lie inside the operating frequency range. To solve this problem, a bandpass pre–filter of at least fourth order, and with a programmable centre frequency at the fundamental resonant frequency would have been needed. Such a solution would have been expensive. A less expensive alternative was to use an N–path FIR (Finite Impulse Response) filter shown in Figure 11.26. The relevant Z–domain transfer function is shown in expression 11.2. In this filter the frequency response is zero at $f_i = 2if_s/8$ $(1 = 0,1,2,..)$

Some Applications of Mixed Signal ASICs

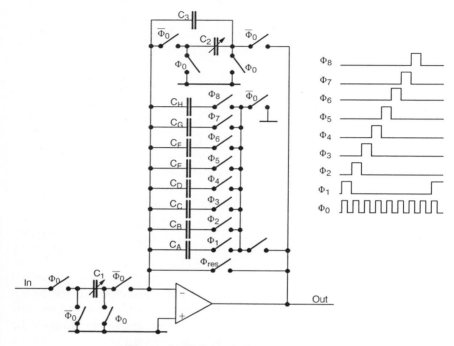

Figure 11.25 Pseudo N–path SC filter

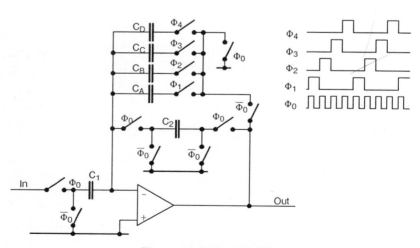

Figure 11.26 N–path FIR filter

$$H(z) = \frac{C_1}{C_2}(1 - z^{-4})$$

11.2

The last block in the receiver section is the tone detector which is shown in Figure 11.27. This is formed by a low pass filtered full–wave rectifier cascaded with a continuous time Schmitt trigger.

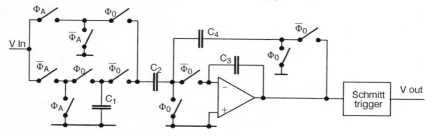

Figure 11.27 SC tone detector

The tone generator produces an eight sample pseudo sine wave signal at the same frequency as the one to which the receiver is tuned. In general sine wave signals are derived from periodic square waves or SC oscillators followed by smoothing filters. A more efficient way to do this in a mixed signal IC is to use a digitally controlled amplifier. In the case of the design under discussion the availability of the eight phases generated for the N–path filter led to the simple solution shown in Figure 11.28. The capacitors C1 and C2 are equal to:

$$\sqrt{2}/2.C \quad \text{and} \quad (1 - \sqrt{2}/2).C$$

respectively, while C3 is equal to C.

The operation of the circuit is as follows. During time slot 1, Φa, Φb and Φd are activated. Therefore the op–amp is in buffer configuration and C1 and C2 are charged to the difference between the op–amp offset voltage Vos and a reference Voltage Vref and C3 is charged to Vos. During time slot 2, C3 acts as a feedback capacitor and C1 injects its charge into the virtual ground. Therefore the first non–zero output sample takes the value

$$\sqrt{2}/2.V_{ref}$$

By following the other time slots in the same scheme shown in Figure 11.28, it is possible to see that the eight values of a pseudo–sine–wave signal with an amplitude of Vref are generated.

The Digital section can be divided into three blocks: a frequency scaler, a clock phase generator and a serial–to–parallel interface register figure 11.29. The frequency scaler is made up of two programmable synchronous counters, with three and nine stages. The three stage counter carries out a prescaler operation for the nine stage counter. Both counters

Some Applications of Mixed Signal ASICs

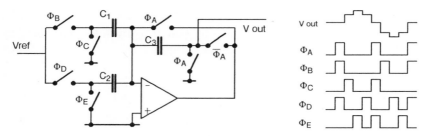

Figure 11.28 SC tone generator

are used when the tone receiver/generator operates in the sub-audio frequency range, while only the nine stage counter is activated if the tones are generated in the audio frequency band. The nine stage counter alone divides the master clock by a factor selected in the range 4–511. The clock phase generator controls the analogue circuit switches. It is a sequential synchronous network composed of two four stage Johnson counters driven by the main clock frequency fs.

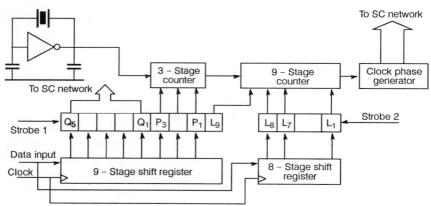

Figure 11.29 Digital section

The last block performs the function of interfacing with the external system and programming the parameters of the tone receiver and generator. In order to reduce the number of pins and hence the external connections to the chip, serial programming was employed at the expense of more complicated microprocessor software. The programming binary word is composed of 17 bits. Of these five establish the Q values of the N–path filter and the other 12 bits define the ratio factor of the counters. These bits are serially charged into two parallel shift registers, of nine and eight stages. They are then transferred into 17 latch stages controlled by two strobe signals. This loading procedure makes the microprocessor operation easier since only the eight least significant bits , which determine the receiver/generator tone frequency, usually need to be changed. The chip was integrated in a 3μm double–poly CMOS process and occupied a total area of 18μm^2.

11.5 Conclusions

This chapter has given brief descriptions of several mixed analogue–digital chips from various segments of the market. It has also considered cases studies of four such chips from the telecommunications and automotive application areas. As stated earlier mixed signal chips are aimed at most segments of the market and as the art of designing these systems develops, the reader find many more examples appearing in the technical literature.

11.6 References

[1]   Garverick S. L, McGrath D. T, Baerisch R. D and Fujino K. " A programmable Mixed Signal ASIC for power metering", IEE, Int. Solid State Ccts. Conf., 1991.

[2]   Volder, J. E "The Cordic trignometric computing technique", IRE. Trans. on Elec. Comp., Vol.EC–8, No. 3, pp.330–334, Sept. 1959

[3]   Ghisio G, Maloberti F, Polito G, and Salerno F, " Knock detection system for spark ignition engines", XXIII FISTIA Congress – the promise of new technology in the automotive industry; Turin, Italy, 7–11 May, 1990

[4]   Hughes J. B, Bird N. C and Soin R. S, " A receiver IC for a 1 + 1 Digital Subscriber Loop", IEEE, Journal of Solid–State Circuits, Vol. SC–20, No. 3, June 1985, pp.671–678.

[5]   Soin R. S, Hughes J.B and Bird N. C, " A Novel approach to the time domain simulation of switched capacitor systems.", Proc. Europ. Conf. Cct. Theory and Design,(ECCTD'85), Prague 1985, pp.713–716.

[6]   Morris S. J, " Mixed mode ASICs for mobile radio systems", Silicon Design, Nov. 1990 (to be published in J. of semi–custom ICs).

[7]   Boser B. E, and Wooley B. A, "The design of Sigma Delta Modulation analogue to digital converters", IEEE J. of Solid State Ccts., Vol. SC–23, No. 6, pp.1298–1308, Dec. 1988

[8]   Sheingold D. H (Ed.), " Analogue–Digital Conversion Handbook", Prentice–Hall, Englewood Cliffs, NJ, pp. 199–206, 1986.

[9] Maloberti F, Tagilani F, Torelli G, "Design of DAC–Drivers for Resistive Actuators.", IEEE, Proc. Int. Symp. Ccts. and Sys. ISCAS, New Orleans, April 1990.

[10] Chiappano G, Colamonico A, Donati M, Maloberti F, " A tunable Switched–Capacitor programmable N–path tone receiver and generator.", IEEE J. Solid State Ccts., Vol.23, No.6, Dec. 1988, pp.1418–1425

[11] Ghaderi M. N, Nossek J. A, and Temes G. C, "Narrow–band switched capacitor bandpass filters.", IEEE Trans. Ccts. and Sys., Vol. CAS–29, pp.556–571, 1982.

[12] Gregorian R, and Temes G. C, "Analogue Mos Integrated Circuits for Signal processing", Pub. John Wiley, New York, 1986, pp.363–387.

[13] Vry M. G and Gerwin P. J, "The design of a 1 + 1 system for digital transmission to the subscriber.", IEE Int. Symp. Subscriber Loops & Services, 1980, p. 36.

[14] Morris S. J and Hague G, "Application–orientated data converters for a CMOS cell library.", Jou. of Semicustom ICs, Vol.7, No. 4, 1990.

[15] Candy J. C , "The use of double integration in sigma–delta modulation.", IEEE Trans. Comms., March 1985.

A Video Communication Application of Analogue - Digital ASICs

P. Senn
CNET Grenoble, France

12.1 Introduction

In the field of telecommunications the development of VLSI technology has allowed the introduction of new services, new systems and more generally a better quality of service to the user. Many of these improvements have been made possible through the development and deployment of digital techniques in place of the more classical analogue techniques for the main signal processing functions. However the signal to and from the end user still has to be in analogue form. Consequently, this transformation in the methods has resulted in an increasing need for fast high performance interfaces between the analogue and digital worlds. There is a new challenge facing VLSI technologies in dealing with these new requirements and allowing complex mixed analogue and digital designs on the same chips. Ultimately the embedding of analogue and digital blocks on the same die may have an economic imperative, but the immediate goal for the design engineer is an improvement in the global function at the circuit level or even at the system level.

This chapter will describe a major example of a mixed analogue–digital circuit in the field of video communications. The first approach will demonstrate that the embedding of three D/A converters in a video processor has economic advantages such as lower packaging costs for the same test complexity as compared with a two chip approach and system benefits such as lower electromagnetic perturbations due to a drastic reduction in the number of external digital buses.

12.2 TV and HDTV applications

New image services and future broadcasting systems such, as the satellite D2-MAC/packet [1] system, currently use digital coding and decoding of the video signal. Digital coding is often necessary to implement complex algorithms in order to reduce bit rate the of video transmitted signal. This concerns all the new image services to be introduced in the ISDN (Integrated Services Digital Network). Digital coding may also be used to improve the picture quality of new broadcasting systems. In the European D2-MAC/packet system (MAC for Multiplexed analogue Components), the picture quality improvement methods adopted, induce the need for a digital processor in the decoder, when an analogue signal is transmitted.

High Definition Television (HDTV) has become a major event in the television arena since the early 1980's. New concepts for production and transmission are being proposed and tested. New standards have already been adopted and it seems obvious that the next challenge for the 1990's will concern the availability of efficient solid state technologies to be able to launch this HDTV concept on the consumer market.

The HDTV signal format adopted in Europe has been established on the basis of the condition that HDTV broadcasting has to be compatible with the MAC/packet family. HDMAC is the name given to the coding system used to broadcast high definition television pictures through conventional channels, based on MAC/packet transmission standards. Table 1 gives the maximum data transmission rate for different standards, starting from the well known 601 CCIR recommendation for todays studio requirement, to future European HDTV production standard. Intermediate standards, such as HDI and HDQ will be used for broadcasting the HDI format will be used for HD-MAC broadcasting [2].

Table 12.1 Maximum data transmission rates for different standards

(Fc = 1/2 Fy)	601 Recomm.	HDI	HDQ	HDP
	625/50/2:1 4/3 16/9	1250/50/2:1 "16/9" 16/9	1250/50/1:1 "16/9" 16/9	1250/50/1:1 "16/9" 16/9
Line frequency	15,625Hz	31,250Hz	62,500Hz	62,500Hz
Pixel frequency	13.5MHz	54–72MHz	54–72MHz	108–144MHz
Samples/line	864	1728–2304	864–1152	1728–2304
Pixels/active line	720	1440–1920	720–960	1440–1920
Instantaneous transm. rate (8 bit/sample	216Mb/Sec	864–1152 Mb/Sec	864–1152 Mb/Sec	1728 –2304 Mb/Sec

It is interesting to note than the bandwidth requirements for HD broadcasting are too large as compared to channel bandwidth used in cable networks or satellite channels, which is lower than 12MHz. As a consequence, mixed-mode broadcasting has been adopted in order to fulfill these channel bandwidth limitations. After a digital time compression, video signals are transmitted in an analogue form, and the sound, synchronization and, data signals in a digital form (duo-binnary form).

The next section will discuss a proposed solution to improve the quality of video restitution in the last stage of a digital TV.

12.3 Concepts for the restitution of video signals

In these future receivers, at the output the processing unit, the video signal will be composed of the luminance component Y and the two colour difference components CR and CB at the output processing unit. The received video signal will in general be digitally coded in accordance with the specifications given in CCIR recommendation 601 which describes digital video studio coding (see table 1). The luminance signal is sampled at a 12.5MHz frequency and the two chrominance signals at 6.75MHz. The three components are coded with 8-bit words. This sampling frequency is equivalent to 702 points per active line in 625-line television systems.

In some applications, i.e. with a very reduced transmission rate such as videotex services, the number of points per line and lines per image may be reduced. The decoder used for the restitution of a digital video signal has to be adapted to a large range of sampling frequencies or equivalent points per line [2].

In a "conventional" decoder, RGB (Red, Green and Blue) are usually reconstructed after cascading three functions: (1) three digital to analogue converters for Y, CR and CB, (2) three analogue filters for anti-aliasing , i.e. eliminating the duplicated spectra introduced by digitization and (3) de-matrixing to obtain R, G and B signals.

To achieve this transformation properly, in accordance with the CCIR 601 recommendation , expensive analogue adjustments are needed. Filters must have a linear phase and a flat pass-band response within the frequency bands (5.5MHz for luminance and 2MHz for CR and CB). In addition, operational amplifiers in this frequency range are expensive. In order to maintain the flexibility discussed previously, the decoder must accept a lower rate input data stream [3]. This means that all the analogue processing has to fit the input sampling rate. Switched capacitor filters are a possible way of implementing the processing part of such a decoder. Figure 12.1 shows a classical architecture using three D/A converters, switched capacitor low pass filters and an analogue de-matrixing block. Figure 12.1 gives the time and also the frequency information concerning the different parts of such decoder.

In order to fulfill the video bandwidth requirement, which is about 6MHz for 601 CCIR recommendation, a number of key points have to be adopted in the design of the filters:

- A very low-sensitive LDI circuit has to be used with an exact synthesis procedure to take into consideration the low ratio between the sampling rate and the input signal bandwidth.

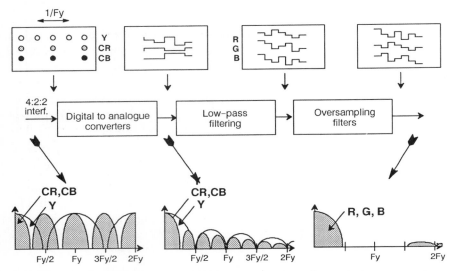

Figure 12.1 Analogue video restitution – Correspondance between number of pixels per line and filter bandwidth

- A double sampling scheme has to be adopted. This technique enables the designer to halve the sampling rate and increase the time slot in which the amplifiers must settle.

- Fast-settling amplifier design.

- Constant group delay in the response.

The constant group delay constraint is the most severe. The classical technique suggests the use of a usual amplitude approximated filter implemented in the leap-frog ladder structure cascaded with all-pass biquad sections. This solution is not efficient and destroys the low-sensitivity obtained by the leap-frog structure when the number of all-pass sections exceeds two. An efficient and original solution has been proposed by including an additional cell in a low-pass elliptic ladder filter [4].

Nevertheless, although some circuits have been fabricated which demonstrate the feasibility of such an approach, it seems more interesting (from an economic point of view) to propose an alternative solution, which uses a more complex digital section and a very simple analogue one. This proposition will be developed in the next section.

12.4 Digital to analogue video processor

Indeed, these functions can be advantageously replaced by digital equivalents (see Figure 12.2). The results to be presented here [6], concern a

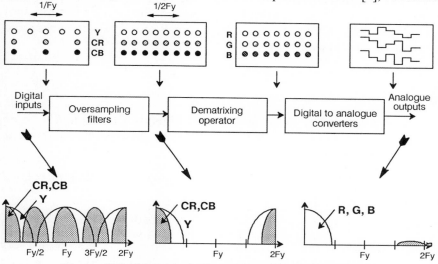

Figure 12.2 Video restitution using digital oversampling

new generation of video processor that combines a more complex and faster digital part with analogue interfaces. The circuit to be presented in the next section complies with the new MAC video specifications. All of the digital and analogue functions have been implemented in a 38mm^2 CMOS chip.

The digital filters perform an interpolation by two for Y and by four for CR and CB. This operation is equivalent to computing one intermediate point between each incoming luminance sample, and three points for the two chrominance signals. In all applications according to 601 CCIR recommendation, such as in a MAC receiver, the output sampling rate for Y, CR and CB becomes 27MHz or equivalent to 1404 pixels per active line in 625-line television systems. After filtering, a matrix operator performs the digital transformation from Y, CR and CB components to R, G and B components. Finally, these components are converted into analogue form by three identical digital to analogue converters. Due to the high output sampling rate, no post filters are required between the converters and final amplifiers.

12.4.1 Digital part

The digital portion consists of two processors. The first increases the sampling rates of the luminance and chrominance input flows. The second is a video matrix operator, which performs conversion of the oversampled flows into the red, green and blue components.

As TV signals are very sensitive to phase shifting, only non-recursive phase linear filters are used. Luminance processing requires a higher performance level, therefore different data paths are used for luminance and chrominance processing.

The circuit is composed of seven filters:

(a) one 19th. order luminance filter,

(b) four different 7th order chrominance filters for doubling the input chrominance sampling rates. The CR and CB are multiplexed on the same data path,

(c) two second order filters for computing the last interpolated chrominance samples.

12.4.1.1 Luminance filter

The architecture of the filter is presented in Figure 12.3. A structure very close to the so-called transposed direct structure of digital non recursive filters is used.

The selection of the coefficients takes into account the sin(x)/x attenuation caused by the Digital to Analogue Converters (DAC's). To achieve a high throughput rate, a parallel organization of the filter is required and to simplify the architecture the following two methods were adopted:

(1) Minimisation of the number of multiplications by choosing a half-band filter. This architecture is well adapted for oversampling a signal, by taking into account the fact that every second input sample is zero.

(2) Simplification of the coefficients [5] by minimisation of the "ONE's" in the coefficient of the filter and use of signed bits:

$$- 0.0001111 = - 0.0010000 + 0.0000001$$

The transfer function of the filter with

$$Z = e^{j2\pi f}/F_s \qquad \boxed{12.1}$$

where Fs denotes the sampling frequency is

$$H(Z) = C0 + C1(Z^1 + Z^{-1}) + C3(Z^3 + Z^{-3}) + C5(Z^5 + Z^{-5}) +$$
$$C7(Z^7 + Z^{-7}) + C9(Z^9 + Z^{-9}) \qquad \boxed{12.2}$$

The incoming and outgoing data flows are 8 bits wide, with values in the 00 to FF range. The internal width of the registers is 16 bits and a carry save structure for the adders has been adopted (Figure 12.3). The time critical

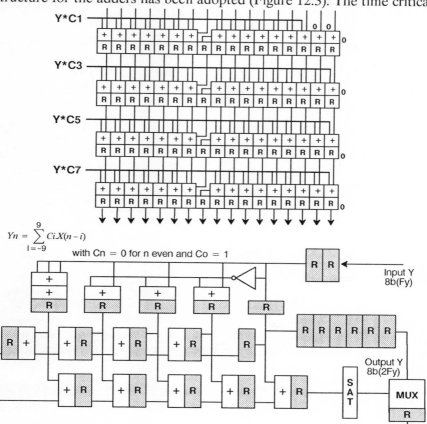

$$Yn = \sum_{I=-9}^{9} Ci.X(n-i)$$

with Cn = 0 for n even and Co = 1

Figure 12.3 Luminance filter - block diagram and data path implementation

part is a two 9-bit words addition in a time interval of 2/Fs, due to the half-band architecture. For the same reason, and in order to save pipe-line registers, a carry select adder organization has been implemented to compute the first multiplication (coefficient C1). To optimize the S/N ratio, no internal truncations occurs and the 8-bit rounding is performed at the output of the filter.

12.4.1.2 Chrominance filters

In a MAC decoder, the colour filter cut-off frequency is the result of a compromise between the colour resolution and an equal noise visibility on

A Video Communication Application of Analogue–Digital ASICs

luminance and colour. Four different filters approximate Bessel functions with a cut-off frequency of respectively 1, 1.3, 1.6 and 2MHz. The real time noise measurement on a MAC test line can be used to select in real time one of the four colour filters available in the circuit.

Figure 12.4 shows the architecture of the chrominance filters. As the two

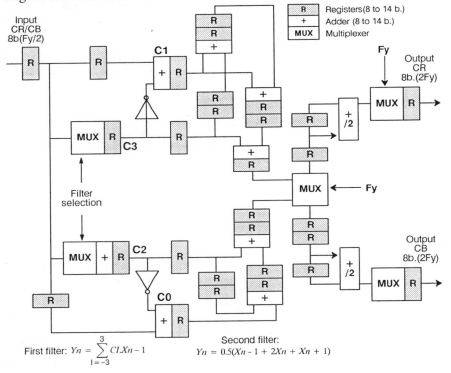

First filter: $Yn = \sum\limits_{I=-3}^{3} CI.Xn - 1$

Second filter:
$Yn = 0.5(Xn - 1 + 2Xn + Xn + 1)$

Figure 12.4 Chrominance filters – block diagram

signals CR and CB have the same specifications, they share the same arithmetical parts for the first seventh order filter. Due to the high output sampling rate (2Fs), the last interpolators are not multiplexed, but are implemented on two identical blocks.

The transfer function of the first filters with

$$Z = e^{2j\,\pi f/CFs/2)} \qquad \boxed{12.3}$$

is

$$H(z) = C0_i + C1_i(Z^1 + Z^{-1}) + C2_i(Z^2 + Z^{-2}) + C3_i(Z^3 + Z^{-3}) \qquad \boxed{12.4}$$

Four different sets (i) of coefficients C_{ji} are selectable, depending on the desired cut-off frequency. All the coefficients are different from zero, but by taking into account that every second input sample is zero, the two crominance signals are multiplexed on the same data path. In order to keep a sample rate of $Fs/2$, the corresponding even and odd coefficient computations are exploded on two different blocks (the upper and lower block in Figure 12.4).

Considering that $C0_i + 2C2_i$ and $2C1_i + 2C3_i$ are equal to 1, the selection of one of the four filters can be performed relatively simply. As an example (see Figure 12.4).

$$Y(n) * C_i + Y(n + 2) * C3_i = Y(n) * (1 - C3_i) + Y(n + 2) * C3_i \qquad \boxed{12.5}$$

It is therefore sufficient to only modify $C3_i$, for example. The same organisation is implemented for the lower port ($C0_i$ and $C2_i$) of the filter.

The last filters are two second order interpolators. With

$$Z = e^{j\pi} f/Fs \qquad \boxed{12.6}$$

the transfer function is

$$Z(z) + 0.5(Z + 1 + Z^{-1}) \qquad \boxed{12.7}$$

with also every second input sample equal to zero.

The width of the internal registers is 14 bits. The two output signals are saturated to 8 bits, and as for the luminance part, no internal truncations are authorized.

12.4.1.3 Matrix operator

This part performs the following digital transformation according to CCIR recommendation:

$$\begin{bmatrix} R \\ G \\ B \end{bmatrix} = \begin{bmatrix} 1 & 1.37 & 0 \\ 1 & -0.698 & -0.336 \\ 1 & 0 & 1.73 \end{bmatrix} \begin{bmatrix} Y \\ CR-0.5 \\ CB-0.5 \end{bmatrix}$$

The four multiplications, are carried out in ROM's, use the above coding (see Figure 12.5). To obtain the 1.37 (CR - 0.5), multiplication, two 16-word ROMs and one adder were used. The first ROM codes 1.37(CRM-0.5), the second ROM codes 1.37(CRL), with CRM and CRL being the most and least significant nibbles of the 8 bit CR bus. The same ROM and adder organization is used to compute the three remaining multiplication procedures.

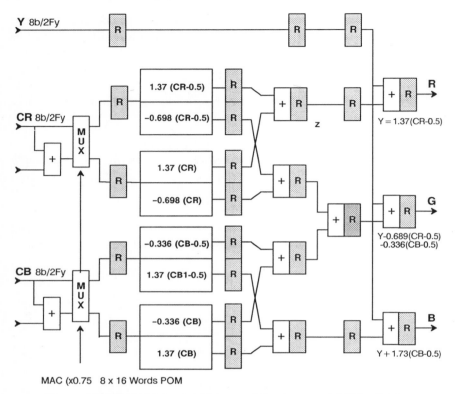

Figure 12.5 Dematrixing operator architecture and 1.333. division

Due to the colour specificity of the MAC satellite transmission, (CR-0.5) and (CB-0.5) are multiplied by 1.3 after the R, G, B matrixing. To compensate this extension, (CR-0.5) and (CB-0.5) must be divided by 1.3 at the decoder. This function is approximated by a 3/4 multiplication and an external command ensures to suppress it if the 601 de-matrixing recommendation is required for other applications. Due to the output noise round-off spectrum density of the chrominance filters, this operation has to be implemented between the oversampling filters and the matrix operator.

12.4.2 Analogue part

12.4.2.1 D/A converters

In order to fulfill the electrical and power consumption specifications, a voltage output architecture has been adopted for the D/A converters. A current output DAC is more power consuming and the output voltage swing is dependant on the external loads. This last constraint is not easily compatible with a low dispersion between the three R, G and B outputs.

Each of the three converter is based on a resistor string used as a voltage divider. An on-chip voltage reference is shared by the three strings. A classical band-gap generator and a power stage have been implemented in order to obtain a 815mV reference voltage . This value corresponds to the standard 700mV transition of a video signal from black to white (levels 16 to 235). The reference voltage generator is directly connected to the three DACs. In order to reduce the diaphony between the DACs, this reference voltage is also connected to an on-chip capacitor. Each converter receives 8 bit parallel data (respectively red,green and blue) which are converted into a one-out-of-N code by an appropriate logic circuitry which activates NMOS gates. A voltage from the reference voltage tap is fed into the analogue output. The impedance adaptation is performed by an op-amp connected as a voltage .follower. It can drive a 75Ω/20pF load. The schematic of the converter is shown in Figure 12.6, and discussed further, below.

(a) Resistor network and switches. This is the heart of the converter, on which the performance depends; particularly the settling time and differential linearity. The string is made of N^+ poly-silicon with a 8Ω typical resistance between taps (sheet resistance of poly-silicon: 25Ω/□). Special attention was given to the design of the string, especially to the design of the corners, which could have induced some deterioration in the linearity. A 0.15lsb nonlinearity for each corner was obtained.

   In order to minimize any gradient effects, the string is folded 16 times. This results in a smaller area required by the string, hence any degradation due to the inhomogeneity of the poly-silicon and any thermal effects is reduced. Another advantage of the folded string is the reduction in parasitic capacitances resulting out of the fact that an analogue output node is charged by only 16 junction capacitances instead of 256. Consequently, the settling time is reduced.

(b) Logic circuitry. Each tap of the string is connected to a NMOS switch. These switches are activated by the 4MSBs through a column-decoding block. This block consists of 16 AND gates. Each output activates 16 contiguous switches. Therefore, 16 contiguous voltage taps are addressed and 16 analogue lines are charged to the corresponding voltage.

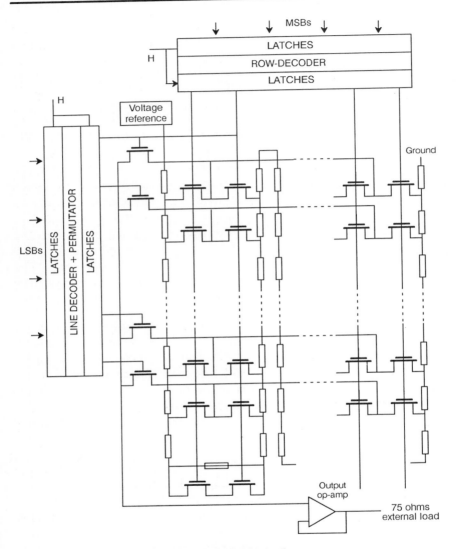

Figure 12.6 DAC block diagram

One of these 16 voltages is selected through 16 row-switches whose gate are driven by the 16 outputs of a row-decoding block. Row and line-decodings are simultaneous.

Latches were introduced between the outputs of the two decoding blocks and the switches. Differences in decoding times are thus eliminated. This is the key to obtaining a low glitch level.

(c) Impedance buffering. The analogue output needs to be buffered for two reasons. First, it cannot drive a large capacitive load without deterioration in dynamic performances. Second, it cannot drive a 75Ω load without deterioration in linearity. Indeed, at the output node (after line-switches) there is an equivalent voltage generator whose internal impedance Ri is a function of the input code and resistor string.

12.4.2.2 Output amplifiers

Each of the three DAC outputs is connected to an operational amplifier used as a unity-gain buffer to drive the (75Ω/20pF) external load. The buffer characteristics must fit the required settling time and linearity error of the DAC. The schematic of the amplifier is shown in Figure 12.7. It consists of a preamplifier gain stage (MN1-MN4) and (MP1, MP2) and a power gain stage (MP3, MP4 and MP5, MN5) which drives the output load.

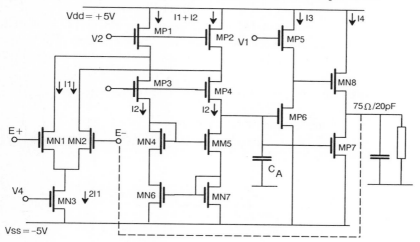

Figure 12.7 DAC output amplifier schematic

The preamplifier is a single pole folded-cascode differential stage [6,7] with a 52dB voltage gain. This gain is ensured by the large W/L ratio (60/3) of the MN1 input transistors. MP2 cascode transistors are small in size (10/3) in order to reject the high frequency pole resulting from the ft of these devices. The C_A compensation capacitor fixes the unity-gain phase margin of the overall amplifier. This capacitor has been made using a double poly-silicon process and has been optimized to obtain the minimum closed-loop settling time. The bias current I2 (150μA) ensures a slew-rate compatible with desired transient performance.

The power stage involves an intermediate level shifter (MP3, MP4) and a common source buffer (MP5, MN5). On account of there being only a positive output voltage, the usual crossover stage is not used. MP5 has been sized to provide a 1mA output quiescent current in MN5 in order to obtain a sufficient gain for a 0 volt output. All devices have a minimum gate length of 3μm and 2.4μm for P channel and N channel respectively, to reject the high frequency poles induced by these devices. A low resistive load and body effects induce a –6dB gain loss at this stage. This gain reduction implies a large voltage swing at the preamplifier output node and compromises the use of a low voltage power supply. Furthermore, the output slew-rate is lowered.

The typical DC gain is 46dB . The unity gain frequency is 50MHz and the phase margin is 65°. The DC gain is sufficient to linearise the DC transfer curve of the output stage and retain a closed loop linearity error <0.1% (0.25LSB for the 8 bit DAC) . The optimal settling time is 25nSec and the slew-rate is 80 V/μSec for a 1V output swing into 75Ω/20pF load. The amplifier quiescent power is 35mW with a 5 volt power supply. The device area is 0.04mm². The amplifier characteristics are summarized in Table 2.

Table 12.2 Typical performance of DAC ouput amplifer

Voltage gain into 50 Ohms/20pF	46dB
Unity gain bandwidth	50MHz
Phase margin	56°
Supply voltage	±5V
Quiescent power	35mW
Unity gain configuration	
Output swing into 50 Ohms/20pF	1.5V
Settling time (1V output swing)	25nSec
Slew rate	80V/μSec
Linearity error	<0.1%

In order to minimize the effects of process parameter variations on threshold voltage, channel length, a controlling block for the output op-amp was introduced (see Figure 12.8). Its settling time is closely related to its quiescent currents. Since the current in a MOS transistor is proportional to $(V_{gs}-V_t)^2$, a small variation in V_t causes a large variation in settling time.

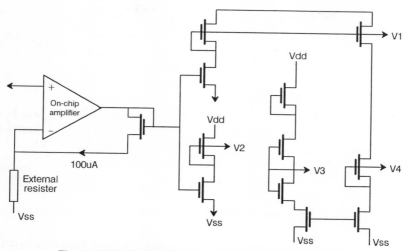

Figure 12.8 Controlling block of the output amplifier

The principle of this block is to keep the op-amp sensitive stage current generators (V_{gs}–V_t) constant. This is performed by generating a current which remains constant with the variations of V_t and ΔL. To achieve this, an N-channel transistor N_0 is included in the feed-back loop of an op-amp. Thus, if the open-loop gain is sufficient, the loop current is constant. The aim is then to connect N_0 with another NMOS transistor, both forming a current mirror. Finally, each MOS current generator of the output op-amp is connected with a MOS transistor of the controlling block, these two MOS acting as a current mirror.

These results place the converter within 8 bit specifications in terms of S/N ratio, linearity and settling time for video applications. The characteristics and typical features of the converters, including output amplifiers, are the following:

settling time - 35nSec
integral linearity - 0.2 lsb
glitch energy - 30nSec.lsb
harmonic distortion -50dB for a 200kHz output signal
diaphony - 75 dB for a 200kHz output signal
45dB for a 5MHz output signal

12.4.3 Experimental results

The architecture of the whole chip is shown in Figure 12.9. An optional intermediate input/output port has been implemented between the matrix operator and the DAC's. This port is used for videotex applications, where

A Video Communication Application of Analogue–Digital ASICs

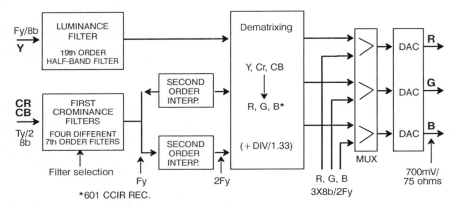

Figure 12.9 Architecture of the digital–to–analogue video processor

TV pictures and alpha-numeric characters, defined with their R,G and B components,have to be mixed. This complementary port also allows the circuit to be tested simply, by separating the digital and analogue parts. Table 3 summarizes the main features of the chip.

Table 12.3 Chip freatures

Process	2μm CMOS process double Poly–One metal
Total device count	40,000
Chip size	38mm²
Supply voltage	±5V
Max clock frequency	30MHz (>40MHz for digital part)
Power dissipation	500mW at 27MHz
Package	40 pin DIL

The layout has been carefully designed over the entire chip in order to minimize the parasitic electrical couplings between the digital blocks and analogue cells. To minimize these effects, on-chip capacitors have been implemented for decoupling certain critical nodes, such as the common internal voltage reference and the DAC power supplies. Simulation and experimental results show that on-chip capacitors of about 200pF on VDD and 700pF on VSS decrease the value of parasitic frequency components (3Fs, 4Fs,...) by a ratio of more than 10dB. The efficiency of the capacitors is partially cancelled by the bounding and package inductances (about 10nH per pin).

The N-well technology used and the negative power supply needed for the analogue part, requires the substrate to be connected to -5V. In order to comply with the speed requirements, the digital part runs between -5V and 0v. Translators have been implemented in the input/output ports in order to maintain an external TTL compatibility.

This circuit has been tested in an experimental MAC decoder. An enhanced picture quality has been observed during fast transition using this approach, compared with a "conventional" analogue decoder. This could be explained by the real linear phase response of the filters.

12.5 Analog-to-digital video processor

Figure 12.10 shows the general architecture of a HDMAC coder and decoder. The input signal originates from a camera working on 1250 lines, 50Hz, 2:1 interlaced with a 16/9 aspect ratio. The sampling frequency is 54 MHz, resulting in 1440 active samples per line for the luminance signal. The required bandwidth reduction is achieved in the encoder by a spatio-temporal sub-Nyquist sampling method, adapted to the picture content [9 - 13].

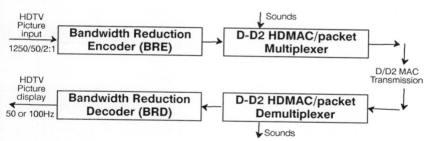

Figure 12.10 Schematic block diagram of a HDMAC coder/decoder

The D-D2-HD-MAC/packet system delivers an HD picture compatible with the conventional (e.g D2-MAC/packet) decoders. The new system requires a 12MHz base-band to recover the whole HD quality. Nevertheless, it allows the use of a reduced base-band transmission channel, down to 5MHz, compatible with a standard D2-MAC/packet system, for example (but without high definition).

The decoder is a complete slave of the coder. This means that in the decoder no complex functions, such as motion detection, take place, in order to reduce the cost of the HDTV receivers. As explained earlier, to recover a picture quality as close as possible to the original, special care has to be

A Video Communication Application of Analogue–Digital ASICs

taken with the limited bandwidth of the transmission channel and the interface circuits. The acquisition function has to guarantee the maximum transparency of the channel (satellite, cable, ..), at the lowest possible cost.

Figure 12.11 gives a possible architecture of the acquisition part of a HD-MAC receiver. The most critical part concerns the A/D converter and the adaptive 1/2 Nyquist filter [8-9].

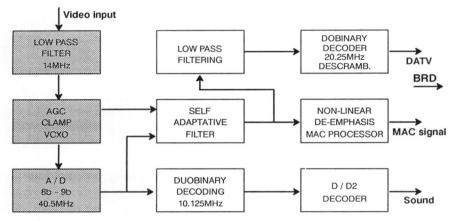

Figure 12.11 Architecture of the acquisition block of a HDMAC decoder

It has been demonstrated that an automotive equalization, to reduce inter-sample interference resulting from echos or linear attenuation of the transmission channel, is a good trade-off between the overall quality and hardware complexity.

As shown in Figure 12.10, the D-D2-HDMAC/packet multiplexer has to make up the base-band signal in a D-D2-MAC/packet form. Some functions are not specific to HD applications (sound encoding, scrambling access control system, time division multiplexing,..).Other functions are specific to HD-MAC, such as Nyquist filtering, non-linear pre-emphasis,.. Indeed, as explained previously, the coding of HD-MAC, based on sub-sampling, implies that the global transmission channel preserves the independence between the consecutive samples, at 20.25MHz (13.5 x 2/3 (MAC time compression). This condition is verified if the equivalent base-band channel meets the first Nyquist criterion, at the vicinity of 10.125MHz. A 10 % roll-off Nyquist filter, equally shared between the transmission and the reception has been adopted. This filter, at the transmission level, is made of a 31 tap transversal filter sampled at 40.5MHz.

Figure 12.12 shows the principle of an automatic equalizer, using a reference signal inserted into the line 624 [17]. The algorithm being processed in order to compute filter coefficients may be the well known Gradient Algorithm (Least Mean Squares, or LMS, Algorithm). This algorithm has to be computed in differed time on a window of 240 consecutive damaged data samples of the input signal. The same 240 samples are used during the entire computation of one set of coefficients (they have to be stored in a looped memory). Typically, this process makes the algorithm converge after 1.5mSec at 40.5MHz (input sampling frequency).

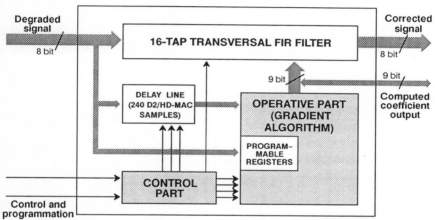

Figure 12.12 Acrhitecture of a HDMAC adaptative equaliser

A noise reduction technique is inserted into the transmission chain in order to reduce the noise sensitivity of the transmission chain caused by an increased base-band (12MHz), as compared with D or D2-MAC signals. Special care has to be taken in the design of the Phase Locked Loop (PLL) in order to guarantee good stability (jitter lower than 2nSec).

HDTV systems require many performing converters. The most critical specifications will concern the encoders, where a sampling rate of up to 144MHz will be used. For this equipment, monolithic or low cost solutions are not required. On the other hand, the decoder will use less critical A/D converters (40.5MHz - 9-b with an input bandwidth of about 12MHz) but with a high level of integration. In this field, the trend is to use multi-step conversion algorithms. Multi-step architectures approach speeds of full parallel designs without a large die area and power consumption. CMOS technologies have proved their ability to achieve such prototype converters with approaching performances (less than 10mm^2 in 1μm CMOS process)

[14-15]. On the other hand, bipolar technologies have proved their ability to implement such converters, but with a higher power consumption [16]. In order to minimize the complexity of the decoder, the Analogue to Digital Converter (ADC) will include Analogue Automatic Gain Control (AGC) and clamping functions. All the additional functions will have to be digitally controlled, including equalization, in order to guarantee the HDMAC quality.

The acquisition part will certainly be the most complex one, due to the immersion of critical analogue parts (A/D converters, PLL, AGC) with complex digital parts (adaptive filtering, non linear corrections,...). BiCMOS processes, with analogue possibilities, now seem to be the best solution for a mono-chip integration. However, if new efficient CMOS A/D converter architectures are proposed, an 0.7 or 0.5μm CMOS process may be a good alternative to a BiCMOS approach.

12.6 Conclusion

This chapter has outlined the incorporation of both analogue and digital circuits in a single chip to meet stringent technical and economic requirements of a video communications application. Such a hybrid approach is becoming very important for many applications in the telecommunications area, such as modems, radio–telephones and ISDN equipment in general. To obtain the maximum benefit from this approach, analogue and digital parts can not be synthesised and analysed separately but a unified approach will be found to be the best. This will result in the development of many specific mixed mode techniques in the future.

12.7 References

[1] M. Mathieu and G. Dvuic "The D2-MAC/packet : A new concept for consumer terminals," in ICCE (Chicago,Il), 1985, pp. 114-115.

[2] P. Senn "Specific Circuits for HDTV Applications" in ESSCIRC '90, Grenoble -France - September 1990.

[3] A. Leger, "Implementation of fast discrete cosine transforms for full videotex services and terminals," in IEEE Globecom (Atlanta, GA) 1984, pp. 333-337

[4] P. Senn, J. Chan Yan Fong and M. S. Tawfik "Concepts for the restitution of video signals using analogue circuits", ISCAS'88, Helsinky - Finland - pp 1935-1938.

[5] P. SihoanI and A. Benslimane, "Design of optimalfinite wordlength linear phase FIR filters: new applications," in ICASP (San Diego, CA), 1984.

[6] Andre Abrial and all "A 27MHz Digital to analogue Video processor", IEEE Journal of solid-state circuits, vol. 23, N⊠6, December 1988.

[7] Paul R. Gray, "MOS Operational Amplifier Design-A tutorial overview" in IEEE j. of Solid-State Circuits, vol SC-17, n⊠6, December 1982, pp. 969-982.

[8] J. M. Fournier, A. Abrial, A. Lelah, P. Senn, J. Bouvier, "A CMOS 120 mW 8-bit voltage output video DAC using a current-cell matrix," in ESSCIRC (Bad Soden, RFA), September 1987, pp. 105-108.

[9] R. Boyer, J.M. Colaitis and D. G. Brook "European HDTV Compatible System" Proc. on European Conference on the Technologies of High Definition Television - Grenoble 89 - France -

[10] J. P. Arragon, J. Chatel, J. Raven and R. Storey "Instrumentation of a Compatible HD-MAC Coding System using DATV "Conference proceeding IBC 88 -Brighton -pp 57-61

[11] F. W. P. Vreeswijk, F. Fonsalas, T. I. P. Trew, C. Carey-Smith and M. Haghiri "HD-MAC Coding of High Definition Television Signals "Conference proceeding IBC 88 -Brighton - pp 62-65

[12] T. I. P. Trew and O.J. Morris "Spatially Adaptative Sub-branches for HD-MAC "Conference proceeding IBC 88 -Brighton - pp. 66-69

[13] J. P. Aragon, F. Fonsalas and M. Haghiri "Motion compensated Interpolation Techniques for HD-MAC" Conference proceeding IBC 88 -Brighton - pp. 70-73

[14] M. Ishikawa, T. Tsukahara "An 8b 40MHz Sub-ranging ADC with pipelined Wideband S/H", Proc. of IEEE International Solid-state Circuit Conference - New-York - USA - February 1989.

[15] A. Matsuzawa, M. Kagawa, M. Kanoh, K. Tatehara, T. Yamaoka and K. Shimuzu "A 10 b 30MHz Two-step Parallel BiCMOS ADC with internal S/H" Proc. of IEEE International Solid-state Circuit Conference- San Francisco- USA - February 1990.

[16] R. Van de Plassche and P. Baltus "An 8b 100 MHz Folding ADC", Proc. of IEEE International Solid-state Circuit Conference - San Francisco - February 1988.

[17] S. Maginot, F Balestro, C. Joanblanq, P. Senn and J. Palicot "A general Purpose High Speed Equalizer" proc. of IEEE CICC - Boston - USA - June 90

High Level Simulation and Hardware Description Languages

R. Cottrell
LSI Logic Limited. Sidcup, Kent, England

13.1 Introduction

The most important simulation tool used in digital design is the logic simulator; in analogue design it is the circuit simulator. Indeed, Circuit simulation is used in digital design, primarily for the detailed analysis of relatively small cells. Digital designers also use simulators operating at higher levels of abstraction, but these are usually directly compatible with logic simulators. It is not surprising, therefore, that there is much interest in developing tools which offer logic and circuit simulation linked together. This topic is the subject of another chapter of this book; it is sufficient here to note that it is a technically demanding exercise to develop such a simulator.

However, circuit-level analogue/gate-level digital mixed-mode simulation will not address all the simulation requirements of mixed-signal ASICs. Circuit-level simulators are only useful in general for circuits containing up to a few hundred transistors. They are therefore not suitable for high-level simulation of complex mixed-signal systems, even if they are only applied to the analogue portions of the system. New methods are required which address these requirements.

13.2 The need for high level simulation

13.2.1 ASIC Design verification

Recent trends have shown a rapid increase in the number of integrated circuit designs involving both digital and analogue circuitry, and market forecasts predict an even more marked rise in the future. This is a natural part of the trend to ever higher levels of integration which yields greater functionality, smaller size, lower power dissipation and higher reliability.

Many of these mixed-signal integrated circuits will be application specific, as is indeed the case for purely digital circuits. This is an inevitable consequence of the high level of integration. For ASICs, time to market is almost always more crucial than designing the most silicon efficient implementation, and so using larger and large building blocks, along with logic synthesis and high-level synthesis tools is becoming increasingly attractive. The days when integrated circuits were designed at the transistor level are long gone.

The importance of time to market is no less for mixed-signal circuits than for purely digital. However, analogue design is a much more complex task than digital, and design automation tools for analogue and mixed-signal design are still in their infancy. This means that design timescales for mixed-signal ASICs are typically much longer than for their digital counterparts, and whereas only 10% of a typical circuit may be analogue, it may well absorb 90% of the effort.

There is therefore an enormous challenge to the electronics CAD industry to produce tools to aid, and effectively de-skill, the design of mixed-signal ASICs. This should not be seen as a threat to the jobs of highly skilled analogue circuit designers; it seems demand for their skills will always exceed the supply. This CAD effort needs to be in parallel with an effort by ASIC vendors to produce libraries of re-usable analogue cells, which are well-characterized and robust. These cells will need to be at ever higher levels of abstraction; today it may include operational amplifiers, digital-to-analogue converters, etc. Tomorrow, it may include complete ethernet interfaces and such like.

The mixed-signal ASICs being designed today are worthy of the term "system on a chip". As an example, the LAD310 series of BiCMOS mixed-signal ASICs from LSI Logic can include up to 100 000 raw digital gates and 10,000 transistors for analogue circuitry. One typical design recently completed incorporated about 20,000 digital gates, four 8-bit analogue-to-digital converters, one 6-bit digital-to- analogue converter, and some extra signal conditioning circuitry. Such circuits present a real challenge to the designer and CAD tool developer alike.

The first and most important emphasis of CAD tool development must surely be design verification. The worst fate to befall an ASIC design project is to go through several prototype iterations before obtaining good silicon. Mature tools are already available to verify integrated circuit layout against a circuit netlist, and to extract parasitic component values from the layout; tools are needed to verify the netlist in the first place, and analyze the effects of the parasitic components. Much of this needs to be addressed by simulation, in conjunction with static electrical rule checking. To this extent, mixed- signal ASICs are no different to their purely digital counterparts, however verifying mixed-signal designs is much harder.

Logic simulators are used to verify the function and timing of digital ASIC designs. With mixed-signal ASICs, there are other additional parameters which required verification, such as bandwidth, noise performance, and cross-talk. Even basic functional simulation of an entire mixed-signal ASIC is a major challenge, although such simulation is now possible using the high level analogue simulation techniques discussed in this chapter. Parametric verification must still be performed in the main on smaller blocks using a circuit simulator. The parametric verification of entire circuits remains a formidable challenge. Solutions to this problem would make a significant contribution to the de-skilling of mixed-signal ASIC design.

13.2.2 Design analysis

Perhaps the most obvious application of simulation is design analysis: to try out an idea to make sure it works. Circuit simulation is widely used for this purpose, however there are circuits for which it is impractical. Complex high order phase-locked loops are examples of such circuits. They often involve clock dividers and phase detectors which are best modelled in a logic simulator, and while phase-locked loops might not contain vast numbers of components, they typically require very long simulation runs. As the mathematics which is used in their design does not usually account for such effects as the non-linearities of a digital phase detector, simulation is highly desirable. The use of high- level analogue simulation and modelling can make such simulations much more tractable.

13.2.3 System level simulation

There is increasing interest in high-level simulation of systems early on in the design cycle. This has become more practical with the development of behavioural simulators. In particular, the development of VHDL, its adoption as an IEEE standard and its wide acceptance has resulted in much increased activity in this area. VHDL simulators are now available which are claimed to implement 100% of the standard language.

One reason for early resistance to behavioural simulation has been the lack of a route from behaviour to structure, other than by manual design. However logic synthesis, which synthesizes a network of logic gates from a set of boolean equations, is now widely used, and high level synthesis is beginning to appear. This makes high-level simulation a much more attractive proposition.

Systems, even more than ASICs, are not entirely digital, and their description and simulation would be made much easier if analogue functions could be used. Indeed, very early in the design cycle it may not be clear which sections of a design will be implemented digitally, and which using analogue techniques. Thus system level design gives a strong motive for providing methods for high-level mixed-signal simulation.

13.3 Methods for high level simulation

13.3.1 Compatibility with digital simulation

This chapter concentrates mainly on high level analogue simulation, but it must be remembered that it is really mixed-signal simulation which is being considered. It is therefore vitally important that any methods developed for high level analogue simulation should be compatible with digital simulation. This is a problem because of the fundamentally different methods employed by the circuit simulators normally used by analogue designers, and logic simulators. The issues involved in linking two such simulators are addressed in another chapter.

There are two principal alternative methods for high level mixed-signal simulation. One is indeed to link together a circuit simulator and a logic simulator, and to provide high level modelling capability within the circuit simulator. This capability can be provided by macro-modelling or by a special- purpose language. The other alternative is to develop analogue models which operate within the event- driven paradigm of digital simulators.

13.3.2 SPICE macro-models

The most well known method for providing high level models of analogue functions is to connect together circuit simulation primitives to construct a mathematical model. The primitives used are usually the various controlled sources (voltage-controlled current sources, etc.) and simple passive components. An example macro-model is given in Figure 13.1. This model incorporates a single pole and finite gain, but is otherwise a perfect operational amplifier.

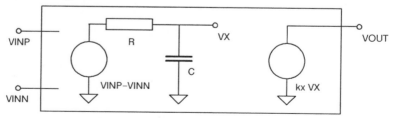

Figure 13.1 Macro-model as a simple operational amplifier

This method suffers from two main disadvantages. Firstly, only analogue functions can be modelled in this way, or more accurately only circuit simulation models can be developed. This is a disadvantage when modelling a function such as a digital-to-analogue converter (DAC), where it would be desirable for the digital inputs to interface directly to the logic simulator, while the analogue output interfaces to the circuit simulator. It would be perfectly possible to construct a circuit simulation macro-model of the DAC, but the conversion of the digital input signals to their circuit simulation equivalent would have to be handled outside the model by the software interfacing the two simulators.

The second disadvantage is that the modeler is limited to using the available simulator primitives. Even if the simulation algorithms are capable of performing a particular calculation, there may be no access to the capability using the standard primitives. In this case, the only recourse is to write software models for new primitives, which is likely to be error prone and tedious.

13.1.1 Special purpose languages

An alternative approach to macro-modelling is to use a special purpose modelling language which gives access to the circuit simulation algorithm itself. This effectively allows new primitives to be developed more easily than in a conventional circuit simulator. As a circuit simulator is essentially a program for solving simultaneous non-linear differential equations, the language must enable such equations to be described. If the language can also give access to the event-driven logic simulator, then it will be much more powerful, and allow effective models of mixed-signal functions such as DACs to be constructed more easily.

The problems with using a circuit simulator on very large circuits arise from the large number of equations which need simultaneous solutions. Using a special purpose language, the behaviour of large functions can be described using many fewer equations than would arise from device-level models, thus extending the usefulness of equation-solving simulators to much larger circuits.

An example of such a language is MAST, the modelling language of the Saber simulator produced by analogy. This indeed allows the description of non-linear differential equations, in addition to event-driven digital models. Saber itself combines circuit simulation and event-driven algorithms within a single simulator; in addition it has been linked to other logic simulators. A simple example of a MAST model is given in Figure 13.2. This model also incorporates a single pole and finite gain.

```
element template op4 inp inm out = gain, pole, offset

electrical inp, inm out
number gain, pole, offset
{
    val v pvout
    var i iout

    values {
        pvout = pole * v(out)
    }
    equations {
        i(out) += iout
        iout: v(out) = (gain * v(inp) - v(inm)
            + offset)) - d_by_dt (pvout)
    }
}
```

Figure 13.2 MAST Model for a simple operation amplifier

13.2.1 Event-driven analogue models

In this case, analogue models are written according to the event-driven paradigm of digital simulators, rather than attempting to combine two different simulation algorithms in a single program. A modelling method needs to be adopted, such as that described in [1], and the models can be written in a conventional behavioural language such as LSI Logic's proprietary Behavioral Specification Language (BSL), or a standard language such as VHDL.

The main advantage of event-driven analogue models is that there is no need to interface two fundamentally different simulators; indeed the mixed-signal simulations can be performed using a standard behavioural digital simulator without modification. The main disadvantages are the difficulty of writing the models, and the problem involved in handling complex interactions between cells. The method is most useful at higher levels of abstraction where interactions between cells are less important, and the ports of a cell can clearly be defined as inputs or outputs.

The method has been used successfully within LSI Logic in two areas. Firstly, simulation of complete complex mixed-signal ASICs has been made possible by the development of behavioural models for analogue cells in an ASIC library. This permits functional verification of the entire ASIC, which would be impossible using only a circuit simulator. It does not remove the need for circuit level simulation of smaller portions of the ASIC to investigate more complex cell interactions.

Secondly, a library of abstract mathematical functions such as two-pole filter sections, multipliers, etc., has been developed to help designers to perform"what if" design studies, and build high-level models of systems. The availability of this library means that the designers do not have to become involved in the task of writing software, where a new abstract model is required it is provided by modelling specialists. This library has been found particularly useful in the design of complex phase-locked loops.

When implemented in VHDL, this method has particular appeal for system-level modelling. With a library of abstract analogue functions available, it would be possible to construct very high level models of systems at an early stage of development before deciding how to partition and implement the design.

13.2.2 A method for event-driven analogue modeling

The method described here is one possible way to model analogue functions in an event-driven simulator [1]. The main problem is the representation of analogue waveforms, which are continuous in value and time. The fact that they are continuous in value is not important, provided that the language being used supports real numbers of sufficient precision. The continuous time waveforms are represented in this method by a piecewise linear approximation, with changes in the gradient of the piecewise linear waveform corresponding to events in the simulator. This is illustrated in Figure 13.3. At each event, a model outputs both the present value of the signal, and its projected gradient over the next piecewise linear segment. A suitable VHDL type definition to implement such an analogue signal is given in Figure 13.4.

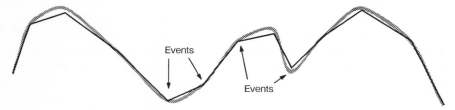

Figure 13.3 Piecewise linear waveform approximation

The accuracy of this waveform approximation is controlled by tolerances, a relative tolerance, and separate absolute tolerances for voltage and currents. It is the responsibility of each model to keep its output waveforms

```
type analogue_signal is
    record
        AS_Value: REAL
        AS_Gradient: REAL
        AS_Time_Set: TIME;
    endrecord;
```

Figure 13.4 VHDL type definition for analogue signal

within the specified tolerances. This tolerance mechanism provides the basis for timestep control. Timestep control is performed separately for each model. The models decide on the time allowed between events on their output ports to keep the waveforms within tolerances. Thus the timesteps are different at different parts of the circuit, and for the same part of the circuit at different times. This provides a significant improvement in efficiency over using some kind of global timestep.

An example of a VHDL model of a 2-input analogue adder is given in Figure 13.5. This model uses a package called analogue which provides type definitions for analogue_signal and a function called present_value which returns the present value of an analogue_signal as a real number. This model, being capable of working exactly, does not need to perform the tolerance-based timestep control which is required in more complex models such as multipliers and filters.

13.6 Example simulations

13.6.1 Event-driven simulation of an on-chip analogue to digital converter (ADC)

A successive approximations ADC is a good example of an on-chip mixed analogue-digital function which can benefit from high level simulation. A block diagram of the ADC is shown in Figure 13.6. Note that the output from the DAC is in the form of a differential current. This is subtracted from the input signal, which is converted into a differential current, using current summing junctions. Behavioural models are used for the DAC, voltage to current converter, and current comparator. The SAR and latch are purely digital models, built up from standard gates available in the appropriate technology library. Results from the simulation are shown in Figure 13.7 and Figure 13.8. The former shows the DAC output currents superimposed in the input current. These waveforms show clearly the operation of the successive approximation converter, and also the errors which arise in such a converter when a sample and hold circuit is not used on the inputs. The latter shows the differential comparator input currents along with some of

```
use work. analogue.all

entity adder is
    port (a,b: in analogue_signal;
            z: out analogue_signal:
end squarer;

architecture behaviour of adder is

begin
    process (a,b)
    variable avar, bvar, zvar: analogue_signal;
    begin
        avar := a;
        bvar := b;
        zvar.as_value := present_value(avar)
                + present_value(bvar);

        zvar.as_gradient := avar.as_gradient
            + bvar.as_gradient;

        zvar.as_time_set := now;
        z <= zvar;
    end process;
end behaviour;
```

Figure 13.5 VHDL model of analogue adder

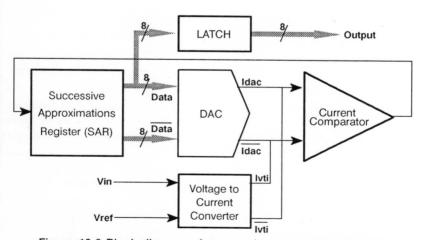

Figure 13.6 Block diagram of successive approximations ADC

the digital signals. The bus called SAROUT represents the output of the SAR, with DIGITS being the output of the latch. Once again, the errors due to the lack of sample and hold circuitry are evident: the two comparator input currents should be identical at the end of each conversion cycle.

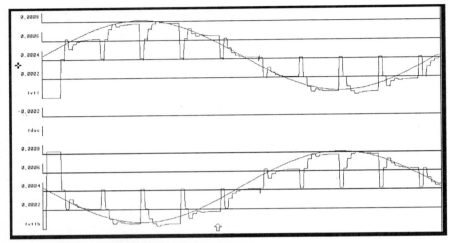

Figure 13.7 DAC Output currents overlaid with input current

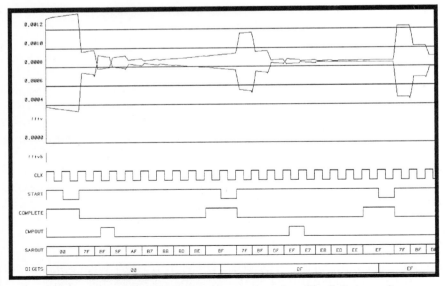

Figure 13.8 DAC Comparator input currents with digital results

13.9.1 Event-driven simulation of an abstract phase-locked loop

A block diagram of the phase-locked loop is given in Figure 13.9. It is a very simple loop built out of abstract mathematical functions. The filter is a single pole filter with a pole at 2.44kHz and a zero at 19.2kHz, which thus operates as an integrator for frequencies in between the pole and the zero. The VCO has a gain of 320kHz per unit input, and a centre frequency of 15.68MHz. The phase detector is mainly digital in operation, but has an analogue output giving three possible output levels: -2.5V (reduce frequency), 0V and +2.5V (increase frequency). The input stimulus used is a single frequency FM signal with a carrier frequency of 1MHz and a 10kHz signal with a modulation index of 0.2. The locking transient obtained from simulation is shown in Figure 13.10. The "thickness" of the signal is due to breakthrough of the input frequency because of the simple nature of the PLL. The 10kHz modulation of the input frequency can clearly be seen in the waveform once frequency lock has been obtained.

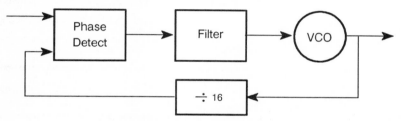

Figure 13.19 Block diagram of PLL

13.12 Future developments

13.12.1 Improved circuit simulators

There will always be a requirement for circuit simulators in mixed-signal ASIC design. It is likely that they will improve in the future, both in terms of speed and their ability to handle larger circuits. The majority of research over the last 10 years or so in circuit simulation has been aimed at improving their performance in the simulation of large digital MOS circuits. An example of these developments is the technique known as waveform relaxation [2]. A similar research drive is now required aimed at analogue circuits incorporating both MOS and bipolar transistors. There is some evidence that this is beginning to happen, as examplified in [3].

13.12.2 VHDL enhancements

VHDL is currently an IEEE standard (IEEE 1076-1987). Under the rules of the IEEE, all standards must be re-balloted every five years. As a consequence, VHDL will come up for re-standardisation in 1992. It is likely

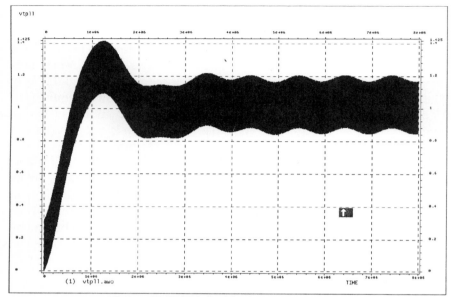

Figure 13.10 PLL locking transient results

that some changes will be made to the language at this time. Some of the changes will be to remove ambiguities, but it is also possible that the language will be enhanced and extended. One area of enhancement which has been considered is support for analogue and mixed analogue-digital modelling, but it now seems unlikely that any significant changes will be made in this direction in 1992. However, there is a possibility of a new standard, possibly 1076.1, being developed over the next few years for VHDL with analogue extensions.

With VHDL as it is stands is it possible to use for event-driven analogue modelling, which has been proved to be useful in high-level system design. The problems which remain are really implementation issues, principally the provision of an adequate library of mathematical functions. These can, of course, be written in VHDL, but could be more efficient if provided by the simulation vendor. The definition of a standard package of mathematical functions would be very useful.

However, VHDL is not presently capable of being used for analogue modelling at lower levels, or where there is a high degree of interaction between blocks, in which case it is necessary to use circuit simulation methods. It is in this area where extensions would be provided in a standard such as the possible 1076.1, specifically time-domain analogue modelling.

The MAST language from analogy represents the sort of functionality which would be required; of course the syntax would be very different to retain compatibility with present VHDL. The aim of the standard would be to define the meaning, or semantics, of the analogue behavioural descriptions, without precisely defining the simulation method to be employed. This would permit standard behavioural descriptions of analogue functions, without constraining future developments in simulator technology.

The main difference between present VHDL and the proposed extension is that the present standard permits code to be written which is run in the pseudo-concurrent environment of an event-driven simulator, whereas analogue behaviour needs to be described in the form of differential equations in continuous time, which need to be solved numerically by the simulator. The two key concepts required in the new language are nodes or equipotentials, which would permit the interconnection of analogue blocks while ensuring the implementation of Kirchoff's laws, and relations, which would allow equations to be described for solution by the simulator. These equations would need to incorporate a derivative operator with respect to time. However, the most difficult problem will be defining the interfaces between the event-driven time of present VHDL and the continuous time of the proposed analogue extensions.

13.12.3 Conclusions

There is a very strong requirement for high level mixed-signal simulation in the design of complex mixed-signal ASICs, for design verification, design analysis and system level modelling. The key issues in such simulations are high level analogue modelling, and compatibility between analogue and digital simulation. The two main alternative approaches are to link together a circuit simulator and a logic simulator, providing some method of high level analogue modelling in the circuit simulator, or to develop high level event-driven models of analogue functions in the context of a digital simulator. There are advantages and disadvantages to each approach.

13.13 References

[1]   R. A. Cottrell,"Event-Driven Behavioural Simulation of analogue Transfer Functions", EDAC 90 Conference, Glasgow, Scotland, 12-15 March 1990, pp240-243.

[2] E. Lelarasmee, A.E. Ruehli and A.L. Sangiovanni-Vincentelli,"The Waveform Relaxation Method for Time Domain Analysis of Large Scale Integrated Circuits", IEEE Trans. on CAD of IC and Systems, Vol. 1, no. 3, July 1982, pp131-145.

[3] D. Patrick and C.Lyden, "An Event-Driven Transient Simulation Algorithm for MOS and Bipolar Circuits", EDAC 90 Conference, Glasgow, Scotland, 12-15 March 1990, pp230-234.

MAST is a registered trademark of analoguey, Inc. Saber is a trademark of analogy, Inc.

Chapter 14
Circuit-Level Analogue/Gate-Level Mixed-Mode Simulation

K. G. Nichols. A. D. Brown
University of Southampton, England
P. F. Kilty. GenRad UK Ltd. Fareham

14.2.1 Introduction

This chapter is concerned with the principles and practice of mixed-mode simulator development, its application potential, and limitations. The prohibitive cost of accurate circuit-level simulation of large, mostly digital, circuits provides the impetus for the development of such simulators.

The designer of a data or signal processing system partitions it into analogue or digital parts as appropriate to the signal abstraction, continuous or discrete, required of the task to be performed by each part. Simulation is an essential aid to design but, conventionally, different types of simulator have been required for analogue and digital circuits. Interaction between the analogue and digital parts of the circuit, particularly when signals are passed round closed loops of such parts, makes independent simulation of the parts less meaningful. The need is to integrate the simulators into a mixed-mode environment which properly supports the interfaces between analogue and digital parts of the circuit.

Another consideration is that although a digital part implements an algorithm for processing discrete-valued signals, the physical realisation of the part processes continuous-valued signals which, at the higher level of abstraction, are viewed as discrete valued. Thus it might be necessary to simulate a digital part as an analogue part, particularly when it is to be operated near its limits of performance or when its performance is sensitive to external interaction. A part implementing a continuous signal processing algorithm cannot, however, be abstracted at higher level as a digital part.

For simulation, a system part is envisaged as a circuit comprising interconnected devices. The degree of abstraction of the models used to represent the devices normally determines the type of simulation performed. Thus, when the devices are viewed as resistances, capacitances and transistors the type of simulation is termed circuit-level. The (nonlinear) electrical behaviour of these devices is related to their physical construction and signals in the circuit are analogue voltages, charges and currents which are continuous functions of time. The modelling can be extended to include the (parasitic) resistances and capacitances of the interconnecting electrical nets which, in a simpler model, are equipotential nodes. This type of simulation is the most accurate available to the circuit designer and also the most costly computationally. Modeling is always a compromise between adequate representation of the electrical behaviour of a device and the computational cost of evaluating this behaviour throughout the simulation period. Even so, circuit-level simulation is rarely used for circuits exceeding a few hundreds of transistors in size - roughly, the maximum size of a leaf cell in a hierarchical design.

When discrete signal abstraction is appropriate, as for a digital part, the latter is still envisaged as a system of interconnected devices but the models used to represent the devices are higher-level abstractions than for circuit-level simulation. When these models approximate signal voltages as one of two discrete values, viz. 0 and 1 (*true* and *false*) and specify delays between input and output signals of a device, the type of simulation is termed *logic-level*. A delay is the interval between a signal transition, 0-to-1 or 1-to-0, at the input of the device and the corresponding transition at its output. A signal is discontinuous during a transition in that its value between 0 and 1 is undefined.

When situations arise in logic simulation where the value of a signal is either 0 or 1 but which is unknown, the additional signal value X is used to represent this. It may be that the signal is constant at X or that the exact timing of a transition is uncertain and so the signal is X in the interval of uncertainty. In other situations, the output of a logic device goes into a high-impedance state which does not affect the voltage of the node to which it is connected; such an output is represented by the value Z. An output value of 0, 1 or X is more dominant than Z and will overwrite the latter when connected to the same node. It is possible to define other logic values but they are not useful in mixed-mode simulation.

Logic simulation is approximately four orders of magnitude less costly computationally than circuit-level simulation, for the same transistor count, and so the cost of the latter can normally be expected to dominate in mixed-mode simulation.

In the sections which follow, suitable interfaces between analogue and digital parts of a system are developed together with methods of synchronisation of analogue and logic simulation times. Examples of mixed-mode simulation are also presented.

14.2.2 Analogue and logic nodes

Before a circuit can be simulated in a mixed-mode environment, it is necessary first to classify each of its nodes (nets) as either of type analogue or of type logic. A logic node has no connection to an analogue device. Any node with a connection to an analogue device is of type analogue notwithstanding its connection also to logic devices. For example, the circuit fragment shown in Figure 14.1(a) depicts a simple latch, comprising

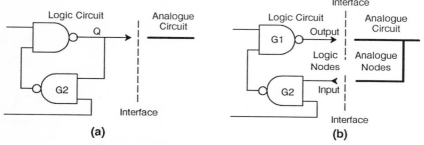

Figure 14.1 Analogue node output of latch

two NAND gates, with an output drive to an analogue part of the circuit, the analogue net is drawn as a bold line. The node Q is an analogue node. It is tempting to classify Q as a logic node and to interface it to the analogue circuit. This would be wrong because the logic connections between g1 and g2 enable correct latching even when the analogue circuit loads Q to the extent of suppressing g1 output transitions and hence the operation of the latch. Simulation would not then expose the malfunction. This pitfall can be avoided by splitting the logic node Q into separate input and output logic nodes and interfacing each to the same analogue node as shown in Figure 14.1(b). The interface has been moved from the position of the broken line in (a) to that in Figure 14.1(b).

Another example is shown in Figure 14.2(a). The classification of the nodes (nets) as logical or analogue (drawn bold) is shown in Figure 14.2(b) together with the positions of the interfaces. Again logic input/output connection A to the analogue transistor has been split into separate input and output nodes before interfacing to the analogue node.

14.2.3 Analogue to logic input interface

The cross interactions of an analogue node interfaced to an input logic node are twofold. First, the input of the logic device loads the analogue node in that the device draws current and, second, the signal on the analogue node is mapped to a logic value on the logic input.

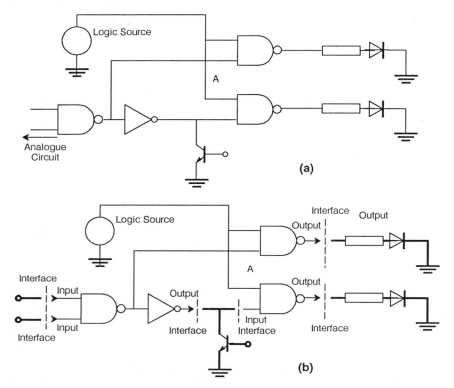

Figure 14.2 A mixed analogue - digital circuit

The effect of the input current into the logic device can be represented by a suitable model of the device on the analogue side of the interface. In an extreme situation, this could amount to a complete transistor-level model of the device and its analogue simulation but then this simply moves the interface problem elsewhere. A compromise is to represent the input by an input resistance chain and an input capacitance as shown in Figure 14.3; the input current to the logic device is i. In the simplest model, *Rupper, Rlower* and C would have constant values but this is a poor representation for most logic devices and so, more generally, these values would be specified as nonlinear functions of the analogue node voltage v. Since these functions have to be handled by the analogue simulator, it is as well to make sure that they are continuous and have continuous gradients, with respect to voltage over the domain of input voltage values.

The interface arrangement described in the last paragraph correctly determines the analogue input voltage v even when i becomes very small (a high-impedance, or weak, analogue drive). Thus, for example, a TTL logic

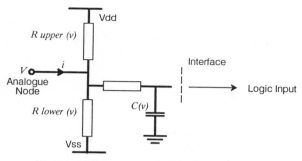

Figure 14.3 Analogue-to-digital interface

device interface would have *Rlower* with a large value compared to *Rupper* so that an open-circuit input (Z) causes v to go high, i.e. a logic 1. But, for a CMOS logic device, both *Rlower* and *Rupper* would be large so that, for an open-circuit input, *C* would retain its charge and the logic input value would be sustained at its previous value.

The mapping of the analogue signal to a logic value across the interface amounts to a loss of information because the domain of continuous signal values is mapped to a value in the domain of values 0, 1 and X (as explained in the previous paragraph, in effect, a Z value also maps to one of these values). This is inherent to the higher level of signal abstraction in logic simulation. It is usual to define two thresholds vo and v1 such that an analogue signal v maps to a logic 0 when v falls below v0 and to a logic 1 when v rises above v1. This map is illustrated in Figure 14.4 for (a) v1 → v0 and (b) v1 ← v0. Note input signal transitions occur only when the input voltage crosses a threshold in the appropriate direction. The logic input waveforms for the two cases are similar in form but the transitions occur at different times relative to the analogue input waveform; whether (a) or (b) is appropriate depends on the logic device or, more specifically, its technology.

The real difficulty with the mapping of the foregoing paragraph arises when the analogue input signal does not behave in a 'logical' manner; see Figure 14.5. Strictly, the logic signal transitions are then as shown by the solid-line waveforms of (a) for v1 → v0 and (b) for v1 ← v0. More usefully, mappings of the input to the logic device are achieved by taking into account the sign of the time derivative of the analogue input as well as the previous value of the logic input. Knowledge of the latter was implicit in drawing the waveforms (a) and (b) of Figure 14.4. Thus, the mapping table for (a) v1 → v0 is, where v- is the value of v at the previous analogue time point,

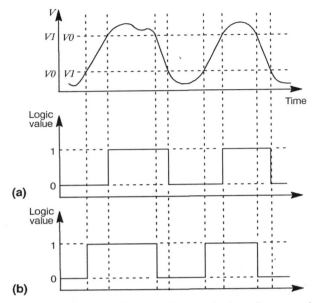

(a)

(b)

Figure 14.4 Analogue voltage-tologic value mapping

Table 14.1 Analogue input

		$V0 \leftarrow V \leftarrow V1$				
		$V \leftarrow V_0$	$V\text{-} \leftarrow V$	$V\text{-} = V$	$V\text{-} \rightarrow V$	$V_1 \leftarrow V$
Previous	0	0	0	x	x	1
Logic	1	0	x	x	1	1
Value	x	0	x	x	x	1

This table has been used to construct the broken-line transitions, indicated in Figure 14.5(a), to the unknown value X. Note the possibility of X as a previous logic value generated by the table. Similarly, the mapping table for (b) v0 ← v1 is

Table 14.2 Analogue input

		$V1 \leftarrow V \leftarrow V0$				
		$V \leftarrow V_1$	$V \leftarrow V\text{-}$	$V = V\text{-}$	$V \rightarrow V\text{-}$	$V_0 \leftarrow V$
Previous	0	0	0	x	x	1
Logic	1	0	x	x	1	1
Value	x	0	x	x	x	1

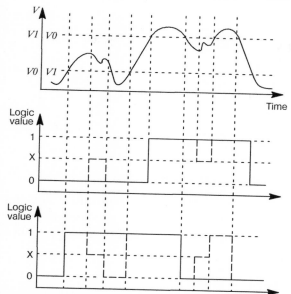

Figure 14.5 Abnormal analogue input waveform to logic output

Since the two tables have the same pattern of entries, just one can be used with different column interpretations for (a) v1 → v0 and (b) v0 → v1.

Another form of non 'logical' behaviour of the input waveform occurs when v remains continuously, although changing only monotonically, between the two threshold levels for too long a period. A useful interface model would map v to X unless v transversed the inter-threshold region within a specified time.

Input transition timing errors occur in practice because analogue simulation time points seldom give rise to analogue input voltages which coincide with v0 or v1. Thus in Figure 14.6, while at first v remains < v1 (and the logic signal remains at 0) at each analogue time point, ultimately a point is attained for which v1 < P v and the 0-to-1 transition occurs. Almost certainly, at this latter point, v1 < v by an amount dependent upon the slew rate of the input signal and the precision of analogue simulation which limits the numerical discretisation. For input waveforms behaving in a 'logical' manner, slew rates will be high and this transition timing error will be small. If it is not, the X generation mechanisms discussed in the previous paragraph are likely to operate and thus indicate the occurrence of the error.

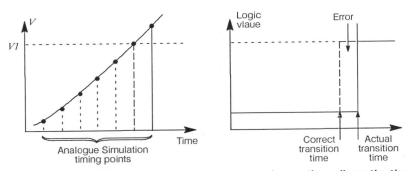

Figure 14.6 Transition time error caused by analogue time discretisation

Note, finally, that the analogue-to-logic input interface is essentially uni-directional. The values of the analogue components of the interface depend on an analogue voltage, determined by analogue simulation, and this voltage is mapped to a logic value.

14.2.4 Logic to analogue output interface

Like the analogue to logic-input interface, the logic-output to analogue interface is uni-directional. The loading of the logic device output on the analogue node is modelled by components, in the analogue environment; see Figure 14.7; Vdd' and Vss' are constant voltages possibly offset from the

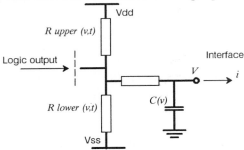

Figure 14.7 Logic output to analogue interface

supply voltages Vdd and Vss. Again, as for the input interface, the values of these components may be nonlinear functions of the analogue node voltages but, additionally, the two resistance values are also functions of the time t. The output logic transition controls the time dependence of these resistances and different logic transitions, e.g. 0-to-1 and 1-to-0, determine different time functions. In a simple modelling situation, the resistances might be independent of v. A 0-to-1 output transition might change the upper resistance from a maximum to minimum value, and the lower one from a minimum to a maximum value, both (say) linearly, over a period initiated by the transition as shown in Figure 14.8. For a 0-to-Z or 1-to-Z

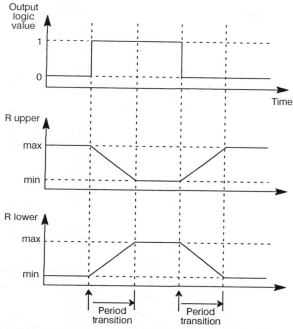

Figure 14.8 Output transition controlling interface resistances

transition, both resistances would be changed to maximum value. An X value has no counterpart in the analogue environment and a transition to this value cannot be mapped realistically. In this situation, almost certainly a design error, an arbitrary decision as to the resistance values would have to be made, e.g. change each to mid-range value.

The transition time and the new logic value are the data passed across the interface, the latter to determine which, or both, of the resistances are to be changed. The component values and the 'period' are essentially analogue characterisation information of the interface. The voltage of the analogue node, as a function of time, is determined by the analogue simulation and will, of course, depend on the resistance and capacitance values. But, in order to minimise timing errors, it is desirable to perform an analogue simulation at the transition time which, of course, is passed across the interface to the analogue simulator. Further analogue time points during the period will be determined automatically by the precision/time-step control mechanism of the analogue simulator.

The discontinuities of time derivatives of the resistance functions are not significant in analogue simulation; indeed, the 'period' might be made zero provided the output capacitance is non-zero. Even so, as for the logic input interface, it is as well to make sure that the resistance functions have continuous gradients with respect to v in order that the analogue simulation is resilient.

14.2.5 Synchronisation of analogue and digital simulation

In section 14.3, it was explained how an analogue signal crossing a threshold generates a transition on a logic input. A small synchronisation error occurs when an analogue voltage at an analogue simulation time point does not coincide precisely with the threshold; generally this is so; see on left of Figure 14.9. In section 14.4, the communication of the time of a logic output

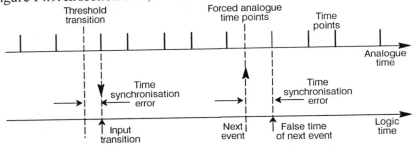

Figure 14.9 Analogue - logic time synchronisation

transition, as a forced time point for analogue simulation, was detailed. When the system is visualised as a number of logic circuit islands embedded in an analogue circuit, the logic simulator can be invoked as a function to return the least next-event time of all the embedded logic circuits. The use of a function in this way avoids a synchronisation error which would arise if logic simulation were delayed until the next analogue time point; see Figure 14.9 on the right hand side. This synchronisation error is likely to be more significant than the other as output slew rates might be smaller and hence time points more widely spaced. The fact that the logic circuits are not directly interconnected is of no consequence provided they are arranged to have common input and event lists. Further, the efficient selective trace techniques can be employed in simulation.

When the next analogue-simulation time point precedes the returned next-event time, the analogue simulation is performed and the logic simulation function invoked at this time; see Figure 14.9. The only processing is of analogue-to-logic input transitions. This might generate a revised next-event time as the returned function value. Otherwise when the returned next-event time precedes the next analogue-simulation time point, the logic simulation function is invoked to process the (possibly) significant list of logic events at this time. Note that, the returned times are mostly those of internal events rather than output events which are less numerous. Even so, in general, many analogue-simulation time points will normally occur between logic events. Accordingly, the total computational cost of the logic simulation will be insignificant, compared to that for the analogue simulation, except when the logic part of the system is exceptionally large.

Alternatively, any system can equally be visualised as analogue circuits embedded in logic circuits. Output transitions of logic elements, and of embedded analogue circuits, schedule future events in the list of the embedding logic circuit. Figure 14.10 shows the event list (OR function of

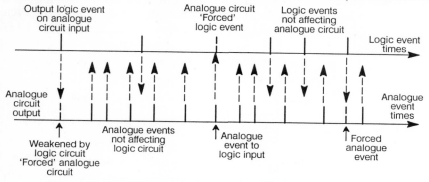

Figure 14.10 Event scheduling, analogue circuit within logic

logic and analogue events) corresponding to an analogue circuit embedded in a logic circuit. A logic event on a node, using the selective trace principle, is listed for all gate and sub-circuit inputs in the fan-out list of the node. When this list includes an input to an analogue circuit the latter is awakened, if latent, by the event, otherwise a time point is forced in its simulation at this instant. Each time step in the analogue circuit is an event in the embedding circuit but creates no events in the fan-out list of an analogue output unless that output crosses a logic threshold and thus effects a logic transition.

A disadvantage of the foregoing scheme is the inefficiency caused by the need to stop analogue simulation after one time step in order to process other sub-circuits at this event time. It would be more efficient to continue simulation of the analogue circuit, either until it causes an output event, or until the next event scheduled in the logic circuit. When there is more than one embedded analogue sub-circuit, however, unless all can be simulated concurrently, the next-event time cannot be known during the simulation of any particular analogue sub-circuit. It might then be necessary, on occasions, to abandon the costly analogue simulations already carried out beyond the ultimate next event.

A more detailed, algorithmic, discussion of these issues follows.

Digital simulators use instantaneous changes in the values of logic variables at discrete instants of time as an abstraction of the actual behaviour of digital circuits. These changes are evaluated in a simulation cycle in which all the changes at the same time instant take place. This cyclic model is also referred to as the time wheel of the digital simulator.

The digital simulator performs the instantaneous value changes in a structured sequence within the cycle. In many digital simulators the cycle may be considered as having essentially two steps, or modes (Figure 14.11).

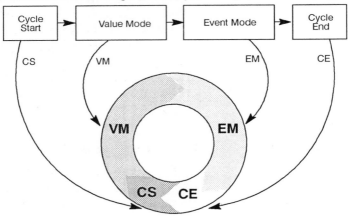

Figure 14.11 The digital simulator time wheel

These are firstly when values are updated, causing element evaluations (value mode), and secondly when events occur (event mode). Whereas element evaluations may affect the values of logic variables at current time through zero delay, events, such as "when clock rises do event" are allowed to affect future values only. When the event mode is completed the digital simulator has both evaluated all activity at current time, and can say when the next activity in the digital domain will occur.

The analogue simulator works in continuous time which it may or may not discretize, transparently to the digital simulator. Obviously all of continuous time lies outside the digital simulator cycle because the digital cycle maps into a single instant in the continuous time domain (Figure 14.12).

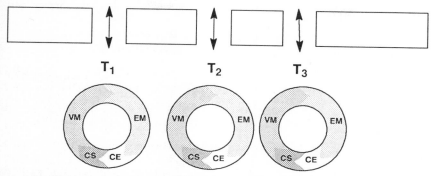

Figure 14.12 The digital time wheel as an instant in continuous time

It will be assumed for the moment that both simulators only move forwards in time. Therefore the analogue simulator will integrate forwards from the end of one digital simulator cycle (which has set up the values at the start of the integration) to the beginning of the next digital simulator cycle. Digital values feed forward naturally into the analogue period, and analogue values feed forward naturally into the next digital cycle. The digital cycles may occur at irregular intervals (Figure 14.13).

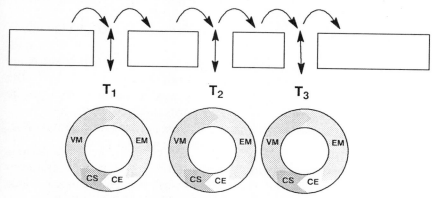

**Figure 14.13 Values flow from the analogue domain
to the digital domain etc.**

The digital simulator can see into the future, for example at the end of any digital simulator cycle it knows what events lie in the future for the digital domain, except for those that the analogue simulator will create (Figure 14.14). The analogue simulator cannot see into the future. It will create inputs back to the digital simulator only by integrating up to that time.

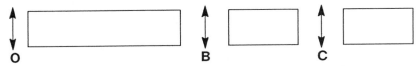

**Figure 14.14 Digital: when cycle O is completed,
event for cycle B and C have been scheduled**

Thus there is no problem for the digital simulator in telling the analogue simulator about the time of the next digital simulator event so that it knows how long to integrate for. The analogue simulator can simulate up to this time as long as it does not itself create any digital simulator schedules.

With only one analogue domain the analogue simulator may integrate up to the time when it creates an input propagating back into the digital domain, or up to the next digital event, whichever is the earliest (Figure 14.15).

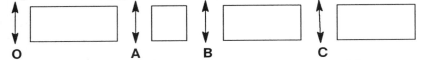

**Figure 14.15 Analogue simulator creates events at A
and hands back to the ditital simulator**

However if there is more than one analogue domain, the analogue simulator must either synchronize those domains itself so that they appear as a single domain, or closely synchronize with the digital simulator.

In the latter case, the analogue simulator must integrate each of its analogue domains in turn only up to the next possible digital cycle. It must then examine the values feeding back to the digital simulator domain to see if any digital simulator schedules are required. If not, it can continue up to the next possible digital simulator cycle. Otherwise it must call back to the digital simulator to schedule the value changes, and then must hand back control to the digital simulator so that it can execute the next digital cycle, after which control is returned to the analogue simulator.

With both simulators only moving forwards in time, the analogue simulator gives back control for any digital cycle. They work in a synchronized or lock step mode, in which the two simulators synchronize whenever there is a signal flow across the interface between the simulators. The synchronization occurs whenever there are events in the value mode of the digital simulator. Synchronization of this type is therefore an implicit function which can be supplied by the digital simulator.

The situation is very different if either simulator can backtrack, for example the analogue simulator. In this case it is not necessary to have the integration periods tightly coupled to all the event times in the digital simulator. It is possible to free the dependence of the analogue domain from value changes in the digital simulator not impinging on the analogue domain.

This is a case for using an additional "external" mode added to the end of the digital cycle (Figure 14.16). Although no schedules after value mode can

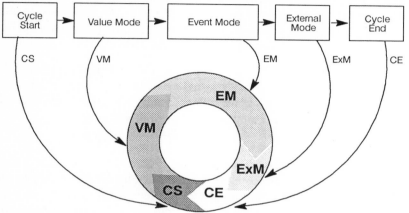

Figure 14.16 Extra mode to allow external scheduling of wake-up by analogue simulator

influence current time within the digital simulator domain, until the schedules in events mode are executed the digital simulator cannot predict the time of occurrence of the next value mode. However using the external mode means that the analogue domain can only influence the future of the digital simulator domain.

The strategy is as follows;

Consider the circuit initialized at time zero (ignoring for the moment how it was initialized).

At time To within the external mode in the digital simulator the analogue simulator is given control. It integrates to a time Ta, taking any path it chooses, including backtracking. Time Ta is either when it has a schedule for digital simulator, or some future time if the analogue simulator does not find a schedule. If there are any value changes, the analogue simulator causes the digital simulator to schedule these changes (at value mode at time Ta). Then, whether or not it has any changes, it asks for a call back at time Ta (external mode), and returns (Figure 14.17a).

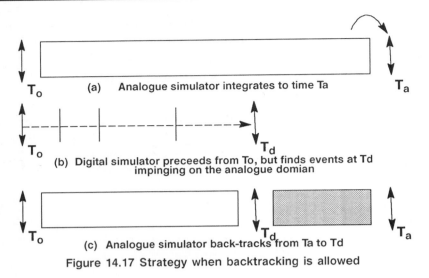

Figure 14.17 Strategy when backtracking is allowed

The digital simulator now proceeds from time To. If it finds values impinging on the analogue domain it automatically creates a call to the analogue simulator in external mode at the relevant time, say Td (Figure 14.17b). Now the analogue simulator is effectively event driven from any input-terminal value changes. Otherwise if no values impinge on the analogue simulator the digital simulator proceeds to time Ta.

When the analogue simulator is called back, it is either time Ta (no value changes) or time Td (value changes).

If the time is Ta, then the cycle from To is simply repeated and the changes scheduled by the analogue simulator at Ta, value mode, will already have taken place.

If the time is Td, then the changes and the call back at time Ta no longer apply. In this case the input terminal value changes from the digital simulator will have already taken place at the value mode at time Td and the changes scheduled by the analogue simulator for Ta in value mode have not taken place.

The integration from To to Ta no will longer apply because of the changes at Td impinging on the analogue integration. The analogue simulator must either re-integrate from To to Td, or find some other way of back-tracking to Td (Figure 14.7c), and it must first undo the schedules at Ta. One way to do this is by simply rescheduling the actual values on the analogue simulator output terminals, since this usually has the effect of overwriting any existing schedules. The schedules themselves will die out in the digital scheduler since they do not change the values.

Thus the analogue simulator must always have the capability of undoing the last integration, and effectively cancelling call backs and future value changes it has scheduled.

The great advantage of this algorithm is that the time steps of the analogue simulator integration can be chosen by the analogue simulator domain itself. Furthermore the analogue simulator is affected only by value changes impinging on it. The two domains synchronize only when they exchange information.

It is also possible to have separate analogue instances with this algorithm. If there is direct coupling between the analogue instances this must be handled transparently by the analogue domain. Otherwise the coupling must be handled via the digital domain.

It is not even necessary to coordinate the synchronizations between the instances. Each instance is allowed to integrate ahead to its own Ta. The digital simulator calls back each instance independently with any schedules for it, always calling the instances in strict time sequence in the order of the schedules. Any analogue instance which is called for a value change schedule will always backtrack to the current digital simulator time. Thus the analogue instances will naturally synchronize independently with the digital domain.

Initialization of the circuit presents a different problem. The analogue domains must set up output values on the assumption that the circuit has been in the initialized state for an infinitely long time. Effectively this means that input values propagate across to outputs in some known way. Thus the initialization functionality is implicitly different from the integration functionality. Furthermore the propagation of these values, repeatedly if necessary, in zero delay is required by the digital simulator to initialize the circuit. This effectively means driving the analogue domain repeatedly from input terminal value changes during the instant of initialization of the digital simulator.

Note that if the analogue simulator needs to iterate signals to a converged value without moving time forwards (for example to initialize the circuit), it can only do so by operating within the value mode of the digital cycle in addition to the external mode.

The back-tracking mode is clearly inefficient. It is the analogue domain that is most expensive to simulate, and clearly it is better never to have to throw away any analogue simulation by back-tracking. It would be more efficient to put the onus on the digital domain to have to back-track.

An alternative is to be more intelligent about the impact of future digital events back onto the analogue domain. After all it is the possible impact of future events onto the analogue domain that prevents the analogue domain from proceeding past the next digital event (unless it is prepared to back-track). However the digital domain can predict the propagation time of any event in terms of its possible impact on any part of the analogue domain. It can do this by static analysis before the simulation starts, stamping each node in the digital domain with the propagation time. If the digital domain then bases the time of the next future event on the earliest future time that a future event can impact the analogue domain, then the analogue domain can integrate for far longer. This has the benefit of greatly reducing the number of synchronisations.

14.2.6 Examples of mixed-mode simulation

Two examples of mixed-mode simulation are presented. The first demonstrates what can be achieved in simulation of a practical circuit and the second was designed to investigate precision of simulation and to assess simulation costs. For both examples a prototype simulator, based on the principles discussed in earlier sections, was used. The simulator was able to process analogue circuits with embedded logic sub-circuits.

The first circuit is an analogue-to-digital converter shown in module form in Figure 14.18. The 8-bit logic counter is composed of 4 master-slave JK

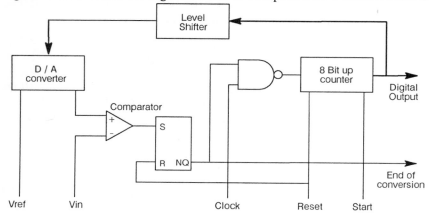

Figure 14.18 Modular description of analogue converters

flip-flops each comprising 8 NAND gates. The digital-to-analogue converter module is a weighted-resistor type of a device using complementary CMOS analogue switches; a simple model is used for the operational amplifier. Thus this module is analogue. The comparator is also represented by a simple analogue model. The level shifter is a small analogue circuit and the flip-flop on the comparator output and the clock-signal transmission gate are both logic circuits. The logic parts of the circuit have a gate count of 41 and there are 45 analogue components in all. Figure 14.19 shows the simulated output of the D-to-A converter module at

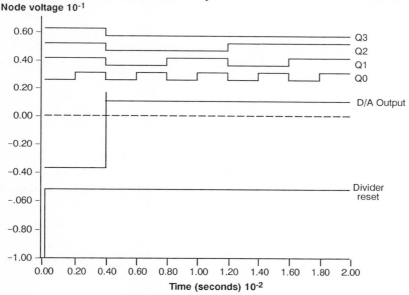

Figure 14.19 Start of conversation process

the start of the A-to-D conversion process; the Q's are the counter outputs. Figure 14.20 shows the simulation results at the end of the conversion process. Note that the vertical scales of these last two figures are different. The apparent uncertain timing at the end of the process, exhibited in the last figure, arises because of the insensitivity of the comparator as demonstrated in the simulation result shown in Figure 14.21; note the significant expansion of the display in the vertical direction.

A module diagram of the second example circuit is shown in Figure 14.22. It represents a variable word-width full adder (including carry propagation) with registered output. The latter is fed back as an input to the adder. The adder comprises 32 transistors per bit of word length and the register 24. A further inverter is used in each bit feedback path making a total count of 58 transistors per bit for the circuit as a whole. The example has been simulated (a) totally at the circuit level (as a precision and cost reference),

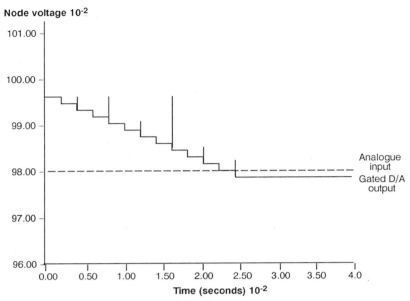

Figure 14.20 Simulation results of end of the conversion process

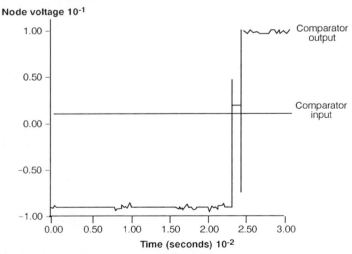

Figure 14.21 Simulation results of comparator input and output

(b) with the adder at circuit level and the register at logic level, (c) with the adder at logic level and the register at circuit level, and (d) totally at the logic level (as the other extreme of cost reference). Values of n range from 1 to 8. A typical result of simulations is shown in Figure 14.23. The dotted waveform depicts the register clock signal. The response displayed is that of

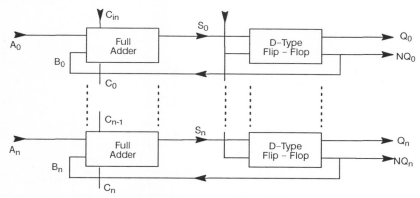

Figure 14.22 Modular diagram of Adder register

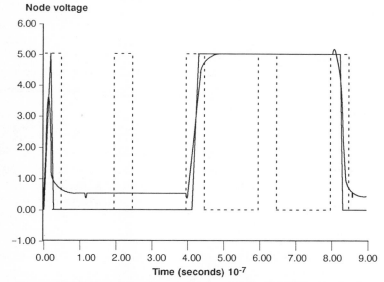

Figure 14.23 Reocision comparison – bit 3 of register output

the bit 3 output of the register (bit 0 is msb), waveforms for cases (a) and (b) virtually coincide as do also those for cases (c) and (d). The result indicates the kind of precision possible with mixed-mode simulation. An interesting feature, in particular, arises because the clock rate is such that the ripple carry from bit 2 arrives too late for the register to be able to latch the correct bit 3 input in time for the '1' bit 3 output on the second clock pulse. Figure 14.24 depicts the bit 3 sum output of the adder. The analogue glitch in the middle of the waveforms, for cases (a) and (c), is not apparent in (d), the wholly logic simulation. Its absence is due to, first, an inertial-delay logic

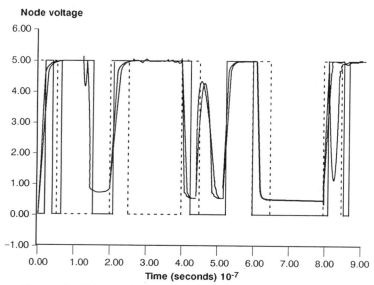

Figure 14.24 Precision comparison – bit 3 of adder output

model and second, failure of the analogue waveform to attain the 0-to-1 switching threshold.

Experimentally observed time complexity for the simulations is shown in Figure 14.25 . As indicated earlier, analogue simulation time dominates. Case (c) shows a slightly lower simulation time than might be expected from an estimation based on transistor count. All that can be claimed for these results is a time complexity slightly worse than linear with n.

14.2.7 Conclusions

The principles on which mixed-mode circuit-level/logic-level simulation is based have been described. The principal areas of new development concern the interfaces between circuit-level and logic devices, in particular, the mapping of signals across those interfaces and the loads reflected onto analogue nodes by logic devices. The feasibility of mixed-mode simulation has been demonstrated by example and questions of precision and cost of simulation addressed.

A number of second-generation mixed-mode simulators have been described in the literature, e.g. Acuna (1989) and related commercial products are now becoming available. It remains to be seen whether these products will be received more favourably than earlier ones by the design community.

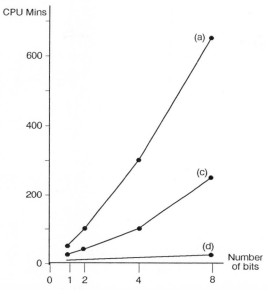

Figure 14.25 Experimentally obsreved relationship between CPU time and complexity

14.8 References

[1] Acuna E, Dervenis J, Pagones A, Saleh R, "iSPLICE3: A new simulator for mixed Analogue/Digital circuits" Custom Integrated circuits Conf. Digest May 1989, pp. 13.1.1 – 13.1.4

14.9 Appendix

14.9.1 Lock step algorthim

```
1. Lock step, many domains

instantiate()
{
    ext_synchronize(); /* specify synchronized mode
                          for this instance */
    instantiate_domains(); /* set up domains */
}

activate()
{
    int time_now = ext_abs_int_time();
    int time_next = ext_next_abs_int_time();
    int actual_steps;

    /* read all input terminals into
       private structures */

    setup_input_values();
```

```
    /* integrate to either time_next or to
       next schedule produced */

    actual_steps = integrate_domains
                      (time_next-time_now);

    /* guaranteed to be called back immediately */
}
integrate_domains(time_steps)
int time_steps;
{
    int answer = NO;

    for (delta_time = 1; delta_time <= time_steps;
                     delta_time++)
{
    for (domain = 0; domain < total; domain++)
    {
        answer |= integrate(domain); /*
                returns YES if caused a schedule */
    }

        if (answer) return delta_time;
    }
    return time_steps; }
```

14.9.2 Back-tracking, many domains – algorthim

```
instantiate()
{
    instantiate_domains(); /* set up domains */
}

LOCAL time_o; /* the last time synchronized */
LOCAL time_d; /* the current digital simulator time */
LOCAL time_a; /* the future time that the analogueue
               simulator has reached */

activate(instance, reason)
int instance;
int reason;
{
    /*** the back-tracking algorithm ***/

    int actual_steps;

    /* get the time that the digital simulator has
       reached */

    time_d = ext_abs_int_time();

    /* check if duplicate call at this time,
       or if call cancelled */
```

```
if (time_d == last_time || (reason == call_back &&
time_d == time_cancelled))
{
    return;
}
last_time = time_d;

/* check if already integrated past time_d,
   if so must re-integrate from time_o */

if (time_d < time_a)
{
    cancel_schedules();        /* at time_a */
    time_cancelled = time_a; /* stash time so that
    call back can be ignored */

    actual_steps = integrate_domains (time_d -
    time_o); /* re_integrate to time_d */
    update_o_from_a(); /* new values always _a
    values after integration */

    time_o = time_o + actual_steps;    /* update the
    time to which we have synchronized */
}
    else if (time_d > 0) /* i.e. not the
                            first call */
{
    update_o_from_a();    /* the normal case after
                            no backtracking */
    time_o = time_a;
}

/* read all analogueue input terminals */

update_inputs_now();

/* integrate, stopping if a schedule is created */

actual_steps = integrate_domains(coarse_time_steps);
/* recreating new _a values */

/* update time_a */

time_a = time_o + actual_steps;

/* schedule call back at time_a */

ext_sched_after(actual_steps);
}
```

Chapter 15
Testing Mixed Analogue & Digital Circuits

Peter Shepherd
University of Bath, England

15.1 Introduction

With the recent improvements in processing technology which have made available reliable CMOS and BiCMOS realizations of both digital and analogue circuit components on the same chip, there has been a large increase in the number of mixed-mode circuits manufactured. The details of the processes and design techniques have been described earlier in this book. The majority of the circuits being realized are designed for specific functions and are manufactured in relatively small quantities, and hence fall under the category of application specific integrated circuits (ASICs). The increase in access to the processing lines via the silicon broker system, with the resulting decrease in price, has widened the availability of these mixed mode ASICs to many more engineers, both in industry and academia. It is expected that this trend will increase in the coming years, with about a 3-fold increase in manufacture of these circuits predicted over the next five years.

There remains the problem, however, of testing the circuits. In general, the testing of ASICs is largely left up to the designers themselves. The manufacturers will do a probe test at the wafer level, so faulty chips may be eliminated before they are packaged. This test may be based on a test pattern devised by the designer, or it may be a test on a 'drop-in' circuit of the manufacturers design, in order to check for any global defects. The use of designer-specified test patterns by the vendors is fairly acceptable for purely digital and even purely analogue chips, although strict guidelines for the form of the test are often given by the vendor. However, in the case of mixed mode ASICs, the manufacturer is unlikely to perform any functional test.

The design of suitable testing routines is therefore very much the responsibility of the ASIC designer, and in recent years the concept of 'design for testability' has emerged as a vital part of the overall design procedure. This has been primarily led by the digital field, where very large scale circuits (10s to 100s of thousands of transistors) are now available for custom design. The problem of successful testing of such circuits is immense, so many tools and techniques have been developed to aid the designer to optimise the testability of his circuits. These include built in self test (BIST), scan path designs, PRBS and signature analysis, etc. The approach to analogue circuits is a lot less clear, however. The diversity of analogue functions available is very wide, and the performance of the circuit can be measured in many ways. For example, measurement of the circuit response in the time or frequency domain, temperature variations, offset voltages, leakage currents, stability conditions (Bode plots), etc., may all be of interest to the designer. To be able to thoroughly test an analogue circuit to ensure conformance with all of these specifications is a very time consuming task, with a large test equipment requirement. With the incorporation of both analogue and digital circuitry into a single chip, the testing task is exacerbated. To add to the difficulty there are circuit blocks which have characteristics of both modes, e.g. A/D and D/A converters and switched capacitor circuits.

The testing task facing the mixed-mode ASIC designer is therefore a very complex one, and at present there is no universal solution to the problem. This chapter presents some of the tools currently available to the engineer to aid him, and discusses some of the techniques which may be developed in the future to provide some form of unified test approach for mixed-mode circuits. Currently, the only practical approach is one of 'divide-and-conquer', whereby the circuit is partitioned somehow into separate analogue and digital blocks, and the characteristic test approaches applied to each. The problem here is that the timing associated with the separate blocks often needs to be coordinated and controlled centrally. The first part of the chapter describes how digital signal processing (DSP) can be used to simulate the analogue test routines, so that central control of both test functions is maintained and overall timing patterns can be synchronized. The second part of the chapter looks at some theoretical approaches which have been proposed to avoid partitioning the circuits, but to apply some form of unified testing technique. These techniques include DC fault dictionaries, digital modelling of analogue blocks, logical decomposition, artificial intelligence (AI) techniques and transient response testing.

15.2 Mixed-mode testing by partitioning and the use of DSP

Currently the standard and practical approach to the testing of a general mixed-mode IC is by partitioning the circuit into separate blocks of digital and analogue circuitry and then application of mode specific testing to each. This partitioning process will involve the placing of extra probe pads, and possibly the inclusion of some multiplexing or BIST circuitry. This enables the circuit to be switched between its normal mode of operation and some test or self-test mode. All this leads to a greater overhead for the chip. Since the cost of a chip is almost directly proportional to its area, and increased area may lead to decreased yield, this approach inherently adds a cost burden to the design. The designer must balance the amount of overhead he adds to the chip design, with the improvements in testability resulting from this. In addition, since few ASICs are functionally tested by the vendor, the burden of design for test, and the testing operation itself lies with the designer and he must also possess analogue and digital test equipment of sufficient quality to test his design. As we shall see, modern automatic test equipment (ATE) is capable of performing both tests in a controlled and synchronized way via DSP, but such test equipment can be very expensive.

Given that the partitioning approach, though far from ideal, is perhaps the only currently practical one, what tools and techniques exist to ease the task and cost?

Powerful digital test equipment is now widely available at relatively modest cost which can be used to test those types of circuit blocks. In addition, many design-for-test techniques have been well developed and are readily available to the digital designer. The analogue side is perhaps not so well developed. As both the need and the processes for large scale analogue or mixed signal ICs have only recently been developed, so both the equipment and techniques have lagged behind their digital counterparts.

However, even given available test equipment for both circuit modes, there remains a further problem with the partitioning approach. Often the correct performance of a mixed-mode circuit will depend on the overall timing of the signals within the circuit and synchronization of events. If the two modes are completely partitioned from each other, it can be difficult to ensure this correct timing and synchronization. Even with controllable analogue instrumentation, it is difficult to ensure synchronization due to cable delays and instrument settling times.

It is to overcome this problem, as well as having other advantages, that the use of DSP based testing has recently been developed.

15.2.1 Introduction to DSP techniques applied to testing [1]

Digital signal processing, when applied to analogue or mixed-mode testing, goes further than simply control of analogue instruments and subsequent processing of the results. The DSP equipment actually simulates the action of the analogue equipment, and therefore substitutes it. This has several advantages. Firstly, both the simulated analogue and the digital test equipment are under the control of a central processor, so that timing of clocks, signals and measurements can be synchronized accurately. In addition the software simulation of certain analogue functions can be more accurate than the instrumentation being simulated. This is because many of the disadvantages of such instrumentation such as non-linearity, noise, drift, aging, improper calibration, settling time, thermal effects, etc., can be reduced or eliminated completely. The DSP based test can often be performed faster when using low-level coded floating point algorithms.

A further point about the use of DSP is the application to non-deterministic testing. In testing a purely digital circuit, the outputs are examined under a particular input stimulation. The sequence of 1s and 0s are checked against a known, expected output sequence. If any of the bits differ from this expected sequence, the device under test is deemed to have failed and the test **need** not continue. This is termed deterministic testing, the precise nature of the expected output has been completely determined before the test starts. DSP is of no direct value in such a case, as there is no overall 'signal' to be processed during or after the test - it is just a direct comparison of signals. However, when analogue blocks are involved there is an acceptable variation in the output signals, such that the output may differ from the ideal expected case, without being deemed to have failed. This is termed non-deterministic testing. There is an acceptable error associated with the test output, but for this error to be properly evaluated in order to determine its acceptability or not, the whole signal or output waveform must be considered. It is here that DSP plays an important role.

The development of analogue test techniques has progressed from the manual test bench which comprised some form of stimulus (DC., single or swept frequency AC, noise, etc.) on an input port, and some form of analogue meter (DC., rms, peak, etc.) on one or more of the output ports. This is the basic setup for a transmission test. There are also parametric tests which may be performed on a single port (e.g. leakage current, offset voltage, impedance, etc.).

With the introduction of minicomputers in the 1970s, a degree of automation was possible. In this case the control of the analogue instrumentation was largely taken out of the hands of the human operator. In addition, the computer could aid with the data logging and processing of results. This often improved the accuracy and speed of the test. However, the analogue source and measurement equipment was still at the heart of the system, with the disadvantages outlined previously still inherent.

The application of DSP has taken a further step forward with the elimination of this analogue equipment and replacement by an emulation of these units. Such a transmission test system is illustrated in Figure 15.1. The emulation of some of these blocks is now described.

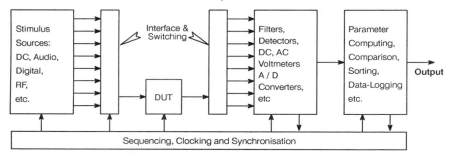

Figure 15.1 DPS Emulation of analogue transmission test

15.2.2 Emulation of analogue instruments via DSP

In order to be an acceptable substitute for their analogue counterparts, the DSP-based instruments must be able to emulate or electrically imitate the performance of the instruments, without having the drawbacks associated with the analogue circuitry. In order to do this, it requires two facilities. Firstly it requires some form of computer programming to simulate the analogue function. For speed, this programming should be at a low level of machine coding. Secondly there should be an interface so that the DSP based instrument can communicate with the analogue device under test. These interfaces are generally based on analogue-digital (A/D) or digital-analogue (D/A) converter circuits, depending on the instrument function.

For example, Figure 15.2 shows the block diagram of a DSP based waveform synthesizer. The input to the instrument is a string of digital data which relates to the desired time sequence of the waveform. This time sequence, or waveform vector, is transferred into a local memory bank and then at the correct time is passed to the D/A block which outputs a sequence of analogue levels. These are in a 'sample and hold' form, so the sequence is passed through a reconstruction filter to derive a continuous, band limited analogue waveform.

The reverse instrument is shown in Figure 15.3 which illustrates a waveform digitizer. The analogue signal is to be sampled, so that each analogue level at a particular point in time can be converted to an equivalent digital word. The sampling rate places a restriction on the bandwidth of the signal being sampled (Shannon's theorem [2]). Without this limit there will be spectral

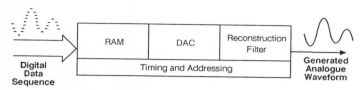

Figure 15.2 DPS Emulation of Waveform Generator

leakage from the higher frequency sample-generated 'image' signals into the signal of interest, causing distortion of that signal. This effect is termed aliasing. The anti-aliasing filter band limits the signal to within half the sample frequency to avoid this effect. The remainder of the block is the reverse of the synthesizer block.

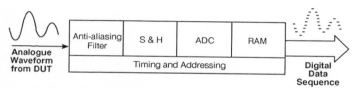

Figure 15.3 DPS Emulation of Waveform Digitiser

It was mentioned earlier that DSP-based testing is generally more accurate and often faster than their analogue equivalents. This is due mainly to the particular structure and circuitry of the analogue instruments. For example in the measurement of voltage, the analogue signal is passed to some form of detector which converts the signal magnitude into an equivalent DC. voltage which can be displayed on a meter. For DC. measurement this conversion is an entirely linear process, and can be performed fairly accurately. With AC. signals the detection involves at least one form of non-linear function (e.g. square law, rectification, clipping, log function, etc.). The poor accuracy of analogue circuitry to reproduce non-linear characteristics means that analogue instruments are inherently less accurate for AC. measurement than DC. measurement. The fact that a good ADC can be much more accurate than an AC. analogue instrument, provides the basis for improved accuracy via DSP. In addition, the settling time of the analogue instrument to provide the smoothed DC. output means that the DSP approach of sampling and computer calculation can often be a faster solution.

These factors mean that for successful hardware emulation, the ADC process must be placed ahead of the detection system, the latter being performed by a computer algorithm.

For example in the measurement of an rms voltage, the traditional analogue instrument will first square the incoming signal, then pass it through an averaging filter to extract the mean. The square root of this value is then taken to provide the rms. The averaging filter is essentially a low pass filter, the time constant of which must be much greater than the period of the signal under test. Therefore the settling time can be relatively long. This is unimportant in human operated instruments as the settling period is still much shorter than human reaction time. But in automated measurements the settling time can be prohibitively long. In a DSP based instrument this filtering action is replaced by a timed integration. This derives directly from the mathematical definition

$$V(rms) = \sqrt{\frac{1}{P} \int V_{in}^2 dt}$$

15.1

where P is the period of integration. The square and square root functions can be performed algebraically by the processor with much greater accuracy than the equivalent analogue circuitry. The filtering function is eliminated and the settling period is replaced by a shorter, controllable, period of integration. However, as the signal processing operations are being performed on a vector of sampled points, the integration cannot be performed as a continuous function, and must be replaced by a summation as

$$V(rms) = \sqrt{\frac{1}{N} \sum_{I=1}^{N} V(I)^2}$$

15.2

It is impossible practically to extract sufficient information from a signal by only sampling one signal cycle. In general the signal is sampled over several cycle periods, and for correct measurement this test interval, called the unit test period or UTP, must contain a whole number of cycle periods (M). In addition, the number of samples taken within the UTP must also be a whole number (N). The relation between M, N and any clock periods associated with the device under test have to obey certain rules for the test sampling to be valid and accurate.

For example, consider an 8-bit ADC (0-255 levels) which is to be tested by applying a 1kHz sine wave as an input signal. Let us sample at a rate of 8000 samples/sec (well above the Shannon limit), and apply the test for 50 cycle periods. Therefore we have :

Ft = 1000Hz   UTP = 50mSec
Fs = 8000Hz   M = 50 Cycles
         N = 400 Samples

This would at first sight seem to be a sufficient rate and time to achieve the test. But if the signal and the test points are examined in the time domain, it will be seen that the sample points occur at the same point in each cycle. At most, 8 unique levels of output will be tested, and this test will be repeated 50 times. This is because there is a common factor of 50 between N and M. For the test process to be optimised N and M must be relatively prime, that is to have no common factor other than 1.

One approach to overcome the problem of common factors between N and M is to slightly offset Ft, so that the input signal drifts slightly with respect to the sampling points. If there is no synchronization of the signal and the sampling, this process is similar to random sampling, and given a sufficient test length, sufficient information can be extracted to fully test the device. However, leaving the process with this random element may lead to prohibitively long test sequences.

By synchronizing the input signal and the sampling process, and by insuring relatively prime M and N we can optimise the distribution of the samples over the input test signal. This process is termed coherent sampling. The relationships to insure coherence can be derived from the following equations :

$$Ft = M * \delta \qquad \boxed{15.3}$$
$$Fs = N * \delta \qquad \boxed{15.4}$$
$$\delta = \frac{1}{UPT} \qquad \boxed{15.5}$$

with M,N relatively prime. δ is termed the primitive frequency. Dividing (15.3) by (15.4) gives the fundamental requirement for coherence,

$$\frac{Ft}{Fs} = \frac{M}{N} \qquad \boxed{15.6}$$

Given Fs at 8000Hz and N at 400 samples, this gives a primitive frequency of 20Hz. Therefore for coherent testing, all periodic components should be multiples of δ, therefore Ft should be chosen from the sequence..., 980, 1000, 1020, 1040..., The nearest possibilities to our original 1kHz, while still maintaining coherence are 980 and 1015. Suppose we use 1020Hz, then

Ft = 1020Hz	M = 51
Fs = 8000Hz	N = 400
δ = 20Hz	UTP = 50mSec

and we have optimised the test.

15.2.3 Further Applications of DSP-Based Instrumentation

In the last section we examined the basics of DSP measurement of an analogue signal, illustrating the technique for calculating the rms voltage of an AC signal, and the importance and conditions of coherent sampling. In this section we discuss some of the further applications of DSP to emulate other measurement functions. These include a vector voltmeter, spectral analysis and noise measurement.

The vector voltmeter provides information about the magnitude and phase of a particular frequency component of a signal, (the same information could be in the form of real and imaginary parts). So it combines the analogue functions of an infinitely selective bandpass filter, AC voltmeter and phase meter. We have already seen how timed integration can be used to simulate a filtering function. To apply this principle to derive a vector voltmeter, we must also make use of the correlation function. This is given mathematically as:

$$R(\tau) = G \int_P A(t).B(t - \tau)$$

15.7

This gives a correlation function R, of two time signals A and B. The integration is performed over a measurement period P, and G is a gain or scaling factor. The variable τ is a programmable delay which is introduced to one or other of the signals to offset the timing of one signal with respect to the other. By using two delay lines, associated with each signal, τ can be made positive or negative, i.e. either signal can lag or lead. The hardware equivalent of this function is shown in Figure 15.4. The scaling factor is to insure that R lies between the values +1 and -1. R = +1 indicates the two signals are identical in form, R = -1 indicates that they have identical but inverted form, and R = 0 indicates that they are statistically unrelated at that particular time shift τ. Auto-correlation is when the same signal is applied at both inputs. Auto-correlation always equals +1 at zero delay, and is an even function of τ.

Now we need to consider two principles to understand how correlators can be incorporated into a vector voltmeter. The first deals with two steady state sinusoids of differing frequencies. If P were infinite, the cross-correlation

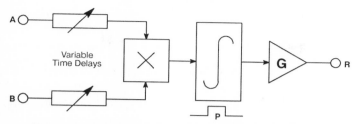

Figure 15.4 Hardware Model of Correlation Function

will be zero. For finite P this condition can still hold, provided that P is such that there is a whole number of cycles of A, and also a whole number of cycles of B. If this is the case, then the cross correlation is zero, irrespective of the relative magnitudes and phases of A and B. We now have the means to our infinitely selective filter. If a signal comprising many coherent frequency components is one input, then by correlating with a reference input frequency at the other input, the correlator will ignore all other frequency components and just give information of the magnitude at the reference frequency.

The second principle deals with two signals of the same frequency but displaced in phase by 90°. In this case, again provided there are a whole number of cycles in the test period, then the cross-correlation is zero, irrespective of the relative magnitudes of the signals and the start point of the measurement. We now have the means of resolving the real and imaginary parts of a frequency component. When two correlators are driven by quadrature reference signals, the output of one represents the real part, and the output of the other the imaginary part at the reference frequency. The magnitude and phase can be determined by the relations :

$$|x| = \sqrt{real(x)^2 + imag(x)^2}$$

$$\boxed{15.8}$$

$$< x = \tan^{-1} \frac{imag(x)}{real(x)}$$

$$\boxed{15.9}$$

This discussion of the vector voltmeter has been mainly illustrative to show how analogue functions may be simulated by DSP techniques and to stress the importance of the coherent sampling condition. In fact the process described above is identical to the mathematical algorithm called the Discrete Fourier Transform (DFT), and it is this algorithm which would be implemented in practical DSP-based equipment. Often though, we wish to determine the magnitudes and phases of many frequency components simultaneously, i.e. to implement a range of vector voltmeters at many frequency points. This is done via an associated algorithm called the Fast

Fourier Transform [3]. This is a mathematical routine which converts time domain data into frequency information and so is the software equivalent of spectral analysis, providing both magnitude and phase information.

It is important to maintain the coherence of sampling, such that all spectral components are whole multiples of δ. By considering all the various harmonics and any inter-modulation products, the UTP can be calculated. All these spectral components will then be orthogonal (i.e. with zero cross-correlations) and so can be uniquely identified by the FFT algorithm. However, if there are any strong spectral components which do not have a whole number of cycles within the UTP, they will not be orthogonal to the other spectral components, the cross-correlations will be non-zero, with the result that there will be a 'leakage' of the signal into other frequency components resulting in erroneous results. This 'spectral leakage' can be minimized by applying shaped time 'windows' [4] to the time data before applying the FFT algorithm. The effect of these windows is to 'smooth' the time domain values near the start and end points of the UTP, to lead to a continuous function across the boundary between two test periods. This attempts to force a periodicity to all the components within the UTP and improve orthogonality.

Another important consideration in the FFT is that the routine is usually based on a decimation process whereby there is a repetitive divide-by-2 operation. This consequently restricts the number of sample points, N, to a power of two. The process then results in $N/2+1$ spectral components (including the zero frequency or DC. component) spaced at the primitive frequency δ.

There exists an inverse algorithm to convert from frequency domain to time domain data, called the inverse FFT. This can be used in DSP test equipment for synthesizing analogue test vectors with specified spectral content.

Multi-tone testing, whereby the input stimulus contains more than one frequency component can be a useful tool to measure inter-modulation distortion, noise, and can also be used as a substitute for a swept frequency source. Provided the stimulating tones are multiples of the primitive frequency, and the test period, P, corresponding to the FFT sampling period, is an integer multiple, K, of the UTP, then there will be $2N + 1$ measured frequencies, which will include all the possible harmonic and inter-modulation frequencies. For example, if the stimulating tones were 3δ and 7δ, then second harmonics would occur at 6δ and 14δ, third harmonics at 9δ and 21δ etc. 2nd order (sum and product) I/M products would occur at

4δ and 10δ, third order I/M products at 1δ, 13δ, 11δ and 17δ, etc. There are certain to be frequencies which are non-harmonic, and no distortion components will fall into these. Any power measured at these frequencies will be due to noise (random and quantization) and so the total noise can be determined by multiplying this power by $2N+1$. The random and quantization components can be separated by using a test period P in which $K>1$. In this case the quantization errors only occur at the frequencies which are a multiple of K. Any non-harmonic components which are not a multiple of K in ordinal value contain only random noise. Any non-harmonic components which are multiples of K in ordinal value contain both quantization and random noise.

A swept frequency measurement can be simulated in two ways. Firstly a series of DFT stimulations and measurements can be performed at a series of frequencies through the desired range of measurement. Alternatively the whole measurement can be done in one go using the FFT and a sufficiently small δ to give the desired frequency resolution. The latter technique has the obvious advantage of speed, but may be slightly less accurate. That is because there are stimulating signal components at all the frequencies of interest, and as the total input power to the device must be limited, the signal to noise ratio of the individual frequency components will be decreased. Alternatively, maintaining high signal levels, the device may be driven into non-linear operation and so give distorted results.

15.3 Other testing approaches

While partitioning and mode specific testing are currently the only practical approaches to mixed IC testing, the technique is far from ideal. The keywords in any test approach are controllability and observability. Controllability is the ability to set the circuit into any particular state (more important in the case of digital circuits where there may be many millions of discrete possible states). Observability is the ability to be able to propagate a potential fault through to a 'probable' node - it is clearly impractical to probe every node on the circuit. Increasing these two factors to an acceptable level involves adding extra circuit 'overhead' dedicated to the testing process. In general there is a trade off between the increased cost in silicon overhead and the reduced testing cost due to the increased controllability and observability. Figure 15.5 shows the general form of the relative costs as a function of overhead. The total cost curve shows a minimum which usually occurs around 10-20% overhead.

In the case of mixed-mode testing by partitioning, the designer is constrained to add overhead so that the boundaries between modes are

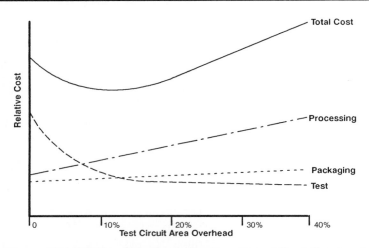

Figure 15.5 Relative Chip Cost as a Function of Test Overhead

controlled and observed - he has no choice in this matter. He may also need to add further overhead to improve the controllability and observability of the digital blocks. The total overhead imposed is unlikely to be near the optimal minimum of the curve in Figure 15.5. Ideally, we require techniques which will 'unify' the testing techniques between the modes, so that the necessity to partition the circuit is eliminated and the designer can select the required overhead to optimise his testing and manufacturing costs.

This section describes a number of approaches which go some way towards this goal. They are not offered as complete solutions or alternatives to the partitioning approach. Each has certain merits and drawbacks, but as the future of mixed-mode testing must lie in a unified approach, it must be from these sorts of approaches, or combinations of them, possibly with some limited partitioning, that a general and successful technique will arise.

15.3.1 DC. Fault Dictionary

The first technique is based on a purely DC. measurement. This is mainly applicable to analogue circuitry, and then only to a limited degree, but with modification may also be applied to digital circuits. The strength of this approach, and the reason for its inclusion in this section, is that it provides a route to selecting the optimum nodes for testing.

The technique described here is based on that proposed in [5], with some modification and improvement [6]. The technique consists of two stages - a pre-test simulation stage to compile the fault dictionary, and a post-test stage to identify or isolate the fault. So the technique can go further than simply go/no-go testing to provide some degree of diagnostic test.

In the pre-test stage the circuit is simulated under fault-free conditions and then including, separately, all the expected catastrophic fault conditions. The voltage values at all the nodes in each case are recorded. These results form the basis of the fault dictionary. The circuit under test is then measured and the results compared with the stored values to determine if a fault is present. When applied to analogue circuits, however, the DC. voltages may have a range of values that are acceptable. Therefore the detection process is established as follows.

(1)   In the case of the fault free circuit each node will have a nominal voltage Vn. Due to fluctuations in the manufacturing process, these voltages may lie within a particular error or ambiguity range $\pm\delta V$. The value of δV should be based on knowledge of the particular process, but may also be based on the resolution of the measurement equipment. Any fault which produces a voltage within this nominal ambiguity set is deemed undetectable.

(2)   The first fault which produces a nodal voltage outside of Vn $\pm\delta V$ forms a new ambiguity set with a range of Vf $\pm\delta V$, where Vf is the calculated faulty voltage at that node. Other faults falling outside the nominal set are checked to see if they lie within an existing faulty ambiguity set. If not a new set is created. If an overlap δVp occurs between two successive sets, then to eliminate it, δVp is divided equally between the two sets with an arbitrarily small guarding voltage to separate the sets.

(3)   To determine which fault is present in the circuit-under-test (CUT), the ambiguity sets are manipulated thus :

(a)   Any ambiguity set that has a single fault within it uniquely defines the fault at that node.

(b)   Ambiguity sets whose intersection or symmetric difference results in a single fault also uniquely defines that fault. In this case all the nodes for all ambiguity sets involved are required.

15.3.1.1 Example

The above algorithm was applied to a CMOS comparator circuit, shown in Figure 15.6. The circuit was simulated under fault-free, followed by 53 fault conditions based on the occurrence of single catastrophic (open and short circuit) faults. The following simulation conditions were applied:

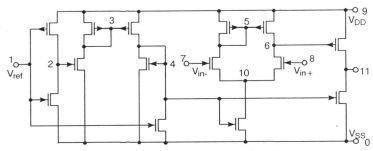

Figure 15.6 Comparator circuit

Vdd = 5.0V, Vss = 0V. With the inverting terminal (node 7) set at -5.0V and the non-inverting terminal (node 8) at + 5.0V, the calculated output voltage (node 11) was 4.97V. In a second simulation the input stimuli were interchanged, the simulated output voltage being -240pV. The faults were then introduced in turn and the simulations repeated to determine all the faulty node voltages. The fault dictionary was constructed as detailed above. The tolerance voltage δV was taken as 300mV. This resulted in 37 of the faults being detected (70% fault coverage), assuming access to all seven internal nodes (including the Vo output node 11). The individual fault coverages of these nodes are summarized in Table 15.1. This shows that 60.4% coverage can be achieved just by accessing node 10. Just accessing the output node 11, which is very often possible, only gives about 40% coverage, not very high. If we look at combinations of the output node plus one internal node, as a compromise, the results are summarized in Table 15.2. This shows that the combinations 5, 11 and 10, 11 both provide 66% coverage, close to the 70% obtainable if all the nodes were accessible.

Table 15.1 Fault Coverage for each node of the comparator

Node Number	No. Faults Detected	No. Faults Isolated	Percentage Fault Coverage
2rcp	3	1	5.7%
3	10	0	18.9%
4	18	2	34.0%
5	29	5	54.7%
6	20	5	37.7%
10	32	3	60.4%
11	21	2	39.6%

To systemize the search for the small set of test nodes to achieve acceptable fault coverage, an heuristic approach has been proposed [6]. We define a sensitivity factor Sn by:

$$Sn = \sum_{f=1}^{F} (Vn - Vf)^2$$

15.10

Where Vn is the nominal voltage of node n, Vf is faulty voltage of node n and F is the number of faults introduced. A high value of Sn implies that that particular node is sensitive to faults and is a likely candidate as a test node.

Table 15.2 Fault coverage for pairs of node including output node

Node Number	No. Faults Detected	No. Faults Isolated	Percentage Fault Coverage
2,11	24	3	45.3%
3,11	28	2	52.8%
4,11	29	3	54.7%
5,11	35	7	66.0%
6,11	25	6	47.2%
10,11	35	5	66.0%

Once Sn has been calculated for all the potential test nodes, the node with the highest value of Sn is considered first to see if it will provide an acceptable fault coverage (possibly in conjunction with the output node(s) which are often readily accessible). If this one is not satisfactory, the next highest Sn can be considered, and so forth.

The DC. fault dictionary approach has a number of drawbacks. While it may identify many catastrophic faults, it does not check for parametric failures, such as gain or bandwidth being out of spec. Faults in capacitors cannot be located. Further, a large amount of simulation and computing must be performed as a prelude to the test. Its main advantage is in the systematic approach to choosing optimum test points within a circuit, i.e. optimising the observability of a circuit block. It may therefore find use in semi-custom circuits where the vendor provides a series of standard analogue macro-cells.

15.3.2 Digital modeling

In the digital modelling approach to unifying the test strategy, catastrophic faults in an analogue module are considered as 'stuck-at' faults in some equivalent digital circuitry. If this can be achieved then the overall mixed-mode circuit can be tested by pure digital techniques. There are two possible approaches to digital modelling, digital logic equivalent, and functional Karnaugh map.

15.3.2.1 Digital logic equivalent

In this strategy [7], an analogue block is conceptually reduced to a combination of digital gates, AND, OR, etc. The test sequences to test such circuits are derived using standard digital techniques. The inputs to the analogue blocks are two-state signals, sufficiently low and high to ensure the equivalent digital switching of the analogue circuitry. An example of an analogue comparator circuit in terms of digital circuitry is illustrated in Figure 15.7 [7].

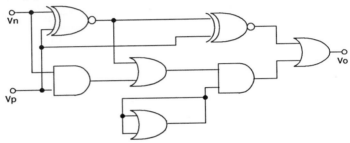

Figure 15.7 Digital equivalent circuit of comparator

Potentially this technique would be a very good approach to a unified test routine. It suffers the same problem as the DC. fault dictionary approach of not fully exercising the analogue blocks, so that gain, bandwidth, etc. specs cannot be tested. The main problem, however, lies in the lack of suitable digital equivalent circuits for most analogue blocks - the comparator circuit is the exception rather than the rule.

15.3.2.2 Functional K-Map

This is a slight variation on the above technique in an attempt to formalize the approach of representing a general analogue circuit by a digital equivalent (as far as testing is concerned). This strategy treats an analogue module as a block (i.e. assumes that access is only available to the input and output terminals), and the DC. response of the module is represented by a Karnaugh map.

The testing algorithm consists of the following steps :

(1)   The fault free circuit is simulated and the response to a suitable set of input signals is calculated.

(2) The circuit operation in step (1) is translated onto a Karnaugh map consisting of 1s and 0s.

(3) The digital function represented by the K-map in step (2) is derived.

(4) The tests necessary to detect stuck-at faults in the digital function in step (3) are then derived.

(5) The tests in step (4) are translated back to DC. voltages applied to the input terminals of the analogue module to be tested.

15.3.2.2.2 Example

This routine was applied to the same comparator circuit as in section 15.3.1. A block diagram of the comparator, the results table for the fault-free simulation and the resulting K-map for an assumed threshold value of 4.0V is shown in Figure 15.8. The digital function represented by the K-map is $Vout = Vp\overline{Vn}$. If the catastrophic analogue faults are considered as digital stuck-at faults, then this digital equivalent function would only require three tests to detect those stuck-at faults, tests T0, T2 and T3.

Test	Code	Vp	Vn	Vout
T0	00	–3.0	–3.0	5 E-14
T1	01	–3.0	+3.0	-2 E-10
T2	10	+3.0	–3.0	4.97
T3	11	+3.0	+3.0	3.26

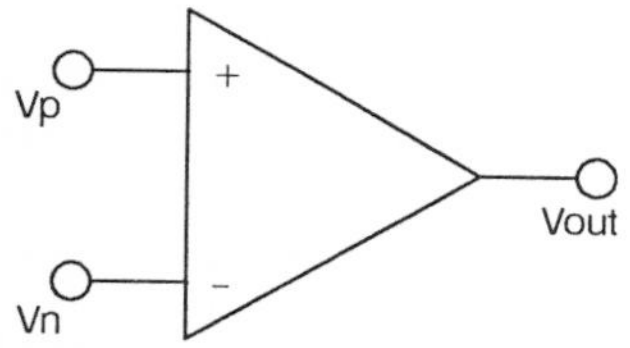

K-Map and Function for Vt = 4.0V

From K-Map : $Vout = Vp.\overline{Vn}$

Required test are : T0, T1 and T3

Figure 15.8 Functional K-Map testing of comparator circuit

To check the validity of this conclusion, the comparator was tested under fault-free and also the 53 catastrophic fault conditions. For each simulation, the four test conditions (T0, T1, T2 & T3) were applied.

To determine which faults were detected by each test, a window is placed around the nominal fault-free output values, in order to represent an allowed tolerance. A fault is detectable if for any test the output voltage falls outside this window. Table 15.3 shows the results of these simulations. Alongside each test is a list of the faults detected by each test, and the total number of faults detected. It can be seen that the test T1 is redundant, as all the fault numbers detected by this test are also detected by the other tests. Therefore the combination of tests T0, T2, T3 will result in the highest fault coverage (70%). In fact the same fault coverage can also be obtained by the test combination T0, T1, T3. If we now define a 'Confidence of Test' as the magnitude of the difference between the nominal fault-free voltage and the voltage generated by a particular fault, then the higher the value of confidence, the more likely we are to detect successfully the fault. If we examine the confidence levels of the two sets of tests, we find in fact that test series T0, T2, T3 has the greater confidence levels, and is therefore the preferred test set.

Table 15.3 Faults detected by tests applied to compatator

Test	Window VI	Window Vh	Faults Detected
T0	−1.0	1.0	1,5,9,12,14,16,19,23,27,28,31,34,39,42,43,50 Tot: 16
T1	−1.0	1.0	12,15,16,19,34,42,50 Tot: 7
T2	3.7	5.0	15,19,20,21,22,32,33,34,42, 50 Tot:
T3	2.0	4.0	5,9,12,13,14,16,17,18,20,21,22,27,31,32,33,38,39,40,41,42, 43,44,45,46,47, 48,49,50,51,52,53 Tot: 31

Although the functional K-map technique is a more general approach to digital modelling and one which can lead to a unified test technique, it still has the drawback of being essentially a DC. test with the problems of not fully exercising the analogue blocks. It may however find some use in future test strategies.

15.3.3 Logical decomposition [8]

Logical decomposition is a technique which has been applied to diagnostic testing of large analogue circuits, and is particularly useful at the board level for detecting faulty chips. It may also prove useful at chip level testing, and because the form of the input stimulus is not critical, may also be applied to digital and mixed-mode circuits.

The method is based on decomposing a network into smaller uncoupled sub-networks. The nodes at which the decomposition takes place include the accessible measurement nodes. The decomposition nodes are termed the D-nodes. Voltage measurements are made at these nodes. Given these voltage readings, the currents flowing into each sub-circuit are calculated or simulated. The summation of these currents at each node is checked against Kirchoff's Current Law (KCL). If the current sum to zero (within some pre-defined tolerance value ϵ), then the node is fault-free and all of the sub-networks connected to that node are also fault-free. If the currents sum to a value greater than ϵ then the node is faulty and at least one of the sub-networks connected to that node contain a fault.

Identification of the faulty sub-network can be achieved systematically by deriving a logical diagnostic function (LDF) for the complete network. The LDF is constructed thus. Each sub-network is associated with a logical variable σ which takes the binary value 1 or 0. $\sigma = 1$ if the sub-network is fault-free and $\sigma = 0$ if the sub-network is faulty. A test Tj is the result of applying the KCL test to a certain D-node. If the particular test is a pass (i.e. Σ KCL $< \epsilon$), then

$$Tj = \sigma j_1 \cap \sigma j_2 \cap \ldots \ldots \cap \sigma j_k \qquad \boxed{15.11}$$

where there are k sub-networks connected to D-node j.

If the test is a fail (Σ KCL $> \epsilon$) then

$$\overline{Tj} = \overline{\sigma j_1} \cup \overline{\sigma j_2} \cup \ldots \ldots \cup \overline{\sigma j_k} \qquad \boxed{15.12}$$

The LDF is defined as

$$T = \overline{Tj} \cap Tj \qquad \boxed{15.13}$$

which means that the LDF is the logical product of the pass and fail tests. The LDF is then minimized using standard Boolean algebra. In the minimized form a σ_i in the LDF represents a fault-free sub-network, while a $\overline{\sigma_i}$ represent a faulty sub-network. If a sub-network is not represented in the LDF, then the applied tests did not fully exercise the sub-network and nothing can be assumed about its state.

Logical decomposition testing can handle multiple catastrophic (open/short circuits) and soft (parametric) faults. It is capable of handling both linear and non-linear networks, although it is best to keep the non-linear elements to a minimum. The form of test stimulus is not critical (D.C., sinusoid, square wave, etc.) so is suitable for mixed-mode circuits.

There are two drawbacks with the technique. Firstly it requires access to a large percentage of internal nodes to get an acceptable fault coverage. At the I.C. level this may lead to excessively high overhead costs. It is, however, suitable for testing at board level to detect faulty chips, and is indeed in use as such. In this case all the chip pins are available for test and the decomposition process is straightforward.

The second disadvantage is in the testing of CMOS circuits. In this case the current flows are very small, particularly into gate inputs, which are often the most accessible nodes. Due to the problem of small tolerance variations, and summing very small values to zero, it becomes impossible to apply practically the KCL test. The technique is therefore only suitable for bipolar based processes.

15.3.4 Artificial intelligence and knowledge based systems

The field of artificial intelligence (AI) involves the use of computers to perform tasks which to some degree emulate the thinking and reasoning processes in humans. In general this involves the computer following a set of rules (which may be expanded and adapted) in order to make decisions. It also involves the use of databases of previous results and experiences on which the decisions can be based. Thus the machine expands and 'learns' more about a system or process as time progresses. Hence these computing systems are also referred to as intelligent knowledge based systems (IKBS).

The role of AI in mixed signal testing is principally one of diagnostic test. It is more suited for testing at the circuit and board level where access to intermediate nodes within the system is more readily available.

There are two main approaches which can be taken in the application of AI to this problem [9, 10]. The first approach involves the automatic generation of a decision tree, based on a library or database of previous measurements and test results. It may also involve data derived from the simulation of circuit faults. Given this data the computer constructs a hierarchy of test decisions and branches depending on the result of the test. Each branch will consist of a group of possible faults identified by that particular test. The decision and branching process is continued until each fault is uniquely identified (where possible). By this process the software guides the test engineer through an optimised series of tests to identify the location of a fault.

This form of 'shallow reasoning' has a number of problems when applied to mixed-mode testing. The database used will be specific to the particular circuit or system under test, so small production runs will lead to small databases and hence incomplete diagnostic information. Feedback paths can alter voltages along an entire signal path, leading to erroneous results. Unexpected or previously undefined faults can also lead to erroneous diagnostics. There is also the ever-present problem of accessing internal nodes to a circuit.

An improved method makes use of the knowledge the engineer will have of the operation of circuit blocks, by employing accurate models for these blocks. The block may be as small as a single component (capacitor, diode, etc.) or may involve blocks or subsystems such as op-amps, ADC/DAC blocks, digital gates or sub-systems, etc. This leads to a 'model-based' reasoning strategy which incorporates into the database not only the knowledge derived from previous testing runs, but also a fault dictionary based on the modelled sub-circuits.

This approach overcomes some of the problems involved with the decision tree approach. The model based data can be incorporated into different circuit implementations, so that each new database does not have to be built completely from scratch. Previously undetected faults can be identified and incorporated into the system. The problems associated with feedback loops can be avoided [9].

In summary, the use of AI based testing techniques can aid the test engineer in performing diagnostic testing, it is not really suited for go/no-go testing. It has the disadvantage of each system being specific to a particular circuit or circuit family, even with the model-based approach. It also suffers the perennial problem of lack of access to internal nodes limiting the degree of diagnostic cover. Hence it is more suited to board testing to identify faulty chips.

15.3.5 Transient response testing [11]

A technique which has the potential to unify the testing of mixed signal ICs is that of transient response testing. The basis of the technique is the application of a digital-like pulse waveform as the stimulating test vector. When a pulse is applied to a linear analogue circuit, the output waveform will consist of the impulse response of the circuit, convolved with the input stimulating waveform. Since the impulse response will be characteristic of the circuit under test, it should be possible to check for faults in the circuit via this form of test.

A mathematical basis has been demonstrated [12] whereby the magnitude and time patterns of the stimulating pulses can be optimised to a particular linear circuit, based on the expected transfer function of the circuit under test, in order to optimise the test pattern. If we now apply this basic technique to mixed-mode ICs, any stimulating pulse train may well be propagated through digital blocks, before being used to stimulate the analogue circuit. Therefore we do not have the ability to tailor the magnitude pattern of the pulses - they must be constrained to the logic rails of the circuit. We still have control over the temporal parameters of the test vector, so we must optimise this pattern to fit a particular circuit under test. In fact there are a number of aspects to this. As the stimulating waveform is a series of digital-like pulses, it can be used to test both the digital and analogue sections of the circuit. Standard digital test techniques can be applied to, for example, path sensitization and propagation of faults to observable nodes. In addition, the optimum test vector could be propagated to the input of an analogue section via preceding digital circuitry.

In fact, theoretically it does not matter whether the digital block precedes the analogue, or vice versa. In both cases the signal remains (largely) in the form of digital-like pulses which will excite a response from each block which is characteristic of that block. The signals may actually pass through several mode transitions and still remain as valid test vectors. Although the concept of transient response as convolution of impulse response with input waveform is strictly only applicable in the linear case, in fact the digital circuits will have a characteristic transient response to a particular stimulus. This characteristic may well be altered by faults within the circuit, and are therefore detectable by this method. The only real restriction on the form of the stimulating analogue signal is that to fully exercise the circuit, the bandwidth of the input signal should exceed that of the circuit under test; i.e. the rise and/or fall times of the stimulating pulses should be sufficiently fast.

The composite response, observed at the external output node(s) of the circuit under test is therefore characteristic of the response of any intervening digital and analogue circuit blocks. Comparison of the expected transient waveform with this measured response provides a basis for go/no-go testing. The analysis of the output transient may be achieved by fourier transformation and signature analysis of the resulting frequency spectrum. A preferred technique is that of a correlation operation between the expected and measured time responses. For a fault-free circuit this will be an auto-correlation function, and will therefore result in a symmetric, even function. Any fault will result in a non-symmetric correlation function and so is easy to identify. Of course, as with any analogue measurement there is some acceptable tolerance on the measured response. This will be dependent on the particular IC process and the circuit itself. So there will be some deviation from the ideal symmetric response which will still be acceptable. The correlation technique is recognized as one which is particularly useful for detecting signals with low signal/noise ratio.

15.3.5.1 Examples

This technique was applied to the same comparator circuit as in sections 15.3.1 and 15.3.2. In this case a fault coverage of 92% was obtained while only accessing the external nodes. This compares with about 70% for the DC. based tests, which also required probing of some internal nodes. To prove that the technique may be applied to mixed-mode circuits, it was used to test a very simple mixed-mode circuit as shown in Figure 15.9 [11]. This consists of a single stage active filter feeding a comparator (the same one as in the previous examples), followed by a digital half adder circuit. The pulse excitation is applied at Vin and the transients are captured at the two digital outputs Vos and Voc. The fault coverage from this circuit is around 85%, again only having access to external nodes.

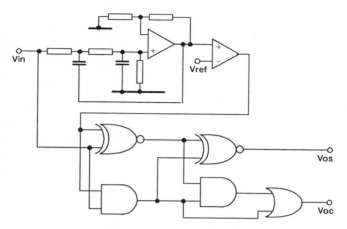

Figure 15.9 Simple mixed-mode circuit

This demonstrates the main advantages of the transient response technique in that reasonable theoretical fault coverage can be obtained by only probing the external nodes. There is no need to partition the circuit into separate blocks. The form of the test vectors is a digital-like train, so it may well be possible to adapt or expand existing digital test equipment to apply the technique. The technique is still in its infancy however, and a number of problems need to be addressed before it can challenge the partitioning approach. It has not yet been proven practically on medium to large scale circuits. It may be that propagating test vectors through large blocks of digital circuitry, for example, will result in a stimulating signal to an analogue block which is not suitable for the required test. The problem of the acceptable tolerance of the output waveform has yet to be addressed. This range will depend on the particular fabrication process and the particular circuit under test, so it would be extremely difficult to apply a general rule for acceptance or rejection. However, initial tests have indicated that catastrophic, and even reasonably large parametric faults tend to cause a large change in the measured waveform and so are easy to detect. It may be that the technique will be suitable for coarse go/no-go testing at an early stage of production.

In summary then, although much work still needs to be done to prove out the method, it perhaps offers the best solution to the unification of mixed-mode testing.

15.4 References

[1] M. Mahoney. "DSP-Based Testing of Analog and Mixed-Signal Circuits", IEEE Computer Society Press, 1987.

[2] C. E. Shannon. "Communication in the Presence of Noise", Proc. IRE, Vol. 37, p. 10, 1949.

[3] W. T. Cochran et al. "What is the Fast Fourier Transform?", Proc. IEEE, Vol. 55, No. 10, pp. 1664-1674, October 1967.

[4] F. J. Harris. "On the Use of Windows for Harmonic Analysis with the Discrete Fourier Transform", Proc. IEEE, Vol. 66, No. 1, pp. 51-83, January 1978.

[5] W. Hochwald & J. Bastian. "A DC. Approach for Analog Fault Dictionary Determination", IEEE Trans. on Circuits and Systems, Vol. CAS-26, No. 7, pp. 523-9, July 1979.

[6] M. A. Al-Qutayri & P. R. Shepherd. "On the Testing of Mixed-Mode Integrated Circuits", Journal of Semicustom ICs, Vol. 7, No. 4, pp. 32-9, June 1990.

[7] D. Brown & J. Damianos. "Method for Simulation and Testing of Analogue/Digital Circuits", IBM Tech. Disclosure Bulletin, Vol. 25, pp. 6367-8, 1983.

[8] A. Salams, J. Starzyk & J. Bandler. "A Unified Decomposition Approach for Fault Location in Large Analog Circuits", IEEE Trans. Circuits and Systems, Vol. CAS-31, No. 7, pp. 609-622, July 1984.

[9] R. Rastogi & K. Sierzega. "A New Approach to Mixed-Signal Diagnosis", Proc. 1990 International Test Conference, pp. 591-7, Washington, DC. September 1990.

[10] T. S. Arslan, L. Bottaci & G. E. Taylor. "An AI Based Approach to Automatic Fault Diagnosis For Mixed Digital/Analogue Circuits", Proc. IEE Colloquium on Design and Test of Mixed Analogue and Digital Circuits, Savoy Place, London, 15th November, 1990.

[11] P. S. A. Evans, M. A. Al-Qutayri & P. R. Shepherd. "On the Development of Transient Response Testing Techniques for Mixed-Mode ICs", Journal of Semicustom ICs, Vol. 8, No. 2, pp. 34-9, December 1990.

[12] H. H. Schreiber. "Fault Dictionary Based upon Stimulus Design, IEEE Trans. on Circuits and Systems, Vol. CAS-26, No. 7, pp. 529-537, July 1979.

Chapter 16

Towards High Level Synthesis of Mixed-Signal analogue-Digital ASICs

Maria Helena Martins, Adolfo Steiger Garcao,
Faculdade de Ciencias e Tecnologia.
Jose Epifanio da Franca
Instituto Superior Tecnico.

16.1 Introduction

The development of methodologies and techniques for the design of mixed–signal analogue–digital integrated circuits is presently one of the most challenging areas of research worldwide. While computer–assisted tools for the design of digital circuits have already attained a considerable degree of maturity, thus enabling the rapid and reliable synthesis of the digital part of mixed–signal application specific integrated circuits (ASIC's), the design of the analogue part is still a very time consuming task due to the very poor support provided by existing computer–assisted tools.

In earlier analogue CAD systems, the automatic synthesis of the circuits was carried out employing simple module generators based on a few standard topologies. Such systems were mostly dedicated to the synthesis of circuits with very regular structures, like switched capacitor (SC) filters and some classes of data converters [1, 7, 15, 16, 24]. Because of the rather limited design solutions which could be provided by such systems, a second generation of analogue CAD tools emerged which, besides providing the automatic synthesis of some circuits, as their predecessors did, also provided optional user–controlled design at different stages of the synthesis process [5, 22].

In recent years, several knowledge–based systems have been proposed with the aim of rendering the automatic design of mixed analogue–digital circuits even more efficient and capable of adapting to the fast technology evolution. Prominent among such systems are Prosaic[2], which is a rule–based system oriented to the stage–by–stage design of operational amplifiers, and IDDAC [6] , OASYS[3] and OPASYN[18] , all of which use knowledge for the topology selection of circuits. Other systems [8] have even introduced an expert manager which handles separate expert systems dedicated to the synthesis of specific classes of circuits.

More recently, mixed–signal CAD systems have evolved towards the development of symbolic tools which not only enable the designer to have a qualitative insight into the key parameters responsible for the behaviour of the circuits [13, 19, 25], but also allow the addition of new topologies in the knowledge base of the system. Such features, however, have been limited to the low levels of design encompassing circuit components such as amplifiers and comparators, and have not yet been incorporated into the higher level design tools that are essential to provide the level of functionality which is so important in modern mixed analogue–digital ASIC's.

This chapter describes an advanced computer–assisted design environment which employs high–level symbolic tools capable of analysing and supporting the synthesis of system architectures defined in terms of specific functional building blocks as well as the architectures of such functional building blocks defined in terms of circuit components. The use of Prolog [23] is highly advisable for the development of such high–level symbolic tools based on pattern matching approaches for the identification of several functional blocks and elements. A strategy for multi–level communication and flow of information is also envisaged in order to efficiently integrate such tools and thus provide the most flexible computer assisted environment for designing mixed analogue–digital ASIC's.

Besides the Introduction, this chapter is organized in four additional sections. Section 2 discusses the hierarchical organisation adopted in our computer assisted environment for designing mixed–signal analogue–digital ASIC's as well as the resulting requirements for the multi–level interfacing and communication mechanisms. There, we shall also introduce the concept of synthesisers which employ a coherent, common framework for the synthesis and analysis of networks implementing different functions, each of which can also be realised using multiple topologies. Essential to such multi–function multi–topology synthesisers is the capability of automatically generating symbolic transfer functions from a given circuit description, which will be described in Section 16.3 for the case of z–domain transfer functions generated from signal flow graphs (SFG's) representing the operation of SC networks. In Section 4 we describe a knowledge–based system for designing at the architecture level SC filters based on multiple methodologies, we also give a detailed example corresponding to a particular methodology based on a cascade of SC biquads. Section 5 summarises the Chapter.

16.2. Mixed–signal design environment

16.2.1 Establishing an hierarchical organization of design

Given the problem of designing a mixed–signal ASIC, the process starts by considering the main functions which are needed to achieve the envisaged functionality. Such functions include linear and non–linear signal processing, signal conversion and control, and can in general be implemented using a variety of techniques and building blocks from purely analogue to purely digital. In an increasing number of applications it is found that a combination of analogue and digital techniques and building blocks usually leads to the optimum solution for realising economically and in the shortest period of time the required functions.

The hierarchical organization established for our design environment is illustrated in Figure 16.1. This comprises the definition of the system architecture at the highest level, and then lower levels of design corresponding to the building blocks (e.g. filters and data converters), circuit components such as the amplifiers (analogue) and registers (digital), and the basic circuit elements, both active (transistors) and passive (capacitors and resistors). The lowest level of design corresponds to the description of the physical dimensions of the basic elements, and their interconnection according to the required circuit components and building blocks.

In the above hierarchical organisation of the design process, it is essential to clearly establish a multi–level interfacing mechanism in order to communicate the specifications and constraints of the design elements employed at the different levels. From top to bottom we communicate the specifications required for the lower levels of design, whereas bottom–up we pass information concerning the constraints of the elements required at the higher levels of design. For example, low level constraints such as the maximum achievable capacitance ratio accuracy in a given process technology can be used advantageously at the higher levels defining the architectures of data converters and filters, and thus allowing the design process to converge much more quickly without the need of going down and back up again several times. It is clearly in this negotiation strategy that the experience of designers plays a key role and where knowledge is much less amenable for structuralisation and unguided automation. The efficient automation of such constraints is probably the most challenging difficulty to be overcome by the generation of systems to be developed in the 90's, but this will certainly be mandatory to achieve the efficiency and wide variety of circuit solutions which are so needed for modern mixed analogue–digital ASIC's.

16.2.2 Design functionality at the architecture level

In the mixed–signal design environment described here, the design methodology at the highest level is illustrated in Figure 16.2. Given a set of

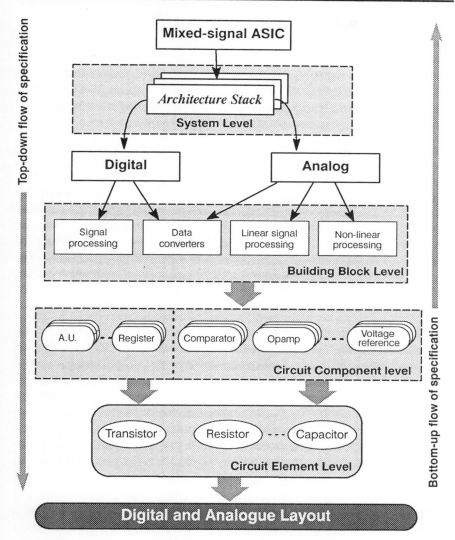

Figure 16.1 Hierarchical organization adopted in our design environment

specifications, the system architecture is characterised in terms of the basic functional building blocks for signal processing and data conversion, whose macro-models are fetched from the corresponding library. Such macro-models are defined by their implementation techniques, and the resulting functional specifications are automatically determined in order to meet the required system specifications.

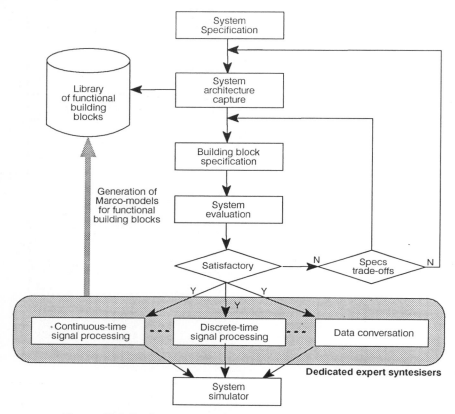

Figure 16.2 Design methodology at the architectural level.

Then, a behavioural system simulator evaluates the correctness of the architecture design as well as the practicality of implementing the required building blocks in the envisaged integrated circuit technology. Whenever the system specifications are not fully met or the resulting building blocks are impractical for implementation, then the design tool can explore new spaces of design, including new specifications of the building blocks for the same architecture and even new architectures yielding new building blocks. This process is illustrated in Figure 16.3, corresponding to the design of an analogue–digital interface system for video signals according to the CCIR recommendation 601 [4, 12]. Here we can see how the adoption of a specific architecture may lead to different types of building blocks (for example, SC and digital decimators) and even different specifications for the same building block (A/D converter operating at either 13.5MHz or 54MHz).

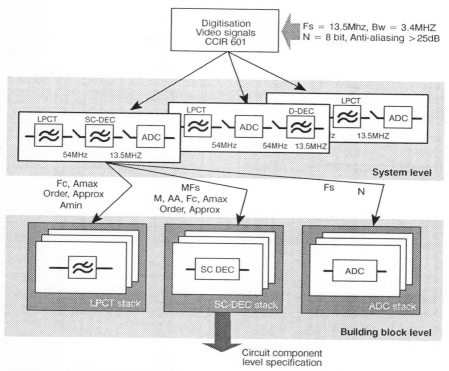

Figure 16.3 Components of the system and functional building block levels for the example of an analogue–digital interface system for digitization of video signals.

Once a given architecture has been established for the system, the specifications of the required functional building blocks are communicated to dedicated expert synthesisers which, in turn, will also carry–out the appropriate design methodology from the architecture level to the lower level where the basic circuit components and elements are designed. After combining and integrating the designs of the various functional building blocks, the system can be simulated either electrically or by using an appropriate mixed–mode simulator.

In order to offer the designer an environment flexible enough to cope with a wide variety of processing functions and design methodologies, it is mandatory to establish highly efficient interfacing mechanisms, whereby the expansibility of the system is achieved through the integration of new high level synthesis modules together with low level design tools. High level synthesis modules comprise not only high level dedicated expert synthesisers but functional simulators and analysers as well. The high level expert synthesisers are developed to have a sufficiently high degree of expertise such that, for given specifications and relevant information on the technology constraints, the most appropriate design methodology can be selected and the most suitable circuit specifications can be determined. Such high level synthesis modules also offer an optional user–guided design mode, in which the designer may be responsible for key design decisions.

16.2.3 Multi–function multi–topology synthesiser of building blocks

As mentioned before, our mixed–signal CAD system incorporates dedicated tools for the synthesis of data conversion circuits as well as circuits for signal processing. Whereas the tool for the synthesis of data converters is extensively described in Chapter 19, here we shall be concerned with the description of a multi–function multi–topology synthesiser suitable for the design of continuous and discrete time signal processing circuits, both analogue and digital.

The basic concepts embedded in the development strategy of our multi–function multi–topology synthesiser for building blocks are described for the case of the dedicated synthesisers employed for designing SC networks for linear signal processing, including filters and multi-rate building blocks. As illustrated in Figure 16.4, such synthesisers are developed to allow the designer to synthesise SC filters, based on both ladder and cascaded biquad methodologies [11] , as well as various circuit topologies for SC decimators and interpolators, with both finite and infinite impulse responses [17, 21] . In all cases the design is based on predefined regular structures which are available in the topology library of the system. The characterisation of those structures consists not only of a description of its elementary components and the way they are interconnected, but also on a symbolic description of their z–domain transfer functions. In those situations where the user wishes to add a new structure to the library he can use the topology editor to obtain the topological description of the structure whose z–domain transfer function is automatically captured in symbolic form using a dedicated generator. The operation of such dedicated generators of symbolic transfer functions is based on two sets of rules which can be established in common for all the functional building blocks and corresponding topologies addressed by the synthesiser. While one set of rules is concerned with the relevant type of variables and functions (for example, s–variable for continuous–time filter functions and z–variable for discrete–time filter functions), the other set of rules is determined by the basic elements which are employed for constructing such functional building blocks. This will be shown in the following Section considering the

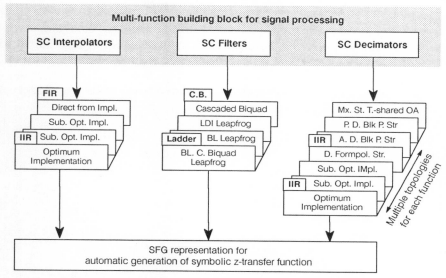

Figure 16.4 Multi–function multi–topology synthesiser for linear signal processing.

automatic generation of symbolic transfer functions suitable for a wide variety of SC networks for linear signal processing.

16.3 Automatic generation of symbolic discrete–time transfer functions

As mentioned before, the automatic generation of the symbolic transfer functions associated with a given multi–function multi–topology synthesiser is crucial to obtain the envisaged flexibility for designing functional building blocks at the architecture level, and achieve the capability for manipulating such structures in a way that currently available tools are not capable of providing. In the example described in the previous Section, the multi–function multi–topology synthesiser is concerned with SC networks for linear signal processing, all of which can be conveniently characterised, both for analysis and for synthesis, using SFG techniques [10]. Therefore, the corresponding automatic generator of symbolic transfer functions must be capable of recognizing the basic SC elements used for building such networks, determining and manipulating the corresponding discrete–time SFGs, and then producing the overall discrete–time transfer function of the circuit.

The generator works in two phases. In the first one, the netlist of a given network is browsed, so that such basic elements as Inverting Switched Capacitors, Non–inverting Switched Capacitors, and Unswitched

Towards High Level sysnthesis of Mixed-Signal Analogue-Digital ASIC's

Capacitors (both in feedforward and feedback branches) are identified, and the z–domain SFG of the network is obtained. Some of the rules employed for the identification of the basic elements in an SC network together with the resulting SFG's are illustrated in Figure 5. In Figure 5a, we can see that,

```
SW(X,Y,Z):-switch(X,Y,Z)switch(Y,X,Z)
scni(X,Y,C):-            /*scni - non-inverting
switched capacitor*/
SW(Inp,X,Fase)
SW(X,ground,Phase),
capacitor(X,Y,C)
SW(Y,ground,Phase)
SW(Y,Outp,Fase)
Phase:=Fase

SC1(X,Y,C):-            /*SC1 -- Inverting
switched capacitor*/
SW(Inp,X,Fase)
SW(X,ground,Phase)
capacitor(X,Y,C)
SW(Y,ground,Fase)
SW((Y,Outp,Phase)
Fase:=Phase
feedbc(inp,Outp,C):-    /*feedbc -- feedback
capacitor */
ampop(Inp,X,Outp):-
capacitor(Inp,Outp,C)
cond(Inp,Outp,C):-     /*cond - unswitched
capacitor*/
capacitor(Inp,Outp,C)
notsc1(X,Y,C)
notscni(X,Y,C)
notfeedbc(X,Y,C)
```

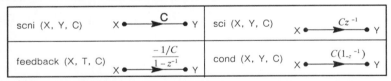

scni (X, Y, C)	$X \bullet \xrightarrow{C} \bullet Y$	sci (X, Y, C)	$X \bullet \xrightarrow{Cz^{-1}} \bullet Y$
feedback (X, T, C)	$X \bullet \xrightarrow{\frac{-1/C}{1-z^{-1}}} \bullet Y$	cond (X, Y, C)	$X \bullet \xrightarrow{C(1_{-z}^{-1})} \bullet Y$

Figure 16.5 (a) Rules for the identification of some basic SC elements, and (b) corresponding SFG's

for instance, an inverting switched capacitor (contained in the outlined box) consists of four switches and one capacitor. The first switch, operating with phase Fase is connected between the input node of the structure and node X. Connected between this node and ground is another switch operating with phase Phase. Between the same node X and node Y is a capacitor with capacitance value C. A third switch is connected between node Y and ground, which also operates with the same phase Phase. Finally, a switch operating in the same phase Fase as the one connected to the input node, is connected between node Y and the output node. Whenever a similar pattern is identified in a given SC network, the generator automatically assigns the SFG given in Figure 5b and which will be used to capture the complete SFG of the network. Figure 5b further illustrates the SFG's corresponding to the patterns of the other basic SC elements also described in Figure 5a[14] .

In the second phase of operation of the automatic generator of symbolic z–transfer functions, after identification of the basic SC elements of the network, Mason's rule [20] is applied to the captured SFG and thus leading to the determination of the corresponding overall z–domain transfer function. This process is illustrated here for the simple example of the SC biquad with multiple damping shown in Figure 6a [9], and whose netlist

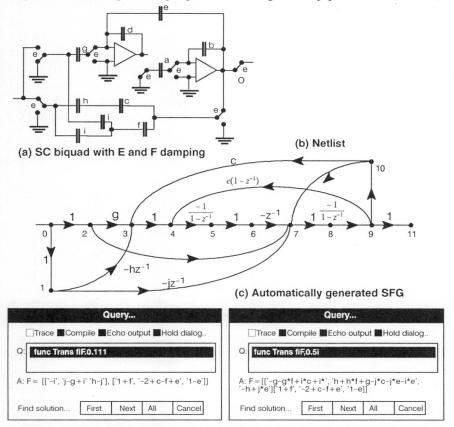

(a) SC biquad with E and F damping

(b) Netlist

(c) Automatically generated SFG

(d) Result obtained by the generator considering the output node taken from the output terminal of the second and the first amplifier.

Figure 16.6 Application of the automatic generation of symbolic discrete-time transfer functions to an SC biquad

description and SFG are given, respectively, in Figure 6b and Figure 6c [10].

The automatic generation of the symbolic z–domain transfer functions of this SC network, considering the output node at the output terminal of either the first or the second amplifier in the network, leads to the results shown in Figure 6c. Such transfer functions are presented as lists of terms for both the numerator and the dènominator polynomials. In either case, the first term corresponds to the coefficient of Z^{-0}, whereas the second and third terms correspond, respectively, to the coefficients of z^{-1} and z^{-2}.

In order to enhance the flexibility of our design environment, the automatic generator of symbolic z–transfer functions can also be used with numerical values for some of the capacitors of the network, while the other capacitors remain represented as symbols. This feature allows the designer to gain insight into the dominant contributions of the various capacitors for the overall performance behaviour of the network, namely the variability of the frequency response against errors of the nominal capacitance values. By precisely determining the effect of the various capacitors in the overall behaviour of a given SC network, the system (and the designer) also captures vital information for optimising the final capacitance values. This is particularly helpful for exploring the merits of new topologies to be designed and evaluated against the background of topologies available in the topology library.

16.4 Knowledge–based system for designing SC filters at the architecture level

16.4.1 Organisation, characteristics and requirements

The knowledge–based system for designing SC filters at the architecture level comprises five main modules, namely a *Specification Translator*, a *Design Manager*, various *Expert Sub–synthesisers* dedicated to the alternative filter design methodologies, a *Knowledge Base* and, finally, an *High–Level Evaluator*. The flow of information among such modules is schematically illustrated in Figure 16.7, and their basic characteristics are briefly summarised below.

After obtaining the desired filter specifications, the Specification Translator is responsible for producing the corresponding discrete time z–transfer function. Then, by taking into account relevant information contained in the Knowledge–Base, and which may concern such topology parameters as sensitivity performance and technology constraints, the Design Manager activates one of the expert sub–synthesisers dedicated to the selected design methodology. The dedicated expert sub–synthesisers are responsible for the architecture synthesis of the topologies associated with a given design

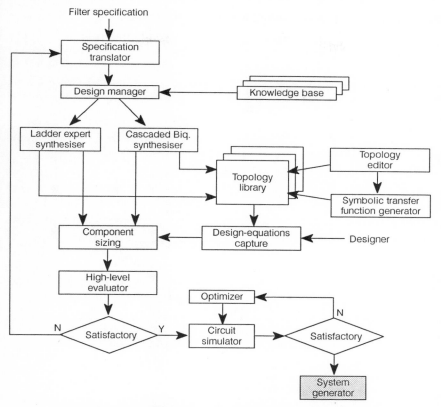

Figure 16.7 Knowledge–based system for designing at the architecture level SC filters with multiple topologies.

methodolgy (in this case, for ladder and cascade biquad design methodologies), and which are based on pre–defined structures residing in the Topology Library. As previously mentioned, the characterisation of the structures contained in the topology library comprises not only their topological description, but a symbolic description of their z–domain transfer functions as well. The addition of new topologies to the library is accomplished through the Topology Editor and the Symbolic z–Transfer Function Generator. Finally, the High–Level Evaluator determines the global values for the capacitance spread, the total capacitor area and the sensitivities of the frequency response with respect to the capacitance ratio errors. In order to illustrate the capabilities of such expert sub–synthesisers employed in our knowledge–based system for designing SC filters, we shall describe next the example corresponding to the methodology based on the cascade of SC biquads.

16.4.2 Expert sub–synthesiser for cascade SC biquads

As illustrated in the schematic diagram of Figure 16.8, the expert

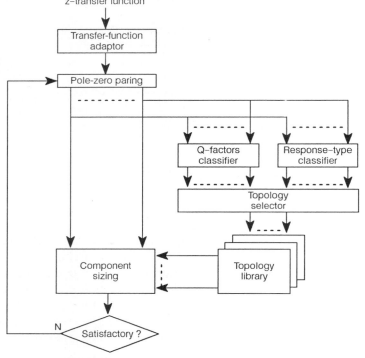

Figure 16.8 Expert sub-synthesiser for cascade SC Biquads

sub–synthesiser for cascade SC biquads comprises a module for fitting the specified z–domain transfer function to the adopted design methodology, i.e. expressed as a product of biquadratic z– domain transfer functions, and another module for designing the SC biquads required for implementing such biquadratic functions. After factorisation, the quadratic functions of both the numerator and denominator are fed into the pole–zero pairing module for determination of the most suitable biquadratic functions for SC realisation. A first cut solution for the initial pole–zero pairing is based on a set of commonly adopted rules for cascade biquad designs [26] .

After the initial pole–zero pairing solution has been determined in numeric form, the expert sub–synthesiser is capable of obtaining relevant information with respect to the resulting values of the pole and zero Q–factors (low, moderate and high) and type of response (low-pass, high-pass, band-pass, band-reject), and which will be used for selecting the most appropriate SC biquad topologies in the topology library. Upon selection of an SC biquad, its z–domain symbolic transfer function is also captured and equated to the corresponding numeric z–domain transfer functions. Then, the capacitance values are determined using a pre–defined set of design solutions which are established to yield reduced capacitor area and low variability of the frequency response with respect to capacitance ratio errors. In case the specifications are too difficult to achieve and lead to circuit solutions which are not practical for realisation, then the designer is advised to adopt an alternative pole–zero pairing strategy, and repeat the synthesis process described before. If the difficulty of obtaining practical designs for the SC biquads persists, then the designer may also try to synthesise a new biquad topology.

16.5 Conclusions

This chapter has described an advanced computer assisted design environment which employs high–level symbolic tools capable of analysing and supporting the synthesis of system architectures defined in terms of specific functional building blocks as well as the architectures of such functional building blocks defined in terms of circuit components. It has presented the general organization of a design environment, including a brief discussion of the strategy for multi–level communication and flow of information which was adopted in order to efficiently integrate such tools and thus provide high flexibility for designing mixed analogue–digital ASICs. We have also introduced the concept of synthesisers which employ a coherent, common framework for the synthesis and analysis of networks implementing different functions, each of which can also be realised using multiple topologies. Essential to such multi–function multi–topology synthesisers is the capability of automatically generating symbolic transfer functions from a given circuit description, and which was also addressed for the case of z–domain transfer functions generated from SFG's representing the operation of SC networks. Finally, we presented a knowledge–based system for designing at the architecture level SC filters based on multiple methodologies, and gave a detailed example corresponding to the methodology based on the cascade of SC biquads.

16.6 References

[1]   Allen, P. E. "A Tutorial – Computer Aided Design of analogue Integrated Circuits", IEEE – Custom Integrated Circuits Conference, 1986, 608–616.

[2] Bowman, R. J. and D. J. Lane. "A Knowledge–Based System for analogue Circuit Design", IEEE Int. Conf on CAD Design, 1985, 210–212.

[3] Carley, L. R. "Automated Design of Operational Amplifiers.", Introduction to analogue VLSI Design Automation., Ismail and Franca ed., 1990 Kluwer Academic Publishers.

[4] CCIR. "XVth Plenary Assembly" XXI–1 Geneve, 1982.

[5] Dasgupta, S, M. M. Mehendale, V. R. Sudershan, R. Jain and N. Subramanyam. "FDT– A design Tool For Switched Capacitor Filters", ICCAD, 1989, 446–449.

[6] Degrauwe, M. G. R., O. Nys, E. Dijkstra, J. Rijmenants, S. Bitz, Bernard L.A.G. Goffart, E. A. Vittoz, S. Cserveny, C. Meixenberg, G. V. D. Stappen and H. J. Oguey. "IDAC: An interactive Design Tool for analogue CMOS Circuits" IEEE journal of Solid–State Circuits, sc–22 6 – December 1987, 1106–1116.

[7] Eaton, G. V., D. G. Nairn and W. M. Snelgrove. "Sicomp: A Silicon Compiler for Switched–Capacitor Filters", 1987, 321–324.

[8] El–Turky, F. and E. E. Perry. "BLADES: An Artificial Intelligence Approach to analogue Circuit Design" IEEE Transactions on Computer–Aided Design., 1989, 680–692.

[9] Fleisher, P. E. and K. R. Laker. "A Family of Active Switched Capacitor Biquad Building Blocks" Bell System Technical Journal., 1979, 2235–22–68.

[10] Franca, J. E. "Switched–Capacitor Systems for Narrow Bandpass Filtering", P.H.D.Thesis, London,1985.

[11] Franca, J. E. "Switched–Capacitor Filters and Multirate Building Blocks", European Conference on Circuit Theory and Design, 1989.

[12] Franca, J. E. and R. P. Martins. "Novel Solutions For Anti–Aliasing Filtering in CMOS Video Interface Systems", VSPC, 1991.

[13] Gielen, G. and W. Sansen. "Automation of analogue Design Procedures.", Introduction to analogue VLSI Design., Franca ed., 1990.

[14] Gregorian, R. and G. C. Temes. "analogue MOS Integrated Circuits for Signal Processing". 1986.

[15] Helms, W. J. and B. E. Byrkett. "Compiler Generation To A/D Converters", Custom Integrated Circuits Conference, 1987, 161–164.

[16] Helms, W. J. and K. C. Russel. "A Switched Capacitor filter Compiler", CICC, 1986, 125–128.

[17] J.E.Franca, R. P. M. "IIR Switched–Capacitor Decimator Building Blocks with Optimum Implementation" IEEE transactions on Circuits and Systems, CAS 37 Jan, 1990, 81–90.

[18] Koh, H. Y, C. H. Sequin and P. R. Gray. "OPASYN: A Compiler for CMOS Operational Amplifiers" Transactions on Computer–Aided Design., 9 2, 1990, 113–125.

[19] Konczykowska, A. and M. Bon. "Automated Design Software for Switched–Capacitor IC's with Symbolic Simulator Scymbal", 25th ACM/IEEE Design Automation Conference, 1988, 363–368.

[20] Mason, S. J. "Feedback Theory – Further Properties of Signal Flow Graphs", 1956.

[21] R.P.Martins, J. E. F. "Infinite Impulse Response Switched Capacitor Interpolators with Optimum Implementation", Int. Symp. Circuits and Systems, 1990, 2193–2197.

[22] Sigg, R. P., A. Kaelin, A. Muralt, W. C. B. Jr. and G. S. Moschytz. "An SC Filter Compiler: Fully Automated Filter synthesiser and Mask Generator for a CMOS Gate–Array–Type Filter Chip", 1987.

[23] Sterling, L. and E. Shapiro. "The Art of Prolog: Advanced Programming Techniques". London, 1986.

[24] Therasse, Y., P. Guebels and P. Jespers. "An Automatic CAD Tool for Switched Capacitor Filter Design: A method based on the generalised Orchard argument", ESSCIRC ,1982, 41–44.

25] Winder, C. L. and R. E. Massara. "A Design Assistant Approach to the Implementation of analogue Integrated Circuits, with Particular Reference To switched Capacitor Filters", Midwiest Symposium on Circuits and Systems, 1988.

[26] Xuexiang, C., E. Sⱥnchez–Sinencio and R. L. Geiger. "Pole–zero pairing strategies for cascaded switched–capacitor filters", 199–204, 1987.

Chapter 17

Automated High Level Synthesis of Data Conversion Systems

Nuno Cavaco Horta, Jose Epifanio da Franca, Carlos Azeredo Leme
Instituto Superior Tecnico, Lisboa, Portugal

17.1 Introduction

Developing analog synthesis techniques and related Computer Aided Design (CAD) tools is currently a hot topic in the CAD world [1], [2], [3]. Applying these techniques to the automatic synthesis of analogue-to-digital (A/D) and digital-to-analogue (D/A) converters is clearly at the forefront of such efforts [4], [5], [6], [7]. This is partly due to the complex design techniques that are so often needed for data conversion systems, specially with high performance, and which are far less amenable for automation than is the case with linear signal processing circuits (e.g. filters) and circuit components (e.g. operational amplifiers and comparators). But, perhaps more importantly, the great deal of attention currently devoted to the automatic synthesis of data conversion systems results from the major role that such conversion systems play in modern VLSI chips combining both analog and digital signal processing functions and the need to efficiently design such chips in the shortest possible fraction of time.

The choice of a particular converter architecture which should be the target for a given automatic synthesis tool is a delicate task [6]. In contrast with, for example, switched capacitor filters, it is very difficult with only one type of architecture to cover a significant range of specifications for data conversion. This is due to the many design trade-offs which relate the required specifications (e.g. conversion resolution and speed) with key targets for integrated circuit implementation (e.g. silicon area and power consumption). As a result of the wide variety of architectures available for both A/D and D/A conversion it is commonly perceived that it would not be feasible to develop a universal tool capable of selecting and optimally synthesizing the most appropriate data conversion architecture to meet a given set of input specifications. Moreover, due to the frenetic activity in this area and the fast evolution of the fabrication technologies, it is likely that such a tool would rapidly become obsolete.

We carried-out a brief investigation of the available data conversion architectures in order to select the most suitable one to meet a broad range of input specifications and perform to a high degree of precision even when considering the non-ideal parameters of the circuit components such as offset, charge injection, noise and component mismatch errors. The data conversion architectures employing binary-weighted capacitor arrays (C-arrays) and the inherent charge-redistribution techniques (multiplying D/A converter and successive approximation A/D converter) [8] fulfill those goals and were therefore chosen as the object architectures for the automation process. A flexible framework for the synthesis of such A/D and D/A conversion systems has been developed to handle not only the complete architecture design and verification process but also to establish an efficient top-down flow of design information to access a variety of lower level design environments including cell libraries, parameterized circuit module generators and circuit level compilers (e.g. [11], [12], [13], [14], [15]). Because of this specificity use for data conversion systems, the tool also includes dedicated lower level compilers which are required for the more specialized circuit components such as the calibration networks and digital control logic. By ensuring such design flexibility at the lower level, the framework described in this Chapter also renders it easier to meet the most stringent specifications for implementation of a data conversion system employing charge redistribution techniques. Various practical examples will be considered to illustrate these design capabilities.

17.2 Data conversion architectures employing binary-weighted C-arrays

Four different architectures for each one of the two possible conversion modes, A/D and D/A, can be synthesized in our framework, and they are all based on binary-weighted C-arrays with inherent charge redistribution operation. The ability to handle both single and segmented C-arrays together with the possibility of including dedicated calibration networks makes it possible to achieve a broad range of specifications.

17.2.1 A/D conversion with single C-array

The basic A/D conversion architecture employs a binary-weighted C-array, a switch array, a comparator, and a digital controller, as schematically illustrated in Figure 17.1. During one full operation cycle the architecture is configured sequentially in the sampling phase, the hold phase and then in the conversion phase. In the sampling phase, the top plates of all capacitors are connected to ground while the bottom plates are connected to V_{in}, with the exception of the most significant capacitor which is connected to V_{ref} to allow conversion of bipolar input signals. Here, the top plates of the capacitors are also connected to both the negative input and the output terminals of the comparator, thus rendering the sampling phase inherently offset free.

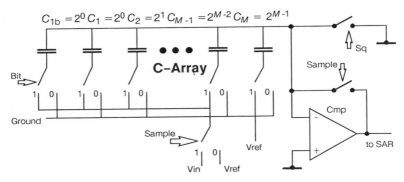

Figure 17.1 ADC architecture with single C-array.

During the subsequent hold phase, the switching operation is controlled in such a way as to disconnect from ground the top plates of the capacitors while the bottom plates are connected to ground. The principle of charge conservation guarantees that the charge in the top plates remain constant and consequently the normal voltage $V_x = -V_{in}/2 - V_{ref}/2$ appears at that point.

Finally, the circuit starts the conversion phase sequentially from the Most Significant Bit (MSB) to the Least Significant Bit (LSB) and carries-out, for the testing of each bit, the operations described in the fluxogram illustrated in Figure 17.2. The testing of each bit consists of connecting first the bottom plate of the corresponding capacitor to V_{ref}, and then evaluate the sign of the new resulting voltage V_x. This is carried-out by the comparator whose output is defined as:

$$V_{cp\_out} = V_{dd}, \quad V_x < 0$$
$$V_{cp\_out} = V_{ss}, \quad V_x > 0$$

and drives the input of the digital controller. If $V_x > 0$, then the capacitor corresponding to the bit under test will remain connected to V_{ref}; otherwise it will be re-connected to ground. This operation is carried-out until all bits have been tested. The converted digital word is determined by the final capacitor connections in the C-array.

17.2.2 A/D conversion with segmented C-array

The basic A/D conversion architecture with segmented C-array is illustrated in Figure 17.3, where the sub-array and the main-array possess, respectively, L and M capacitors for a conversion resolution of $N = M + L$. For high resolution converters, this architecture can achieve a significant reduction of the overall capacitor area in comparison with the area that would be taken by a single C-array to implement the same resolution. The

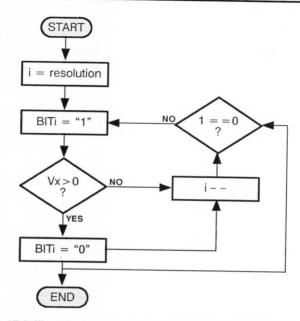

Figure 17.2 Fluxogram representing the ADC conversion phase.

minimum capacitor area for a segmented C-array can be achieved for N even and $M = L = N/2$. The operational amplifier coupling both arrays ensures that the weights of the capacitors in the sub-array are attenuated by 2^{-L} such that the weight of its most significant capacitor is normally half the weight of the less significant capacitor in the main-array. The operations which are needed to carry-out the conversion process are identical to those for the previous architecture with single C-array. However, because of the inversion produced by the active attenuator, the switches in the sub-array are switched in the opposite way of those in the main-array, thus achieving the correct conversion sign.

17.2.3 D/A conversion with single C-array

A parasitic insensitive D/A conversion architecture with a single C-array is shown in Figure 17.4. During the initialization phase, we connect the top plate of the capacitors to the output of the amplifier, for offset free conversion, and connect the most significant capacitor either to ground or to the reference voltage depending on the sign bit. Then, for conversion, we configure the architecture as a multiplying D/A converter. This consists of connecting the most significant capacitor to the amplifier output and connect either to ground or to the reference voltage the remaining capacitors, depending on the values of the corresponding bits of the digital

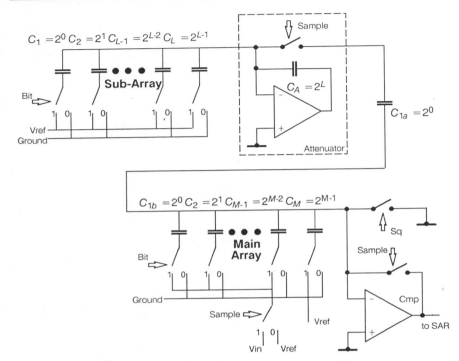

Figure 17.3 ADC architecture with segmented C-array for reducing capacitance spread.

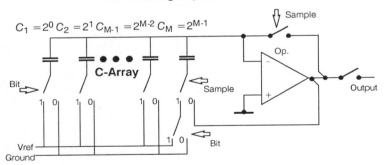

Figure 17.4 DAC architecture with single C-array.

word. In this way, the output digital code corresponding to the converted analog voltage is expressed in two's complement notation.

2.4 D/A Conversion with segmented C-array

In the D/A conversion architecture with segmented C-array, shown in Figure 17.5, we employ again, as in the case of the A/D conversion, an active attenuator for coupling the sub-array to the main-array. As before, the switches of the sub-array are also switched in opposition to those of the main-array in order to ensure that the packets of charge produced in both arrays are consistent with respect to their signs.

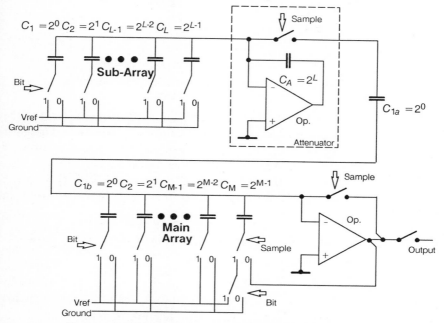

Figure 17.5 DAC architecture with segmented C-array.

17.2.5 Calibration networks

When the conversion resolution increases it becomes mandatory to introduce self-calibrating techniques in order to guarantee the required resolution and linearity. This consists of associating to each relevant capacitor in the binary-weighted C-arrays employed in the architectures the typical calibration network shown in Figure 17.6 [10]. Here, the smallest capacitor can introduce a nominal weight of $\pm 1/4$ LSB, depending on whether it is switched as C_k (positive weight) or in opposition to it (negative weight). The number of capacitors in the network depends on the capacitance value as well as on the accuracy of the capacitor being calibrated.

The calibration cycle illustrated in Figure 17.7, consists of determining the appropriate configuration for the calibration networks such that the main binary-weighted C-arrays can achieve the required capacitance matching accuracy within better than $\pm 1/2$ LSB. Under the control of a dedicated

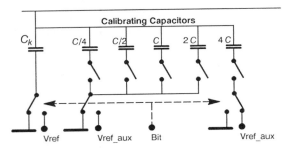

Figure 17.6 Typical calibration network for high resolution data conversion applications.

digital circuitry, this process is carried-out sequentially from the least to the most significant capacitor that needs calibration [10]. For each capacitor being calibrated the residual voltage is determined a sufficient number of times in order to obtain convergence for the calibrating configuration. This cycle is repeated for all the calibration capacitors of each calibration network using the successive approximation algorithm described in Figure 17.2.

17.3 Design methodology and dedicated functional simulation

17.3.1 Design methodology

The design procedure schematically illustrated in Figure 17.8 starts by considering first the required conversion mode, A/D and D/A, and then a set of input specifications which include the resolution, conversion speed, reference voltage and various technology parameters.

The specified conversion resolution and reference voltage are fundamental to determine the required resolution of the comparator which must be able to resolve a minimum voltage of $V_{ref}.2^{-(n+1)}$ to achieve an error smaller than $\pm 1/2$ LSB. The reference voltage is also important for establishing the range of values available for the input signal, $[-V_{ref}$ to $V_{ref}]$, and hence also contribute to determine the speed of the comparator. The conversion speed F_{conv} determines the clock frequency $F_s = F_{conv}(N + 2)$, where the extra two clock cycles represent the time needed to discharge the capacitors, provide offset cancellation and take a new sample of the input signal. Finally, the technology parameters specified at the input can be set either to the values of an already characterized process or to default values if no technology information is available.

Based on the input specifications and available technology parameters, the architecture is defined by deciding first whether to introduce the segmentation and/or the calibration options. The number of capacitors

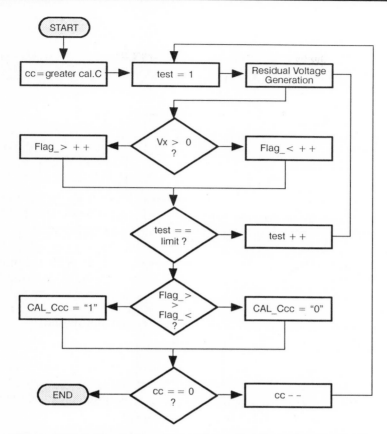

Figure 17.7 Calibration cycle for the calibration of one capacitor.

assigned to the arrays can be automatically determined or, alternatively, can be defined by the designer in order to meet special requirements with respect, for example, to the capacitor area and KT/C noise. The need for the calibration option is highly dependent on the required linearity and expected matching accuracy of the capacitance ratios. For optimizing the design, especially the calibration network which has many design parameters to define [10], we need to simulate the operation of the circuit. This is carried-out by an internal dedicated behavioral simulator which has been written based on the formal description of the data conversion architectures, in terms of the individual circuit components and their specifications, as well as the corresponding conversion algorithms. Statistical models for defining the capacitance ratios and taking into account perturbations such as additive gaussian noise are also included.

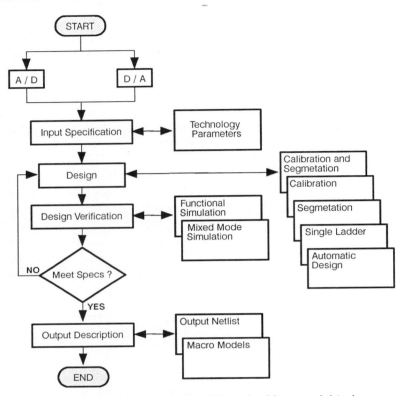

Figure 17.8 Design methodology for binary-weighted data conversion systems.

A structured output netlist is generated when the desired input specifications are met. Here, the synthesized data conversion architecture is defined in terms of the individual circuit components such as amplifiers, comparators, switches, C-arrays, and logic gates, together with their specifications. Such design information is then passed on to a variety of lower level design environments including cell libraries, parameterized circuit module generators and circuit level compilers (e.g. [11], [12], [13]) where the individual circuit components can be designed with great flexibility. This makes it possible to meet the most stringent specifications for the implementation of data conversion systems employing the architectures described before.

17.3.2 Dedicated functional simulation

The built-in functional simulator dedicated to this type of converter architectures is based on the corresponding conversion algorithms and takes into account the non-ideal parameters of the circuit components. For some of these effects, for example capacitance mismatch errors, we randomly generate the values for simulation based on the appropriate statistical information. This allows to verify the functionality of the architecture and its linearity under more realistic conditions.

The functional simulation of the A/D conversion mode to carry-out a Quick Test is illustrated in Figure 17.9. It starts by randomly generating the input digital word that will ideally be obtained if a correct conversion is performed. This digital word is then applied to an ideal D/A converter in order to generate the analog voltage which is needed to test the A/D converter. The converted output digital word is obtained by considering for the various circuit components the realistic parameters previously determined. By comparing the input and output digital words we can therefore determine the resulting conversion errors expressed in LSB's. This simulation and verification can be carried-out for as many input digital words as the user may require, and considering as many sets of component values as determined by their statistical behavior.

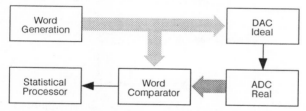

Figure 17.9 Schematic description of the functional simulation of the A/D conversion mode for the quick test.

The functional simulation of the A/D conversion mode to carry-out a Total Test is schematically indicated in Figure 17.10. This differs from the above because now we apply to the A/D converter under test a large set of analog voltages uniformly separated and covering the full input voltage range. By linear interpolation, additional samples can also be inserted between the analog voltages corresponding to two consecutive digital words and hence increase the verification accuracy. The conversion results are stored only if they are relevant to determine the INL (Integral Non-Linearity) and DNL (Differential Non-Linearity) characteristics.

As in the previous cases, the functional simulation of the D/A conversion mode can also be carried-out for quick and total tests, as is illustrated respectively in Figure 17.11 and in Figure 17.12. Here, the input digital words which are either randomly generated, for the Quick Test, or sequentially generated for every conversion code, in the case of the Total

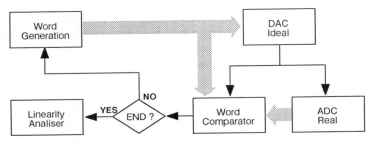

Figure 17.10 Schematic description of the functional simulation of the A/D conversion mode for the total test.

Test, represent both the input of the D/A converter under test and of the ideal D/A converter. By comparing the two converted output voltages and quantizing the result we can express the conversion errors in LSB's and then use this information for extracting both the differential and integral non-linearity characteristics of the converter.

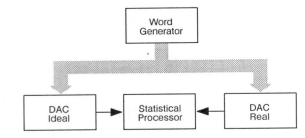

Figure 17.11 Schematic description of the functional simulation of the D/A conversion mode for the quick test.

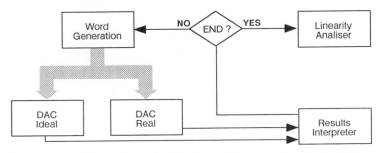

Figure 17.12 Schematic description of the functional simulation of the D/A conversion mode for the total test.

The functional simulation of the A/D and D/A conversion modes to carry-out a Monte Carlo analysis makes use of the Total Test for a wide and representative set of randomly generated errors. This analysis allows a histogram to be generated representing the INL and DNL boundaries for the sets generated as well as the estimation of the most suitable set of errors for the acceptance margin previously specified. This is fundamental for the automatic design mode decisions and for the lower and higher level descriptions.

17.4 Menu driven operation and interfacing

17.4.1 Menu driven operation

This tool can be operated either in a fully design automated mode or in an user-guided design mode. In the latter case the user has access to all the design parameters and options of the tool through a menu-driven user-friendly interface, as schematically illustrated in Figure 17.13. This allows the synthesis and verification of a practically unlimited number of architectural solutions for data conversion systems employing binary-weighted C-arrays.

As shown in Figure 17.13, there are several options available to the user in order to make his designs, tests and circuit descriptions of the chosen architecture. But, firstly, a default design is automatically generated by the tool for a given set specifications, and only then can the designer reach the full potential and flexibility of the Menu World.

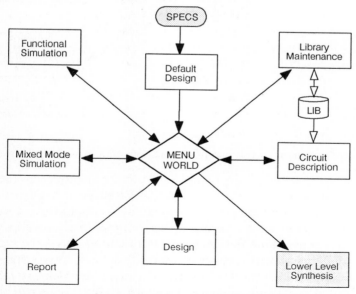

Figure 17.13 Menu world for the user-guided design mode.

In the Menu World the user has the ability performing a range of tasks, these include:

- the generation of alternative data conversion architevtures,

- report generation for designs,

- functional simulation for behavioural verification of designs,

- mixed mode simulations for detailed analysis of performance,

- management of library cells for optimum component selection (if available) for the synthesised converter

- circuit description generation for the design of remaining lower level components.

All such tasks can be performed by manipulating the menu system illustrated in Figure 17.14.

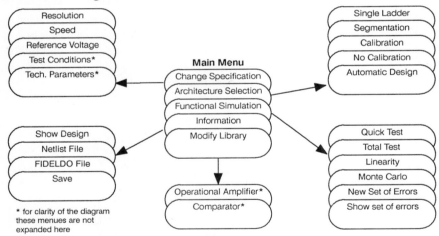

Figure 17.14 Illustration of the complete menu system available in the tool.

In the design block the designer can change the technology parameters, adjust the input specifications, if needed to relax stringent requirements, modify the test conditions and even choose another architecture, if strictly necessary or required for optimizing specific circuit characteristics. The circuit reports provide the designer with information about the current characteristics (including parameter values) of the circuit corresponding to the synthesized architecture. The functional simulation block gives access to the dedicated behavioral simulator which allows the designer to obtain statistical results for the converter, including the integral and differential non-linearity characteristics. There is also a possibility of performing a mixed-mode simulation by generating a description of the circuit to serve as the input file to, for example, the FIDELDO mixed-mode simulator [16]. The library block gives access to, and constantly updates, a library of cells which may be employed in the design process and from which the technology parameters can be extracted for simulation of the data conversion architecture. Finally, a circuit description using the cells in the library can be generated and then passed to the lower levels of the design hierarchy.

17.4.2 Tool interfacing

In order to obtain the full advantages of the hierarchical design approach it is essential to establish efficient tool interfacing mechanisms for transfering the design information, as is schematically represented in Figure 17.15.

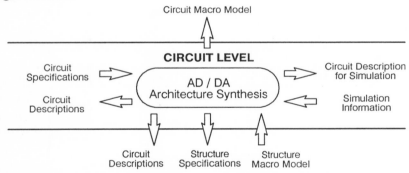

Figure 17.15 Tool interfacing for transferinf the of design information.

At the architectural level, there are two possible tools which can be interfaced in order to render even more comprehensive the design process. One such tool is the mixed-mode simulator FIDELDO, where the analog part of the converter is described electronically whereas its digital part is described functionally. This type of simulation can be used to provide more realistic information on the performance behavior of the synthesized converter. The other tool consists of a cell library database where we can search for the most appropriate circuit components that meet the specifications determined by the architecture synthesis process. This makes it possible to considerably simplify the design path down to the layout level.

Whenever it is not desired by the designer, or not possible, to obtain components in the cell library database, then the tool establishes a top-down flow of information in order to access lower level tools for designing such circuit components. This is the case, for example, for parameterized module generators for ·C-arrays and switches, and silicon compilers for operational amplifiers, comparators and reference voltages.

The bottom-up flow of information consists of generating an appropriate macro-model of the converter whereby we can convey to a higher level of the design information concerning its input and output variables as well as the integral and differential linearity characteristics, including relevant statistical behavior performance.

17.4.3 Examples of output information

In order to analyze the converter performance the user may wish to obtain statistical information about the converter by manipulating the quick test option in the functional simulation menu. The format of such information is illustrated in Figure 17.16, for the example of a 10-bit converter with a random set of capacitance values where the most significant capacitor has a large mismatch error causing an error larger than 2 bits for about 50% of the tested codes.

Conversion Cycle			
Iteration No.	Input Code	Output Code	Error Rate (%)
1007	815	820	
1009	701	704	1 LSB 28.795
1010	665	669	2 LSB 0.000
1011	779	785	>2 LSB 52.004
1013	912	917	
1014	262	263	
1016	300	301	
1017	357	358	
1018	665	669	
1020	122	121	
1021	111	110	
1022	776	782	
1023	862	867	

Figure 17.16: Statistical results obtained with quick test.

A more detailed simulation may be obtained by asking for the generation of INL and DNL characteristics, as will be shown for the design examples described in the next section.

In case the user requires a more complex evaluation of the circuit, then the tool can produce a circuit description file for the mixed-mode simulator

FIDELDO. Such a file consists of one non-critical part which is described in the FIDEL description language and functionally analysed by the FIDEL analyzer, and a second critical part which is described in the typical SPICE electrical description format, and electrically simulated by the ELDO analyzer. The functional and electrical analyzers are both integrated in the same simulation environment, as schematically illustrated in Figure 17.17.

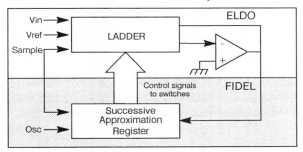

Figure 17.17 Evaluation of a data conversion architecture in a mixed-mode simulation environment

It is also possible to observe and extract the parameter values determined by the tool for the various circuit components comprising the synthesized data conversion system. This is illustrated in the Table 17.1, for the comparator, and in Table 17.2 for the amplifier, of a 12-bit segmented A/D converter with conversion speed up to 250kHz and a reference voltage of 2.5V.

Table 17.1 Example of specifications determined for the comparator.

Input Resolution	Setting time for Comparison	Auto-zeroing Condition		
		Settling Time	Final Error	Loading Cap.
90uV	275nSec	600nSec	350uV	35pF

Table 17.2 Example of specifications determined for the coupling amplifier.

Gain	G B W	Setting Time	Final Error	Loading Cap.
55dB	22MHz	75nSec	25mV	4pF

The macro-model of a synthesized A/D conversion system is illustrated in Figure 17.18, and this may be used for an even higher level of synthesis where such converter is needed as a building block. Such macro-model informs about the worst case INL and DNL characteristics as well as the relationship between the input clock and the conversion speed, and also indicates the input/output terminals corresponding to the reference voltage, input master clock , analog input signal, converted digital word and end of conversion, and also the enable control terminal.

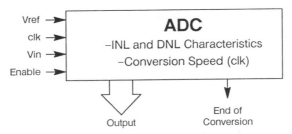

Figure 17.18 Macro-model for an A/D conversion system.

17.5 Design examples

In order to demonstrate some of the features described before, we consider here the design of a data conversion system to meet the specifications given in Table 17.3 for an A/D converter. For each one of the four possible A/D conversion architectures we obtained the basic requirements indicated in Table 17.4 for the binary-weighted C-arrays resulting for the synthesis of the 12-bit A/D converter. Furthermore, considering the selection of the architecture with segmented C-array, we can see in Table 17.5 how the required calibration network would vary as a function of the expected capacitance matching accuracy of the arrays.

Table 17.3 Example of input specifications provided for the synthesis of a data converter.

Conversion Mode	Resolution (bits)	Speed (kHz)	Vref. (V)	Techno.
A / D	8	100	2.5	AMS
A / D	12	100	2.5	AMS

Table 17.4 Basic characteristics of alternative architectures. matching accuracy.

Architecture	C-Spread	Total-C	Calibration	Accuracy
Non-Seg	2048	4096	No	0.02%
Segmented	32	160	No	0.02%
Non-Seg. / Cal.	8192	4104	Yes	Can Vary
Seg. / Cal.	3368	162	Yes	Can Vary

Table 17.5 Calibration requirements as a function of the capacitance matching accuracy of an A/D converter with segmented binary-weighted C-array.

Matching Accuracy	Bits 10 to 12	Bit 9	Bit 8	Bit 7	Bit 6	Bits 5 to 1
0.1%	Cal.	Cal.	No	No	No	No
0.5%	Cal.	Cal.	Cal.	Cal.	No	No
1%	Cal.	Cal.	Cal.	Cal.	No	No
2%	Cal.	Cal.	Cal.	Cal.	Cal.	No

Assuming a capacitance matching accuracy of 0.1%, we have carried-out the behavioral simulation in order to verify the design of the 12-bit A/D converter with segmented C-array, with and without calibration. The resulting INL and DNL characteristics are illustrated in the diagrams of Figure 17.19. For completeness, this figure also shows the linearity results obtained for the case of the 8-bit A/D converter with a single, non-calibrated C-array and considering the same matching accuracy of 0.1%.

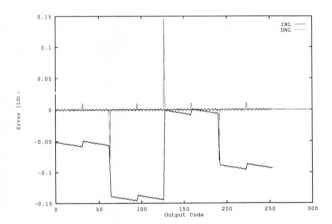

Figure 17.19 (a) 8-bit with single C-array

After a satisfactory architecture design has been achieved, the corresponding description containing relevant information of the specifications for each circuit component may be generated to continue the design process at lower levels. The typical placements which are also automatically indicated for the circuit building blocks employed in an A/D converter are illustrated in Figure 17.20(a) and in Figure 17.20(b), respectively for the case with a single C-array and for the case with a segmented C-array. In the case of the A/D converter with single C-array we

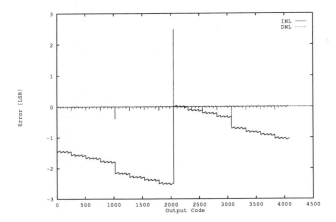

Figure 17.19 (b) 12-bit with segmented C-array

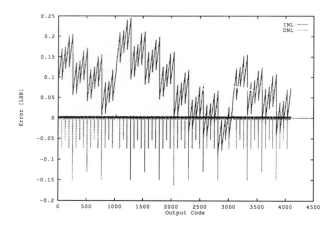

Figure 17.19 (c) 12-bit with segmentation and self-calibration.

Figure 10.19 Illustrations of the INL and DNL characteristics obtained for different architectures of an A/D conversion system assuming a maximum capatitance matching accuracy of 0.1%

can see a squared box representing the C-array, with the box for the comparator sitting on its right, the switch-array on top, the digital cells on the left, and finally the analog pads separated from the digital ones. In the case of the segmented C-array we obtain a similar placement with the addition of a second C-array, an amplifier, and also the generator of the auxiliary voltage reference.

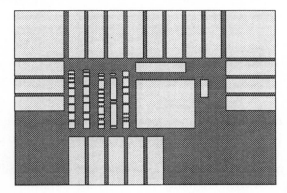

Figure 17.20(a) Typical placement of the circuit building blocks in an A/D conversion system (8-bit with single C-array).

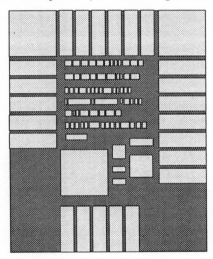

Figure 17.20(b) Typical placement of the circuit building blocks in an A/D conversion system (12-bit with segmented C-array).

17.6 Conclusions

We have considered in this Chapter a flexible framework for the automatic synthesis of data conversion architectures employing binary-weighted C-arrays which can meet a broad range of input specifications, both for A/D and D/A conversion, and perform to a high degree of precision even when considering the non-idealities of the circuit components such as offset, charge injection, noise and component mismatch errors. This included not only the description of the complete architecture design and verification automation processes but also the discussion of an efficient top-down flow of design information to access a variety of lower level design environments including cell libraries, parametrized circuit module generators and circuit level compilers. By ensuring such design flexibility at the lower level, this framework also renders it easier to meet even the stringent specifications for implementation of a data conversion system employing binary-weighted C-arrays.

Acknowledgements: Part of the work described in this Chapter has been developed within ESPRIT II Project 2193 (ADCIS) sponsored by the Commission of the European Community. We are grateful to colleagues at Anacad (Germany and France), CNM and University of Seville (Spain), CNET (France), NMRC (Ireland), and Matra MHS (France), for the useful discussions and suggestions during the development of this work. We also acknowledge our colleagues Paulo Montalvao and Pedro Abramtes who have been involved in the initial phases of this work.

17.7 References

[1] M. Ismail, J. Franca, "Introduction to Analog VLSI Design Automation", KAP, 1990.

[2] P. Allen "A Tutorial - Computer Aided Design of Analog Integrated Circuits", IEEE Custom IC Conference, pp. 552- 555, 1986.

[3] J. Franca, "Present Role and Future Trends of Analog Design Automation", 6th. ESPRIT Conference, Brussels, November 1989.

[4] P. Allen, P. Barton "A Silicon Compiler for Successive Aproximation A/D and D/A Converters", IEEE Custom IC Conference, pp. 608- 616, 1986.

[5] W. Helms , B. Byrkett "Compiler of A to D Converters", IEEE Custom IC Conference, pp. 161- 164, 1987.

[6] G. Jusuf, P.R.Gray, A. Sangiovanni-Vincentelli, "CADICS - Cyclic Analog-to-Digital Converter Synthesis".

[7] S. Sabiro, P. Senn, M. Tawfik, "HiFADiCC: A Prototype Framework of a Highly Flexible Analog to Digital Converters Silicon Compiler", IEEE International Symposium on Circuits and Systems, New Orleans, May 1990, pp.1114-1117.

[8] J.McCreary, P.Gray, "All-MOS Charge Redistribution Analog-to-Digital Conversion Techniques - Part I", IEEE J. Solid-State Circuits, Vol. SC-10, pp.371-379, Dec. 1975.

[9] C. Leme, J. Franca, "An Overview and Novel Solutions for High-Resolution Self-Calibrating Analogue-Digital Converters", Proc. 1989 IEEE Int. Symp.. on Signals, Systems and Electronics, Sept. 1989.

[10] C. Leme, J. Franca, "Minimum Area Analogue-Digital Network for High Resolution Self-Calibrating Converters", Electronics Letters, Vol. 26, No.18, pp.1491-1493, 30th. August 1990.

[11] M. Degrauwe et al., "IDAC: an Interactive Design Tool for Analog Integrated Circuits", IEEE J. Solid-State Circuits, Vol. SC-22, pp. 1106- 1116, Dec. 1987.

[12] D. Garrod, R. Rutenbar, R. Carley, "Automatic Layout of Custom Analog Cells in ANAGRAM", Proc. 1988 IEEE Intrl Conf. on CAD, November 1988.

[13] H. Koh, C. Sequin, P. Gray, "OPASYN: A Compiler for CMOS Operational Amplifiers", IEEE Transactions on Computer-Aided Design, Vol. 9, pp. 113- 125, Feb. 1990.

[14] R. Harjani, R. Rutenbar, R. Carley, "OASYS: A Framework for Analog Circuit Synthesis", 1987, IEEE Transactions on Computer-Aided Design, Vol. 8, pp. 1247-1265, Nov. 1989.

[15] M. Degrauwe et al., "The ADAM Analog Automation System", IEEE International Symposium on Circuits and Systems, New Orleans, May 1990, pp. 820-822.

Chapter 18

Automated Analogue Design

Adoracion Rueda and Jose L. Huertas
Centro Nacional de Microelectronice, Universidad de Sevilla
Avda Reina Mercedes s/n. Edif. CICA

18.1 Design automation for mixed analogue/digital integrated circuits

Analogue IC's and mixed analogue/digital ASIC's are nowadays needed in many different fields such as telecommunications, robotics, data storage, biomedical instrumentation, etc. Since the analogue parts are strongly dependent on applications, the challenge of producing optimal designs for these ASIC's involves problems very different from one application to another, making the problem very unlike those encountered in digital circuits. This increases engineering costs and design time, which are key drawbacks of analogue and mixed integrated circuits.

To catch up with the advances of CAD tools for digital design, the world research community is now making significant efforts to develop new software programs with capabilities for the efficient design of analogue and mixed ASIC's. Attention is being paid mainly to the design automation of the analogue parts of the system, as well as to the development of efficient simulation and modeling tools. The latter must cope with the accuracy requirements of analogue circuits and with the problems derived from the coexistence of analogue and digital circuits on the same substrate [1]. On the other hand, the former must provide ASIC's designers with aids to evaluate trade-offs and alternative designs, to generate specific components in a short time, and to interconnect these components forming the final chip.

From a practical viewpoint, it is frequently invoked the need of developing silicon compilers, computer programs which would be able to transform system level specifications into working silicon. Although the term Silicon Compiler has been coined some time ago, it is rather vague and may be misleading since it is used with several meanings. In a broad sense, an analogue silicon compiler can be defined as a program that automatically translates a high-level description of an analogue circuit (or system) into a physical layout description.

According to the component complexity it is customary to consider different levels of compilation. In what follows we will distinguish between module (low-level) and system (high-level) compilers, depending on their complexity. Usually, both of them coping with the three phases of a design, namely synthesis, schematic sizing and layout. In this terminology, module compilers (frequently called module generators) handle the design of low complexity building blocks like opamps, comparators, etc. On the contrary, analogue silicon compilers at system level have been mainly limited to circuits or systems with a very regular structure, requiring a considerable number of similar building blocks, for example switched-capacitor filters, A/D and D/A converters, and so on.

To save time and computing effort, early compilers resorted to a library of predefined cells. A consequence of this fact is that the resulting circuits are scarcely optimal in terms of area and performance. Of course, this is not a drawback when the main goal is a fast access to silicon, but it is when high performance is demanded. An alternative approach is to provide designers with an Integrated Design Framework, i.e., with a set of cooperative tools performing the same task as compilers. They are not but open compilers, i.e., software pieces whose overall job is the same of that a compiler, but where a formal hierarchy is translated into a hierarchy of programs, and human designer interactions are welcome. In these frameworks, both the expert designer and the novice have a place, the results being possibly distinct in each case although the tool used is similar.

But behind a successful framework there is a great investment in developing input and output formats, translators for each program and data' management methodologies. At this time, open frameworks integrating silicon compilers for regular subsystem, module generators, layout tools for custom cells, general floor-planning and routing facilities, and supporting the addition of user-developed tools as well, are a strong challenge in the digital CAD world [2]. Of course, less mature is this idea for analogue designs. Research efforts in design automation of analogue circuits are just recently receiving increased attention.

The idea of an Integrated Framework can be better understood by means of Figure 18.1. In this Figure we can see a set of programs cooperating throughout a programming backbone. There is a common user interface and several programs grouped into one of two categories, either module or system generators, each one enclosed in the Figure in one of the shaded blocks. System generators are programs able to accept functional specs and select one (or a few) system architecture among different possibilities. Every system generator is devoted to both a design style and an application. They produce specifications for the module generators attached to the framework. This is the main difference with a conventional system compiler. Unfortunately, system generators are improperly called system compilers, creating some confusion. However, since the term silicon compiler is broadly used with both meanings, we will maintain it along this text.

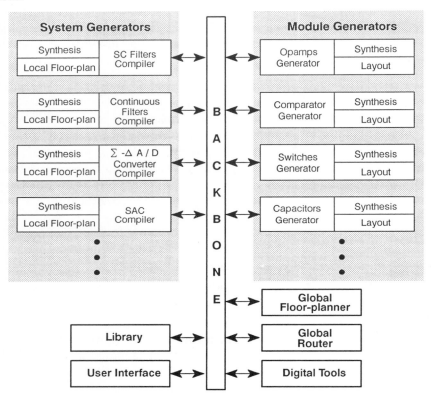

Figure 18.1 Integrated analogue framework

An example would be the case of an A/D converter employing binary-weighted capacitor-arrays; it will receive as inputs the required accuracy, the conversion speed and technology, etc. and generates requirements for opamps (gain, settling time, etc.), comparators (minimum voltage resolution, settling time, etc.), capacitor arrays (capacitor values) and switches (settling time). Once these specs are found out, they are passed to the different Module Generators implied, each one of them handling first the electrical synthesis of the corresponding basic block and afterwards generating the layout of this cell. When all the Module Generators finish their job, their outputs are passed trough the involved System Generator in order to carry out the layout of the system itself. Both a Global Floor-planner and a Global Router are used to assemble different subsystems together in a chip. Since every system design and layout can benefit from different concepts derived from designer expertise, even its "local" layout and floor-planning are specific, the global router and floor-planner being reserved for the final assembly. Ideally, this hierarchical flow might be extended for applications where the level we have called System level could provide the components to a higher level.

Figure 18.1 is a representation of how an Integrated CAD Framework must be, but the reality is somewhat more modest. Although some programs have been developed as standard tools, only a few are compatible with existing environments. There is a tendency to incorporate the new tools into commercial frameworks, all these tools being hierarchically cooperating. But still is quite early for results to come, and this hierarchical integration could be considered a future must for CAD.

After this revision, we are in a position for understanding the organization of this chapter whose purpose is twofold: first of all, we will quickly review the Analogue System Compilers reported in the literature. Secondly, we will go more in depth into the existing (or on-going) Circuit Module Generators.

18.2 A Glimpse at analogue system compilers

We have just concluded that compilers can play an important role in analogue ASIC design, although the degree of automation varies widely depending on the problem complexity. It is agreed that systems with regular structures are particularly amenable to being automatically designed.

The synthesis procedure we call silicon compilation might be divided into two stages: circuit synthesis and layout synthesis. Circuit synthesis relies, on the mapping process of either analytical equations or performance specifications to a circuit topology containing low-level blocks. Specifications are translated to these blocks whose circuit topology is then selected and their elements sized. The output is a netlist and/or a model for verification and layout synthesis. The aim of layout synthesis programs is the creation of a mask layout from a sized device schematic. They are of great importance in analogue compilation due to the heavy dependence of an analogue IC's performance on specific layout details. The overall design process is strongly application-dependent. Dedicated software has to be coded for each class of application.

In this section the most recently reported analogue silicon compilers are briefly reviewed. Following the two kind of systems on which these compilers have been focussed, we will organize this material into SC filter and data converter compilers.

18.2.1 Switched-capacitor filter compilers

A general SC filter compiler methodology can be described as illustrated in Figure 18.2. Automatic translation of global specifications for a SC filter to silicon is a complex process involving three major steps.

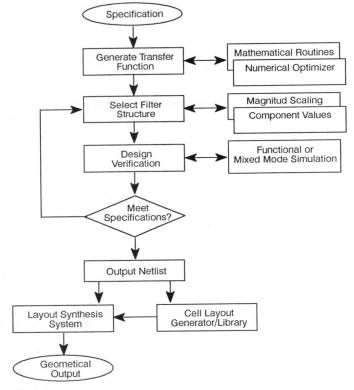

Figure 18.2 A General SC filter compiler methodology

First, mathematical approximation and numerical optimisation are carried out to generate transfer functions. Second, synthesis of structure and component values computation. At this step a design verification process must be provided either through a built-in simulator or by interfacing an external mixed-mode one such as SWITCAP. The third step deals with the layout generation. The usual approach for this purpose is to start by constructing a basic SC block (integrator) and then, blocks are all connected forming the whole filter. The output of this stage should be not only a layout description but and electrical extraction or model for simulation.

Table 18.1 SC Filter compilers comparison

	Filter Structure	Layout Style	Simulator
Helmes, 1986 (CONCORDE, Seatle Silicon Comporation)	- Classical approximation function - Cascade of biquads - LC Ladder Simulation	- Parameterized cells - Analogue channel router	SWITCAP
Therasse, 1987 (VITOLD, U.C. Louvain)	- Equiripple filters - Cascade of biquads - LC Ladder Simulation	- Analogue cell library - Standard cell floor-planning style	DIANA
Eaton, 1987 (SILCOMP, U. Toronto)	- Classical approximation function - LC Ladder Simulation	- Analogue cell library - Parameterized capacitor	SWITCAP
Assael, 1988 (CNET)	- Classical approximation function - LC Ladder Simulation	- Analogue cell library - Capacitor module generator - Fixed floor-plan	SWITCAP SCYMBAL ELDO
Barlow, 1989 (Asahi Kasei Microsystems)	- Classical approximation function with numerical optimization - Cascade of continuous anti-aliasing blocks, single stages, biquads and interpolators	- Module generators for Opamps, switches and capacitors - Fixed floor-plan - Analogue channel router	Built-in SC circuit Simulator

A lot of software has been generated to face some parts of this design process. Some have addressed the approximation and design tasks [3-4] and others the automatic layout generation [5-10] (approach in [10] also includes the automatic synthesis of opamps and switches).

Fairly complete SC filter compilers have also been reported [11-15], leading SC filter compilation to be a mature field. In order to make some kind of comparison amongst these different compilers. Table 18.I the main features of each one related to three points: (a) the approximation functions and the filter structures they are able to handle, (b) the layout style used, and (c) the simulator employed to verify the design.

It can be remarked that the main evolution has been at the layout level. Early SC filter layouts were based on a standard-cell layout scheme for placing and routing fixed height blocks with variable widths depending on the total capacitance corresponding to each block [5-8, 11-13]. These blocks are predefined parameterized bi-quad or integrator sections implemented as abutment of "slices" in the form illustrated in Figure 18.3a. Sensitive signals (summing node connections to opamp negative input) are hidden inside bi-quad sections to ease placing and routing. This standard-cell approach is relatively area-inefficient and its scope of application is limited. But, the main limitation comes out from its technology-dependent character; changing the system to accommodate a new technology means a considerable effort to change each cell.

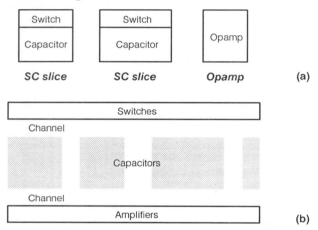

Figure 18.3 Two Different SC Filter Layout Strategies

More interesting layout approaches have been described in [9, 14, 15], all of them relying on a fixed floor-plan strategy consisting of arrays of capacitors located between opamp and switch arrays as indicated in Figure 18.3b. Switches and amplifiers can be either selected from a library or automatically generated, whilst capacitors are always automatically generated.The total area and aspect-ratio of each set of capacitors belonging to the same integrator are computed. Then each one is made of a certain number of unit capacitors placed according to a growing algorithm which minimizes area and signal crossover over sensitive nodes. The result is a technology independent approach allowing much better density and performance.

18.2.2 Silicon compilers for data converters

Automating the design of A/D and D/A converters is rather difficult due to the broad range of requirements encountered in real applications. Accuracy, speed, area, power dissipation, and component matching constraints are the main coordinates of their design space. The trade-offs among them being tightly dependent on the particular application field for which a converter is needed. The result is the existence of many architectures, each one being essentially different to the others and tailored to a specific problem (video, audio, etc). Hence, selecting an architecture appropriate for many different applications seems to be impossible, as well as building a compiler able to deal with different architectures is still far from reality.

Since this fact above is widely accepted, existing compilers for data converters [16-19, 24, 29] rely on the selection of a particular architecture, thus defining the range of applications for which the resulting converters are adequate. The most popular choice are Successive Approximation Converters [16-19], their versatility being invoked as the main reason justifying this election. They do not present stronger advantages or weaker points than other converter types, but they give a good compromise in terms of application wideness, bandwidth, resolution, etc.

Generally speaking, a data converter compiler is structured into the same two synthesis phases we have described in the introduction of this Section, driven by a netlist generator and by a layout generator, respectively, differences existing in the way they are built as well as in their interactions. Resident in a database these compilers store informations about technology, component matching statistics, design rules and floor-planning style. The compiler inputs are system specifications, like resolution, conversion speed, input signal bandwidth, maximum differential nonlinearity, maximum integral nonlinearity, power dissipation, supply voltage, reference voltage, area and aspect-ratio.

The first part of the compiler, the Netlist Generator, synthesizes all the different functional blocks required by the converter. Starting from the system specs, this generator designs every block optimizing it locally and, after assembling all the blocks, predicts or simulates the resulting converter performance. If the result does not meet the given specs, the generator tries to proceed by relaxing some of the user's requirements and repeating the task. The relaxation of specs might affect the total area or the power dissipation but resolution, speed and linearity have to be fulfilled. Then, if after relaxing the specifications that can be considered "flexible" a design cannot be obtained, the user is asked for new specs.

After designing the converter at the electrical level, the layout generator carries out the geometrical design. Layout flexibility is variable for every compiler, it ranges from fixed styles based on standard cells [16,18] to very free ones where passive and active components are customized [17,19].

Going into some details for every reported compiler, CADICS [17] implements an improved one-bit-per-cycle architecture and incorporates a built-in behavioral simulator. It uses OPASYN [37] to generate opamps and comparators,and routers like GLITTER or ART are employed. HiFADDiC [18] is also able to design Successive Approximation Converters, in this case a pipe-lined cycling converter with analogue compensation. Behavioral simulation is carried out by FIDELDO [23] using macro-models for the analogue blocks, and both the floor-planning and the layout are fixed. Library cells are used and assembled in predefined horizontal bands with separated analogue and digital channels routed by an analogue-oriented channel router. The compiler proposed by Allén and Barton [16] and the one in [19], generate charge redistribution successive approximation converters. The former can handle structures using only a binary weighted capacitor array and structures using a combination of a resistor string and a capacitor array. Its geometrical generator is AIDE2, a program to interconnect parameterized cells into a predefined structure. Concerning [19], it is formed by two main blocks, MDAC which is a menu driven architecture compiler, and ALSAC, which provides the converter mask layout. The key features of this compiler are: (a) its capability of synthesizing a broad range of resolution and performance requirements by incorporating different circuit topologies as well as powerful self-calibration techniques, (b) the ability to consider non-ideal issues through dedicated functional simulation and precise layout techniques, and (c) the layout generation, carried out with a new floor-planning strategy [21], and dedicated module generators [22]. We cannot finish this section without making a reference to programs that extend the capabilities of other tools allowing them to synthesize converters [24, 29]. All of them are devoted to sigma-delta converters, the reason being the close relation of the analogue part of this kind of converter to filters, as well as that the large digital part these converters require can be designed by a digital silicon compiler.

18.3 CAD Tools for circuit module generation

In the last section we have highlighted some attempts to make computers face the design problem. They lead to the construction of silicon compilers at the system level for simple classes of applications. A few of them do include the capability for the specific design of the required primitive building blocks. Providing this capability through invoking programs carrying out technology-independent building block designs, is the best way to surmount library shortcomings. These programs, usually called circuit module generators, offer more flexibility than fixed library cells. Moreover, in addition to be more useful to a large number of applications, they are the best vehicle to push performance limits further.

In the light of the previously presented SC filter silicon compilers it can be said that they require of three classes of analogue sub-blocks: switches, capacitors and opamps. They usually are clustered forming a macro-block, such as an integrator. Other analogue sub-blocks such as comparators, band-gap voltage references, etc. have to be considered in A/D and D/A converters. Then, two categories of module generators are required for automated generation of analogue systems. The most simple are those dealing with a flexible and error-free implementation of simple components: resistors, capacitors and switches. The other category corresponds to module generators for complex building blocks: opamps, comparators, etc.

In this section we give due prominence to some of these module generators, remarking on the features making them of current interest.

18.3.1 Module generators for simple components

Successful methodologies for the technology-independent automated generation of simple components have been reported in the context of both SC filter and data converter silicon compilation. The most promising approaches are presented in [9, 22]. The key goal of these tools is providing compact high-accuracy arrays of capacitors and switches meeting some fixed floor-planning style. Accuracy is obtained by considering optimum transistor sizes for switch schematics, and by minimizing, through layout techniques (i.e. parallel connections of unit capacitors), systematic errors in capacitor values or ratios resulting from process variations and mask movements. On the other hand, compactness is achieved using specific layout templates for each type of switch, and by applying flexible capacitor shapes together with efficient unit-capacitor placement algorithms.

The methodology described in [9] for capacitor arrays generation in SC filters has been proven to be efficient in different SC filter floor-planning style [10], as well as extendible to other applications requiring precise matching of several capacitors, such as A/D and D/A converters based on the charge redistribution principle [22]. Resorting to this methodology, capacitor blocks with a flexible aspect-ratio are obtained. The generation process is based on three stages. In the first stage, technology parameters are considered to limit permissible unit capacitor values (i.e. unit capacitor aspect-ratio), and dimensions of each capacitor block (all capacitors belonging to a same integrator) are evaluated to minimize the overall area of the filter. Minimum area is attained following an interesting optimisation algorithm based on two considerations: a) tracking requirements for any two capacitors in different integrators are not needed, so, unit capacitor shapes can be different in each integrator block, b) assuming a fixed width W of the overall filter layout, imposed by the opamps, the dimensioning algorithm looks for a minimum uniform height for each capacitor block such that the sum of the parcial block widths matches W. In the second stage, a symbolic matrix is assigned to each capacitor block showing the relative positions of unit capacitors for each capacitor in the block. In the third stage, symbolic matrices are converted into mask layouts for the given design rules. The implemented algorithm tries to find assignments of unit capacitors in the symbolic matrix such that each capacitor faces the best block side in order to meet routing constraints.

In [22] a similar methodology has been developed for the automatic generation of precise binary-weighted capacitor arrays for A/D and D/A converters. Flexible algorithms are necessary for these converters for two main reasons:

(1) Capacitor area can be large and variable depending on the number of bits, converter architecture and technology optimum unit-capacitor value. This means that modules with variable relative form-factor and area should be made up.

(2) Matching requirements are fundamental for high resolution converters. Thus, each capacitor of the array should be realized as the parallel connection of several identical unit capacitors, and unit capacitors should be equally surrounded. On the other hand, common-centroid or axial symmetric geometries might be recommended.

The capacitor module generator presented in [22] has been developed to meet these two requirements. Input specs accepted by the program are: number of bits, geometrical form-factor for the complete array and unit capacitor value. The main body of the generator is the array synthesis. It is based on a technology independent symbolic synthesis process which generates a symbolic matrix for the array. The process is quite different from that used in [9] because when dimensioning capacitor blocks it intends to meet a given geometrical form-ratio, instead of minimum height and width; on the other hand, all capacitors in the block have the same top plate while their bottom plates are connected to different switches.

Two algorithms have been developed to generate symbolic matrices. One of them gives rectangular symbolic matrices with all the bottom plates of capacitors having access to only one side of the block, and the common top plate having access to any other side. Symbolic matrices with common-centroid geometry are generated by the second algorithm, which places capacitors around a symmetry axis all of them having access to an inner common channel. Figure 18.4 illustrates graphically the generation of these symbolic matrices.

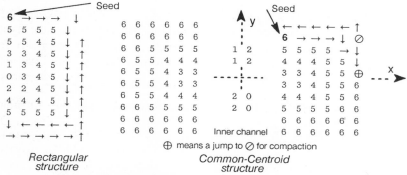

Figure 18.4 Illustrating symbolic matrices generation

Both algorithms are based on the same placement function, whose C description is schematized in Figure 18.5. They start placing a seed for the capacitor to be built (for instance, capacitor 6 in Figure 18.4), and an intelligent grow-up algorithm is then applied to synthesize the symbolic matrix. This grow-up algorithm must insure the connectivity of all unit capacitors, the compactness of the matrix (to avoid empty entries), and the achievement of a given form-ratio.

The layout of the capacitor array is generated in several steps. First of all, a unit capacitor is realized by sizing the capacitor template, which is chosen between those implemented in the database. Second, unit capacitors are placed in their relative positions as given by the symbolic matrix. Connections are then made in two levels, being the type of level imposed by the technology. For a double poly template, bottom plates of unit capacitors belonging to the same capacitor are connected in Poly1. Top plates are connected in Poly2, adding dummy taps to keep constant the capacitor value. In the case of common-centroid structures it is needed to size the inner channel allowing two types of connections, connections among unit capacitors and connections to the output switches. Finally, dummy capacitors are added and a surrounding well is included for isolation purposes.

```
Init-Symb( ) {
for (each capacitor)
     { Place the seed( ) ;
          Look for the last placed unit capacitor ( ) ; {
               Look for the neighbours and initiate one ( ) ; }
     }
     Place dummys ( ) ;
     Generate masks ( ) ;
}
```

```
Look for neighbours and initiate one ( ) {
     for ( each free neighbour ) {
          if ( connection laws are met ) {
               if ( compact the design ) {
                    if ( Form-Factor is met ) {
                         initiate one unit capacitor ( ) ;
                    }
          . . . . . . .
}
```

Figure 18.5 Symbolic capacitor placement function

Concerning switch module generators, we will describe herein that reported in [22] for application to successive approximation A/D and D/A converters based on the McCreary charge-redistribution principle [20]. This switch module generator is addressed towards the automatic design of the switches connected to the bottom plates of the binary-weighted capacitors. Although it is integrated into the ALSAC layout compiler [19], it is flexible enough to be used into other compilers. It realizes the three phases of the design process, namely, topology selection, transistor sizing and layout generation. The generator structure is shown in Figure 18.6. The database is organized in two levels: the representation level including data structure definitions and data processing functions, and the knowledge level which contains information relative to all switch types and switch layout templates. The input specifications are: array type, number of bit, conversion time (or settling time for switches) and unit load capacitance value.

In the synthesis process, the generator distinguishes different types of switches ranging from simple PMOS or NMOS transistor switches to double throw CMOS switches. The variable "array type" in the input specs serves the generator to process one of them. The problem the synthesis block must solve is to appropriately size switch templates by considering the conversion time requirements, which has to be specified as the settling time for the charge redistribution. Since the capacitive load increases with the number of bits, a careful switch sizing is important in order to trade-off speed specifications, and area and power limitations. In other words, making transistor dimensions different for each switch, we have the same

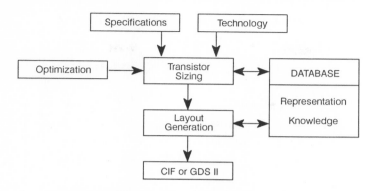

Figure 18.6 Scaled switch generator structure

· RC time constant for each branch, thus saving area and reducing transient currents. This problem has been focused as a worst-case design and the analytical worst-case relationships have been stored in the database knowledge level. Using these equations and the input specifications, initial transistor sizes are obtained, which will be then optimized by considering the most significant associated parasitic capacitors.

Templates for the layout are made up from transistor soft-macros in a predefined strategy for placement and routing. Switch layout is generated according to the previously calculated transistor dimensions, this meaning that the active areas as well as the contacts and wires are sized taken into account electrical requirements. At the final stage, switches are placed fulfilling a predefined arrangement to facilitate the routing. In the current version this arrangement is fixed according to the abstract description for the binary-weighted capacitor array given by the capacitor module generator. Figure 18.7 shows the generated layouts for the switch array (top view) and for the capacitor array (bottom view) corresponding to a 6 bit charge-redistribution array.

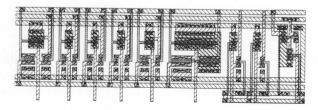

Figure 18.7(a) Layout generated in [22] for scaled switches and capacitor arrays for an ADC (Top View)

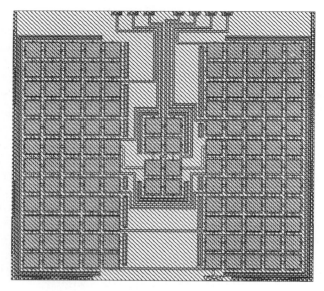

Figure 18.7(b) Layout generated in [22] for scaled switches
and capacitor arrays for an ADC (Bottom View)

18.3.2 Analogue cell module generators

Every experienced analogue designer knows that designing a good analogue cell, an opamp for instance, is a really difficult task. Complicated trade-offs must be solved at each level in the design process in order to best fitting specifications. A right solution involves a deep knowledge of circuit operation together with some iterations with an electrical simulator. Even so, when there are different possible ways (i.e. different topologies or design styles) to meet specs, decisions at the top level are, to some extend, purely random; designers do not explore alternative topologies except when an optimum solution does not result from the first choice. Analogue cell module generators are pieces of software aimed to automate complex cell design. They not only help skilled designers by relieving them from time-consuming routine work, but also allow design exploration and/or optimisation. Besides, they should be able to provide optimum parameter values without interaction with the user, thus being valuable to novice designers.

The global goal of any analogue module generator is to automatically translate block behavioural description to mask layout, normally starting by exploring the design space to obtain near-optimum parameters for circuit schematics; and then, resorting to layout techniques able to take into account matching and crossover restrictions. The overall process is usually carried-out by two software tools strongly related, one of them for the circuit synthesis step and the other for the layout generation step (see for instance, the examples of the pairs IDAC/ILAC [28/45], OASYS/ANAGRAM [33/46], CAMP/SLAM [35/43], etc). In order to better review the different methodologies proposed to cope with both steps, we will discuss them separately. Thus, approaches to make circuit synthesis are going to be considered in first place.

18.3.2.1 Synthesis step

There are three fundamental tasks any cell module generator need to perform for automating circuit synthesis: topology selection, parameter assignment and design verification. A closed loop implementation of the design cycle can be viewed as illustrated in Figure 18.8. It represents a typical manual design procedure. After the first two stages, a first circuit design is obtained from the given performance specs and technology data. Designers use their experience as well as simple equations to achieve this first design (sized-devices topology). Then a simulation is carried out, and circuit performance is compared with input specs and a new design is realized (if necessary) by adjusting circuit parameters or even changing circuit topology. Several iterations are usually made to achieve the final

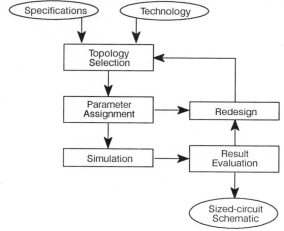

Figure 18.8 Implementation of an analogue design cycle

design. Owing to the difficulties behind the complete automation of this design process, only a few cell module generators have an actual closed loop form, including a simulator. However, recent approaches incorporating designer expertise into intelligent programs have led to successful module generators [25-42], and efficient design assistants [49-54]. Although all the reported module generators fit the global goal for most analogue building blocks, and most of them are based on expert systems, they differ widely in philosophy, i.e. they employ different ways either to capture human knowledge or to handle the design process, differing also in the way they perform design verification.

The circuit synthesis problem can be based on optimisation, systematic synthesis methodologies, or both together in a complementary way. In the latter, synthesis provides a starting point and optimisation tunes it. Automatic synthesis tools incorporating these methodologies can be classified into three categories of approaches, they will be reviewed in what follows.

(a) Optimisation-based approaches:

We bunch into this category tools to make dimensioning (parameter assignment) by relying on a classical optimisation problem formulation. They consider design problems mathematically defined as: determine a design parameter vector such that a multi-objective non-linear programming problem is solved. Usually, they are constructed using a circuit simulator combined with optimisation algorithms [25-27]. They can be successfully applicable to a wide range of circuits, but a good initial guess to converge to an optimum solution is required. These approaches have two main drawbacks concerning operation speed (i.e. the time required to get a working design). One is that they need intensive design knowledge to enter the problem description, although they usually provide specific languages for that. The other is that they are very costly in CPU time because many circuit simulations have to be performed during a design cycle.

(b) Knowledge-based approaches:

We will include into this category, tools that emulate expert designers mainly in the following actions:

(i)   Using knowledge to select topology

(ii) Resorting to a design methodology based on heuristic knowledge together with one of the two following alternatives: use of analytic equations (analysis and design equations) or use of a simulator.

(iii) Applying failure-correction knowledge-based procedures.

These kinds of software tools are based on design expertise obtained from analogue designers. They need to catch knowledge from human designers either to create heuristic or to obtain accurate while simple analysis and design equations, as well as to define a simple design flow for each topology or design style. But once knowledge is stored in the base, users are only interacted within some nodes of the design process and only when the system is not run automatically.

A number of expert-system based approaches have been recently reported [28-36], differing each from the other in the specific methodology to achieve sized schematics. Although they can do some specific optimisation during the design tuning, they face the dimensioning problem according to the acquired experience, usually implemented as a mixture of algorithms and rules. These approaches are currently considered the quickest way to achieve correct complex cell designs from simply specified design requirements.

All knowledge-base systems perform a translation of input specs into sized circuit schematics. However a comparison of the systems can be made according to their different sizing strategies and to their implementation. Since a sizing strategy is strongly dependent on the architecture or circuit topology, we will distinguish two types [35]: Multiple fixed-architectures approaches (e.g. IDAC [28,29]) and sub-circuit assembly approaches (e.g. PROSAIC [30], BLADES [31], OASYS [32,33], An-Com [34], CAMP [35], ISAID [36]).

For the former, one of the circuits stored in the knowledge base is selected (by the user or automatically) according to the input specifications. After that, transistor sizes are evaluated to achieve the required performance. The sizing process relies on developed routines specialized for a given analogue function, which include detailed expertise about the circuit. Since the chosen circuit topology can not be changed during the design process, they are known as flat approaches.

In sub-circuit assembly approaches, non-fixed circuit topologies are used and the design task is hierarchically partitioned into several sub-circuit tasks. Useful decompositions and assignment of sub-circuit goals are based on heuristics trade-offs, although in each system, this is carried out in different ways. These approaches can also be called hierarchical approaches. The main feature of a non-fixed topology approach versus a fixed topology one is flexibility. Non-fixed topology approaches can fit an unlimited number of specifications, and particular solutions to one design problem may be reused in another.

(c) Knowledge-based approaches including mathematical optimisation:

Tools included into this group are devised as an alternative to the above mentioned approaches, their aim being to get the best of each one in order to construct cell generators achieving more optimum cell designs.

To reach design goals designer's knowledge is employed in three ways:

- To guide topology selection.

- To supply a first design (if necessary).

- To create an appropriate search space, as well as analytic models.

The first two actions are usually carried out automatically in the same way tools in category (b) should do. On the other hand, the essential independent design parameters and constraint relations for optimisation are stated externally and then stored in the knowledge base for each circuit topology. For analogue cell design, a first design means either an annotated sized topology or a topology description together with a limited range for transistor sizes and bias current. Optimisation algorithms are applied to minimize some cost function subject to multiple linear or non-linear constraints through interactive adjustment of design variables. Since in analogue cells most performance specifications are related to small-signal behaviour, in each iteration a DC analysis of the circuit is also needed.

Several tools of this type have been reported [37-42]. The majority have been proto-typed only for opamp generation, although they could be applied to other analogue cells. The main differences between them are related to the optimisation strategy. Some make use of analytic equations to evaluate circuit performance [37-39], and others rely on SPICE simulations [40-42]. Furthermore, different search spaces and optimisation algorithms are considered, for instance, a multi-start steepest-gradient descent algorithm as in [37] or a Newton-Raphson like based on an expert system as in [40].

18.3.2.2 Actual implementations

In what follows, the key points of the most representative design tools belonging to the two last categories are highlighted in order to better understand the different philosophies in automated analogue cell generation. In Figure 18.9 the different knowledge based approaches to the sizing problem are shown, and reported module generators are related with there corresponding methodology.

IDAC [28] [29], is the representative design environment of a fixed-architecture approach. It has capability to automatically design a

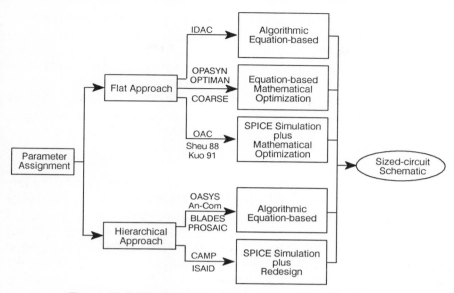

Figure 18.9 Different knowledge-based approaches and syntheses tools for the sizing problem

large number of building blocks (OTA, OPAMP, Comparators, zV_{ref}, over-sampled A/D, etc.), contained in a library of fixed topologies. The key points of IDAC are:

(1) Extensive input specifications. Besides selecting a technology, circuit specs, and the chosen schematic, the user can give some design options (assignment of gain, MOS operation in weak inversion, etc.), as well as information concerning minimum and maximum values of parameters, extreme temperature, etc.

(2) Worst-case distortion analysis is performed (previous to the design) to redefine specifications according to temperature, technology and bias current fluctuations. New nominal values are computed to satisfy the required input specifications over these deviations.

(3) A flat approach design strategy dedicated for each schematic. For this sizing step, IDAC requires general circuit knowledge (simplified design equations) which are incorporated into the "formal description" for each schematic. This is no more than a dedicated routine (i.e. a procedure defined by an ordered set of equations) stored together with the corresponding topology. For large circuits and to permit introducing new modules, the program supports a limited

hierarchy allowing decomposition into sub-circuits and reuse of module generators.

(4) Dedicated built-in analyzers are used to verify the design. They make use of transistor models which are continuous from weak to strong inversion. If some specifications are not satisfied after analyzing, a set of new specifications are defined and the formal description is again executed. For these modifications, IDAC resorts to the stored knowledge common to the family of circuits it is dealing with.

(5) A dedicated optimisation is included as a separated option for each circuit in the library, but it only serves to tune the final circuit, in terms of minimum area, power, etc.

Concerning sub-circuit assembly approaches, the main distinguishing feature is hierarchy, i.e., the circuit design task is hierarchically decomposed into sub-circuit design tasks. Hierarchy has two main advantages. First, reduces the size of the knowledge base as well as the complexity of the rule base. Second, architectures can be changed by modifying or adding sub-circuit primitives. Any improvement in the primitive design procedures benefits a large number of complete circuits. Unfortunately, it suffers from an important disadvantage in analogue design, namely, an optimum design for the complete circuit by accounting for all sub-circuit operation might not be reached. Two main methodologies have been followed to construct systems in this group. One relies only on analytic equations and rules stored in the knowledge base [30-34] and the other uses, in addition, SPICE simulations [35-36].

Although preceded by two significant tools [30,31], OASYS [32,33] is the most representative system of the former type. Its hierarchical structure evolved overcoming the difficulties of an initial classical rule-based version similar to [30,31], being the current design process more algorithmic than rule-based (about double number of steps than rules). The key points of OASYS are:

(1) The synthesis process in OASYS consists of a set of selection/translation steps in an explicit hierarchical structure. Selection means choosing a fixed functional sub-block arrangement (i.e. a design style) from among alternative templates in the database. Translation is driven by a planning mechanism which translates input specs into sub-block specs at the current level.

(2) Knowledge is stored in templates for each design style as a set of rules driving selection, and a plan implemented as a set of rules and design equations to make translation.

(3) Plans supporting translation are static processes involving decision, computation and refinement actions to meet performance specs and to change the plan when some cannot be met. To reach these goals, the planning mechanism must handle rules derived from designers expertise, linear and nonlinear algebraic equations expressing component values in term of performance parameters, technology parameters and previously determined values.

(4) OASYS is provided with failure handling and particular optimisation capabilities by allowing local or global back-tracking in the hierarchy. Thus, failures could involve translation or design style modifications, and numerical optimisation can be implemented in the form of a fixed-point iteration process through different plan steps.

(5) Design refinements are contemplated as a number of additional plans associated to a same design style aimed to obtain designs for minimum power, area, noise, etc.

(6) As IDAC, OASYS is an equation-based expert-system-assisted approach. However, hierarchy used in OASYS greatly reduces the effort needed to create analysis and design equations for new circuit topologies. Moreover, when new device models are needed to accommodate technology evolving, only the lowest-level design styles should be modified while complete plans still remain useful.

(7) To date, OASYS can automatically design about 128 different CMOS op-amp's device-level topologies from 2 fixed high level design styles, and it is progressing towards implementing comparator design styles.

An-Com [34] uses a hierarchical strategy similar to OASYS which also relies on knowledge stored in templates and on top-down planning mechanisms. Apparently it offers the same compilation capabilities as OASYS although only simple CMOS opamps have been used to test the approach. Moreover, An-Com includes a macro-modeling utility whose aim is the verification of decomposition at each level (except at transistor level). This seems to be an alternative to iteration between different levels in the hierarchy.

Knowledge-based approaches including an electrical simulator in the design loop, to deal more accurately with parameter assignment, can be illustrated with CAMP [35] and ISAID [36]. Both are hierarchical expert-system-assisted approaches but as far as SPICE results are used to evaluate and verify design solutions at any intermediate step, the concept of hierarchy is differently exploited. Furthermore, the expert system actions are directed not only to generating a solution, but also to inferring conclusions from SPICE results and then to exploring on space of improvement solutions.

In CAMP, the improvement solutions space is represented by various improvement solutions implemented for each circuit performance parameter. These solutions can be either appropriate adjusting procedures or alternative circuit topologies for any functional block that can be identified in the current circuit. For the latter cases, CAMP is provided with a self-reconstructing capability. It consists of the identification of a portion of the circuit which can be replaced in order to improve a particular performance parameter. Then, circuit reconstruction is realized through equation substitution in the design knowledge base for posterior analysis. As a result, an initially-chosen circuit topology could be altered during the sizing process. A similar improvement procedure is proposed in ISAID [36], where a circuit generator module and a correction procedure module are interactively applied together with SPICE to automatically design CMOS opamps.

The use of SPICE simulations in the closed-loop design cycle makes these two methodologies more CPU time consuming, however, high performance designs can be obtained. In addition, the integration of an efficient expert system within the loop could cut down the number of SPICE iterations, thus offering powerful and reliable design systems.

Knowledge-based methodologies that apply mathematical optimisation are reported in [37-42]. Optimisation in these systems is viewed as a multiple parameter (device sizes and bias current) determination process under multiple design objectives and constraints. Several strategies have been followed to cope with the many degrees of freedom in parameter spaces, as well as with the multiple circuit performance evaluation. Equation-based approaches [37-39] substitute expensive circuit simulation with algebraic evaluation of an analytic circuit model. As far as these analytic models relate circuit performance specifications to several circuit parameters, without an explicit relation with circuit topology, equations-based optimisation approaches are not applicable to hierarchical design methodologies. Analytic models usually consist of: a set of independent design parameters, relationships between dependent and independent design variables, design equations relating circuit performance specs to the design variables, knowledge about the circuit, and general constraints. Owing to the use of simplified equations the final circuit has to be simulated before fabrication.

On the other hand, approaches in [40-42] use multiple SPICE iterations to guarantee the accuracy of the results. They always start from a first design obtained using some knowledge-based approach or selected from a library among those circuits fulfilling the given specifications. After that, the tuning problem formulated as a multiple-objective optimisation problem is solved using SPICE to evaluate performance of each intermediate design.

OPASYN [37] is a compiler for CMOS operational amplifiers that use optimisation for circuit sizing. The outstanding features of OPASYN are:

(1) It operates with 5 fixed circuit topologies stored in the database for CMOS opamps. The first step in the design phase consists in selecting a circuit topology that can meet the given specifications.

(2) The topology selection step is based on a decision tree included in the base. Using a set of rules, the decision process checks nodes of the tree and decides the leaf node or nodes (circuit topology) that better fit specifications

(3) Dimensioning is carried out in OPASYN using mathematical optimisation and a set of simple analytic equations attached to each topology in the base. The steepest descent algorithm is applied to calculate independent design parameters that minimize a design cost function, which is expressed as a combination of functions of the performance parameters. A first-cut design is not required, instead a random grid sampling through the bounded domain of all the independent design parameters is the starting guess.

(4) Designer's knowledge is used to evaluate the number of independent design parameters needed in each topology, as well as to create the design equations. These equations include fitting parameters that have been introduced to obtain more accurate performance evaluation. But the fitting parameters are technology-dependent, they are obtained from SPICE simulations. Thus, updating technology means an effort to create new design equations.

OPTIMAN [38] like OPASYN is a mathematical optimisation-based approach to sizing, which relies on analytic models. Its main differences from OPASYN are:

- It is independent of technology.

- A cost function must be supplied by the user, and it must be expressed as a weighted sum of at most four basic objectives (noise, power, area, Gain Bandwidth Product

- The optimisation algorithm is simulated annealing, and a limited initial move range is provided.

- If deviations are noticed after the final SPICE simulation, the program indicates which characteristics are responsible for the failure.

- Because it is integrated with the symbolic simulator ISAAC [50], it is provided with some capabilities for the automatic modeling of new topologies.

In [39], an alternative methodology to apply optimisation is proposed. The great difference from the previous approaches is that the DC operation point is the design variable, while the transistor sizes and AC characteristics are used as objectives. The aim of the method is to avoid the iterating DC analysis needed in the previous approaches, but it is not clearly stated the real advantages this method would offer.

18.3.2.3 Layout step

Most of the reported cell module generators provide a complete path from specifications to layout. They are usually formed by two separated software pieces responsible for the circuit synthesis and for the geometrical layout, respectively. The layout process is of essential importance in analogue circuits because physical design has a strong influence on certain performance parameters. Our concern here will be layout methodologies specifically targeted to analogue cells; so general placement and routing strategies usually borrowed from digital systems are not considered.

In laying out analogue cells, care must be taken about: control of absolute element values, optimum device matching, and harmful parasitic effects. Hence, tools for automating analogue layout should take into account:

- Topological, electrical and geometrical constraints, such as symmetry, sensitive and noisy node location, device orientations, etc.

- Complex primitive structures whose sizes and shapes can vary over wide ranges, even for the same functional building block when working under different specifications.

In addition, technology and schematic independence should be a common feature of all these tools.

The above mentioned specific layout constraints make layout automation for analogue IC'S much harder than for digital. However, many existing tools adopt digital layout styles that are accommodated to analogue problems. The ways to include analogue constraints range from applying layout heuristics [43, 44] to using algorithmic and procedural techniques [37, 41, 45, 47, 46].

An automatic layout generator consists of a set of low-level cell generators, a placer, a router, and a tool for post- processing and compaction. A general flow-diagram is shown in Figure 18.10. The specific implementation of each step defines the differences between all the reported systems.

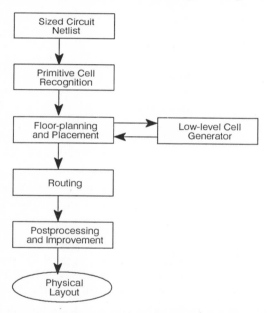

Figure 18.10 General flow-diagram of a cell layout program

From an input netlist (generally annotated with critical constraints), the circuit is partitioned into a set of primitive functional blocks (leaf cells) which can be basic devices (transistor, capacitors, etc) or structures (differential pairs, active load, etc.). Systems have available a set of parameterized primitive cells which are procedurally generated by dedicated generators from given device parameters. For instance, ILAC [45] uses the generator called STUCCO to obtain shape and interconnect information for all possible different realizations of each leaf cell, as well as a geometrical layout in a generic layer for the selected shape. The same capability is included into other layout generators, the exception being [46] where only transistors and capacitors are considered as primitives and both multiple-transistor structures and wells are generated during placement.

Floor-planning and Placement are the processes of determining the shapes, relative positions and orientations of the primitive blocks such that a number of objectives are optimized. This is carried out with the information provided by the leaf cell generators about the different shapes and orientations for primitives, and applying an algorithm of optimisation. Different algorithms and cost functions have been resorted to in analogue layout approaches, being the formers usually borrowed from digital approaches. However, the cost function must include, in addition to the usual area and aspect ratio objectives, specific objectives related to analogue constraints such as symmetry, proximity etc., that are far from common in digital applications.

The most used floor-planning strategy is slicing-tree, which provides a simple representation of the relative position of the blocks and a simple identification of the routing channels, while allowing a fast estimation of the total area. However, since slicing-tree restricts routing to channels and lacks abutment connections, the resulting layout is far from that created by expert layout designers, except when simple low-level primitives with minimum sizes are used. More promising seems the new placement algorithm implemented in KOAN [46], which can deal with complex layout symmetries, dynamic merging and abutment of individual devices, as well as flexible generation of wells and bulk contacts.

Concerning placement optimisation, it is realized in different ways that can be divided into constructive and iterating algorithms. In SLAM [43] a knowledge-based approach handling prioritized layout constraints is followed. In OPASYN [37], each circuit topology in the database is assigned with a slicing-tree that specifies a topological arrangement for the building blocks in the circuit, and the Stochmeyer's algorithm is used to fit the given floor shape with minimum area. In ILAC and KOAN, simulated annealing is employed, it is an iterative general optimisation technique that, in spite of being computationally intensive compared to other techniques, can produce optimum placement for a wide range of design styles and problem types.

Placement and routing are intimately related and interdependent. Routing of analogue circuits has to attack more aggressively capacitive and resistance coupling problems, since parasitic effects will strongly affect circuit performance. Hence, analogue routers devise different types of nets:

sensitive, noisy and neutral nets. They are assigned with different priorities and nets are routed based on its priority. Since analogue cells are small circuits with a reduce number of blocks, dedicated routing has been mostly applied.

Again, different algorithms borrowed from digital circuits [48] have been considered in analogue cell routing, such as line expansion [46], scan-line [45], and switch-box [37]. In general, they adopt the one-wire-at-a-time, method, including cross-talk as a penalty function. Also a routing over devices capability is included in some of then.

The last phase of layout generation is the post-processing and improvement phase, it is aimed to produce the optimum physical layout. Intermediate geometrical layouts generated in the previous phase are converted into the final layout with the process specific layers. Moreover, compaction and interconnection improvements by rip-up re-routing are carried out.

18.3.3.3 Design assistants

Although several software systems for the automatic synthesis of analogue cells have been developed, two important tasks still remain to the human designer. One concerns the knowledge acquisition, which mainly means creating analytical equations and appropriate models for new circuits, as well as storing them in the knowledge base. The other consisting in circuit verification and characterization, which generally means, many iterations with a circuit simulator.

Recently, some software tools directed towards the automation of these two tasks have been published [49-54]. They are usually developed in the context of existing analogue synthesis systems and can be viewed as design assistants, making these systems open in the sense that knowledge bases can be dynamically extendable by the users. Hence, more general and topology-independent design systems are emerging as a reality, coping with new circuits and new topologies for high-performance applications.

Symbolic analysis is a cornerstone for the automatic modeling of analogue integrated circuits. Symbolic analyzers as those in [49- 51] provide a mechanism for the automatic entry of circuit knowledge into the synthesis system, through the automatic generation of symbolic equations describing circuit performance. Symbolic formulas include, gain transfer functions, power supply and common mode rejection ratios, impedances, pole/zero extraction, nonlinear distortion, etc. Although symbolic analysis is not a new idea, only recently has it been efficiently exploited in connection with analogue design systems. In addition to symbolic analyzers, other tools are needed to automatically generate design plans; that is, programs performing equation manipulation in such a way that design parameters will be related to performance specs. The tool reported in [52] has been developed for that purpose; it is integrated in a design system [53] which through the combination of symbolic simulation, numerical optimisation and knowledge techniques, performs as an intelligent and flexible analogue design system.

Concerning tools for automated verification and characterization, they are simulation assistants supporting circuit-specific aspects like test function generation, automated simulation and result evaluation, data sheet generation, etc. The system in [54] provides these capabilities for various design environments exploiting expert knowledge, which means the user defines a specific work-plan as an unstructured set of functions for the circuit characterization, while the work organization and control is carried out by the program.

18.4 Conclusions

CAD tools for analogue (and mixed analogue-digital) circuits are receiving a great deal of attention from both research institutes and companies. Among these tools, a strong commitment is made towards developing practical silicon compilers that can be embedded into existing commercial frameworks. Present compilers are restricted from the viewpoints of design style, architecture and application, although there is an emerging generation which fits better into the structure of software environments.

Much more effort has still to be done, especially covering higher level tools capable of dealing with complex systems. In that sense, standardization initiatives (like ECIP) as well as analogue description languages (like AVHDL) and powerful behavioral simulators seem to be very necessary. Also, efficient performance-driven layout generators are required as part of analogue automatic design tools. The main goal must be to increase designer's productivity since as IC fabrication becomes cheaper and more available, many more chips will be designed and the design cycle has to be reduced as much as possible.

Existing compilers are targeted to "simple designs", i.e., to designs requiring a unique grouping and/or interconnection of ordinary analogue functions. Future automation has to be directed towards two different goals. On one hand, "simple designs" will be extended to a broader class of users, including novice designers without a strong experience in analogue circuits. On the other hand, CAD tools will assist the expert designer in the efficient and rapid completion of "complex designs", i.e., designs pushing the limits of precision and performance for a particular integrated system.

Concerning software advances, object-oriented programming tools and databases will be in the future, since it will be necessary to pay attention to the management of hierarchical object-based descriptions of systems. Concerning new CAD concepts, there is a need for toolthat can capture the designers expertise in an automated way and to store it efficiently into a knowledge base. All of these advances have to be performed in such a way that analogue circuitry will become more closely integrated with digital circuits.

Software integration is a crucial requirement as well. As far as standards for design environments are being developed, efforts for incorporating both system and module compilers into open design frameworks become more and more important. The availability of module generators, system compilers, layout tools for custom cells, general floor-planning and routing facilities, as well as the integration and easy interaction with digital and user-developed tools are essential features for the new analogue CAD software. The last to be considered is the use of the emerging tools as a way to learn analogue design. This is specially true for module generators, dealing with the electrical and physical design phases. Since they are intended to be very interactive, might offer new possibilities for training analogue design engineers.

18.5 Acknowledgements

Authors are grateful to all their colleges in the Centro Nacional de Microelectronica for their continuous support, and to Maite and Angeles for assisting with the document preparation. A. Rueda wishes to thank Alberto Yufera for his valuable collaboration and many hours of fruitful discussions, and Jose M. Quintana for his inestimable aid.

18.6 References

[1] P. E. Allen "A Tutorial: Computer Aided Design of analogue Integrated Circuits". Proc. IEEE Custom Integrated Circuit Conf. pp 608-617. 1986.

[2] D. S. Harrison et al., "Electronic CAD Frameworks". Proc. IEEE pp 393-417. Feb. 1990.

[3] E. Sanchez-Sinencio and J. Ramirez-Angulo, "AROMA; An Area Optimized CAD Program for Cascade SC filter Design". IEEE T. on CAD, Vol. 4, No. 3, pp 296-303. July 1985.

[4] D. G. Nairn and A. S. Sedra., "Auto-SC, an Automated Switched-Capacitor Ladder Filter Design Program". IEEE Circuits and Devices Magazine, pp 5-8, March 1988.

[5] P. E. Allen and E. R. Macaluso., "AIDE2: An Automated analogue IC Design System", Proc. IEEE Custom Integrated Circuits Conference, pp 498-501. 1985.

[6] C. D. Kimble et al., "Autorouted analogue VLSI". Proc. Custom Integrated Circuits Conference, pp 72-78. 1985.

[7] T. Pletersek et al., "High-Performance Designs with CMOS analogue Standard Cells". IEEE I. of Solid-State Circuits, Vol.21, no.2, pp 215-222. April 1986.

[8] G. Winner et al., "analogue Macrocell Assembler". VLSI System Design, pp 68-71. May 1987.

[9] H. Yaghutiel et al., "A Methodology for Automated Layout of Switched- Capacitor Filters". Proc. IEEE Int. Conf. on Computer Aided Design. pp 444-447. 1986.

[10] M. Negahban and D. Gajski., "Silicon Compilation of Switched-Capacitor Networks". Proc. European Design Automation Conference 1990, pp 164- 168.

[11] W. J. Helms and K. C. Russell., "A switched Capacitor Filter Compiler". Proc. IEEE Custom Integrated Circuits Conference, pp 125-128. 1986.

[12] Y. Therasse, L. Reynders, R. Lannoo and B. Dupont., "A Switched-Capacitor Filter Compiler". VLSI System Desing, pp 85-88. Sept. 1987.

[13] G. V. Eaton, D. G. Nairn, W. M. Snelgrove and A. S. Sedra., "SICOMP: A Silicon Compiler for SC Filters". Proc. Int. Symp. on Circuits and Systems" 1987, pp 321-324.

[14] J. Assael, P. Senn, M. S. Tawfik., "A Switched-Capacitor Filter Silicon Circuits" IEEE Journal of Solid-State Circuits, Vol.23, no.1, pp 166-174, February 1988.

[15] A. Barlow et al., "An Integrated Switched Capacitor Filter Design System". Proc. IEEE Custom Integrated Circuits Conference, paper 4.5. 1989.

[16] P. E. Allen and P. R. Barton., "A Silicon Compiler for Succesive Approximation A/D and D/A Converters". IEEE Custom Int. Circuits Conference, pp 552-555. 1986.

[17] G. Jusuf et al., "CADICS - Cyclic analogue to Digital Converter Synthesis" IEEE Custom Int. Circuits Conference, pp 286-289. 1990.

[18] S. Garcia Sabino, P. Senn, M. S. Tawfik., "HiFADiCC: a Prototype Framework of a Highly Flexible analogue to Digital Converters Silicon Compiler". Proc. IEEE ISCAS'90, pp 1114-1117.

[19] C. Leme et al., "Flexible Silicon Compilation of Charge Redistribution Data Conversion Systems". IEEE Midwest Symp. 1991.

[20] J. L. McCreary, et al., "All-MOS Charge Redistribution analogue-to-Digital Conversion Techniques". IEEE J. of Solid-State Circ. Vol.SC-10, no.6, pp 371-385. Dec. 1975.

[21] L. Paris, et al., "Floorplanning Strategy for Mixed analogue-Digital VLSI Integrated Circuits". Proc. EDAD'91.

[22] A. Yufera, A. Rueda and J. L. Huertas., " Flexible Capacitor and Switch Generators for Automatic Synthesis of Data Converters". Proc. ISCAS'91.

[23] H. L. Tahawy, A. Chianale, B. Henmion., "Functional verification of analogue Blocks in FIDELDO: a unified Mixed Mode Simulation Env.". Proc. IEEE, ISCAS 1989.

[24] J. G. Kenney and L. R. Carley., "CLANS: A High-Level Synthesis Tool for High Data Converters". Proc. ICCAD'88, pp 496-499.

[25] W. T. Nye et al., "DELIGHT-SPICE: An Optimization-Based System for the Designs of Integrated Circuits". IEEE Trans. CAD, vol 7, N.4, pp 501-519. April 1988.

[26] J. C. Lai et al., "ADOPT-A CAD System for analogue Circuit Design". Proc. CICC 1988, paper 3.2.

[27] J. M. Shyn and A. Sangiovanni-Vicentelli., "ECSTASY: A New Environment for IC Design Optimization". Proc. ICCAD 1988, pp 484-487.

[28] M.G.R. DeGrauwe et al., "IDAC: An Interactive Design Tool for analogue CMOS Circuit". IEEE J. Solid-State Circuits, Vol. SC-22, n.6, pp 1106-1118. Dec 1987.

[29] M.G.R. DeGrauwe et al., "Towards an analogue System Design Environment". IEEE J. Solid-State Circuits, Vol. 24, n.3, pp 659-671, June 1989.

[30] R. Bowman and D. J. Lane., "A knowledge-based system for analogue integrated circuit design" Proc. ICCAD, pp 210-212. 1986.

[31] F. M. El-Turky and E. E. Perry. "BLADES: An Artificial Intelligence Approach to analogue Circuit Design". IEEE T. on CAD, Vol. 8, n.6, pp 680-692, June 1989.

[32] R. Harjani, R. A. Rutenbar and L. R. Carley., "A Prototype Framework for Knowledge-Based analogue Circuit Synthesis" Proc. 24th ACM/IEEE Design Automation Conference, pp 42-49. 1987.

[33] R. Harjani, R. A. Rutenbar and L. R. Carley., "OASYS: A Framework for analogue Circuit Synthesis" IEEE Trans on CAD, Vol.8, n.12, pp 1247-1266, Dec 1989.

[34] E. Berkcan, M. d'Abreu and W. Langhton, "analogue Compilation Based on Successive Decomposition". Proc. 25th ACM/IEEE Design Automation Conference, pp 369-375. 1988.

[35] B. J. Sheu, J. C. Lee and A. H. Fung., "Flexible Architecture Approach to Knowledge-Based analogue IC Design". IEE Proc. Pt. G, Vol.137, n.4, pp 266-274, August 1990.

[36] C. Toumazou, C. A. Makris and C. M. Berrath., "ISAID-A Methodology for Automated analogue IC Design". Proc. ISCAS 1990, pp 531-535.

[37] H. Y. Koh, C. H. Sequin and P. R. Gray., "OPASYN: A Compiler for CMOS Operational Amplifiers". IEEE Trans. on CAD, Vol.9, n.2, pp 113-125. Feb. 1990.

[38] G. Gielen et al., "analogue Circuit Design Optimization Based on Symbolic Simulation and Simulated Annealing" Proc. European Solid-State Circuit. Conf. 1989, pp 252-255.

[39] P. Heikkila, M. Valtonen and H. Pohjonen., "Automated Dimensioning of MOS Transistors without using Topology-Specific Explicit Formulas". Proc. ISCAS 1989, pp 1131-1134.

[40] B. J. Sheu, A. H. Fung and Y. N. Lai., "A knowledge-Based Approach to analogue IC Design". IEEE Trans. on CAS, Vol. 35, n.2, pp 256-258. Feb. 1988.

[41] H. Onodera, H. Kanbara and K. Tamara., "Operational Amplifier Compilation with Performance Optimization". Proc. IEEE CICC 1989. Paper 17.4.

[42] C. Y. Kuo, L. G. Chen and T. M. Parng., "An Automatic Synthesizer for CMOS Operational Amplifiers". Proc. EDAC 1991, pp 470-474.

[43] D. J. Chen, J. C. Lee and B. J. Shen., "SLAM: A smart analogue Module Layout Generator for Mixed analogue-Digital VLSI Design". Proc ICCD 1989, pp 24-27.

[44] M. Kayal et al., "SALIM: A Layout Generation Tool for analogue ICs". Proc. CICC 1988. Paper 7.5.

[45] J. Rijmenants et al., "ILAC: An Automated Layout Tool for analogue CMOS Circuits". IEEE. J. Solid-State Circuits, Vol. 24.n.2, pp 417-425. April 1989.

[46] J. M. Cohn, J. Garrod, R. A. Rutenbar and L. R. Carley. "KOAN/ ANAGRAM II: New Tools for Device-Level analogue Placement and Routing". IEEE J. Solid-State Circuits, Vol. 26, n.3, pp 330-342. March 1991.

[47] E. Berkcan and M. d'Abreu., "Physical Assembly for analogue Compilation of High Voltage IC's". Proc. CICC 1988 paper 14.3.

[48] B. Preas, M. Lorenzetti. "Physical Design Automation of VLSI Systems". The Benjaming/Cummings Publishing Comp. 1988.

[49] S. J. Seda, M. Degrauwe, W. Fichtner., "A Symbolic Analysis Tool for analogue Circuit Design Automation". Proc. ICCAD'88, pp 488-491.

[50] G. G. E. Gielen, H. Walscharts, W. Sansen., "ISAAC: A symbolic Simulator for analogue Integrated Circuits". IEEE J. Solid-State Circuits, Vol. 24, n.6, pp 1587-1597. Dec.1989

[51] F. V. Fernandez, A. Rodriguez- Vazquez and J. L. Huertas., "Interactive AC Modeling and Characterization of analogue Circuits via Symbolic Analysis". Accepted for publication in analogue Integrated Circuits and Signal Processing. Kluwer Academic.

[52] K. Swings, and W. Sansen., "DONALD: A Workbench for Interactive Design Space Exploration and Sizing of analogue Circuits". Proc. EDAC'91, pp 475-479.

[53] G. Gielen, K. Swings and W. Sansen., "An Intelligent Design System for analogue Integrated Circuits". Proc. EDAC'90, pp 169-173.

[54] D. Milzner, and W. Brockherde., "SILAS: A knowledge-based simulation Assistant". IEEE J. Solid Circuits, Vol. 26, n.3, pp 310-318. Mar. 1991.

Index